Advanced Interferometers and the Search for Gravitational Waves

Volume 404

For further volumes:
http://www.springer.com/series/5664

Massimo Bassan
Editor

Advanced Interferometers and the Search for Gravitational Waves

Lectures from the First VESF School on Advanced Detectors for Gravitational Waves

Springer

Editor
Massimo Bassan
Department of Physics
Università Tor Vergata
Rome
Italy

ISSN 0067-0057 ISSN 2214-7985 (electronic)
ISBN 978-3-319-03791-2 ISBN 978-3-319-03792-9 (eBook)
DOI 10.1007/978-3-319-03792-9
Springer Cham Heidelberg New York Dordrecht London

Library of Congress Control Number: 2014931208

Printed on acid-free paper

Springer is part of Springer Science+Business Media (www.springer.com)

Foreword

Detecting gravitational waves is still a dream. After several years of commissioning of Virgo, and after reaching its target sensitivity, we know that a gain of roughly one order of magnitude is still necessary, so that a new detector called "Advanced Virgo" is in construction. Gravitational waves are not only a unique messenger coming from extreme regions of the universe, but also, it is well known, an experimental challenge. Starting from the very beginning, with resonant bars, if a gravitational signal is detected in 2015, it will crown one century of General Relativity theoretical studies, and half a century of technological developments. The coupling coefficient between matter energy and waves is so tiny, and the amplitude of waves so weak that any spurious physical effect negligible in an ordinary context must be carefully assessed: shot noise, frequency noise, residual seismic activity, thermal noise, stray light, and other microscopic effects are however much larger than the signal if no care.

This particular circumstance has defined since years the matter now called "experimental gravitation." The lessons presented in this book reflect the various fields that proved necessary to allow some hope for a first historical detection. We hope that for the young searchers who took part in this school, it will be some day classical science.

Jean-Yves Vinet
Virgo spokesperson

Preface

Why this book? Whom is it intended for? How should the reader use it? These are the questions a preface is called to answer.

Let us start by pointing out *what this book is not*: it is not an introduction to the theory and basics of gravitational waves (GW), that our reader should have already acquired. Any recent textbook on General Relativity has a chapter on GW generation, propagation, and detection. Besides, there are excellent monographs on GW, from the books of P. Saulson (unfortunately out of print) and M. Maggiore to the reviews of D. Blair and the thorough online overview of B. Schutz and Sathyaprakash [1–5].

However, very little is required to read and fully exploit the contents of this book: the reader should be familiar with the fundamentals of GW detection (not much more than the very basic $h(t) = \delta L(t)/2L$) and be at ease with concepts like interferometers and noise spectral density, signal-to-noise ratio and chirp.

This book *is not* an overview of all current experimental efforts to detect GW: we cover neither Pulsar Timing Array, a promising technique to detect sub-μHz radiation, nor Laser Interferometer Space Antenna, the so-called LISA concept for a space mission that will fly in the early 2030s, placing a Gm long interferometer in a heliocentric orbit to observe GW in the 1 mHz−1 Hz frequency range. This book only deals with the present and the future of ground-based, km-long interferometers, aiming at detecting GW in the 10 Hz–3 kHz range. We did not attempt to cover the advances in data analysis, that would deserve a book on its own [6].

What is then the goal of this book? Its goal is to be a guide for both new and experienced scientists that decide to venture in the field of experimental search for gravitational radiation. It aims to be a handbook for the physicist who wants to join, or is already a member of one of the large collaborations now engaged in the construction of ground based, second generation, GW detectors; Advanced LIGO in the USA, Advanced Virgo and GEO in Europe, KAGRA in Japan. It aims to be a reference book on how to tackle the huge, impressive technological challenges that can intimidate even the experienced experimentalists.

The past of gravitational radiation is well known: it is part of scientists' common knowledge since the papers [7, 8] of 1916 and 1918, almost a century ago, where A. Einstein himself worked out the wave solution to his field equations, passing through the debate of the 1950s on the detectability of these waves to the first attempts of building a detector and the indirect, but indisputable evidence of

GW from observation of binary pulsars [9, 10]. Resonant antennas, where the vibration of a massive cylinder is to be detected against thermal and readout noise, held the scene from the early 1960s, with the pioneering work of J. Weber, to the cryogenic apparata of the 1980s and finally the ultra-cryogenic detectors (Nautilus, Auriga, Minigrail) of the 1990 and early 2000s, when they were finally surpassed, both in sensitivity and bandwidth, by the large interferometers. Interferometers themselves have a long history of technological development: from the first, table-top, seed work of Forward, Miller, and Moss [11] in 1971, they have grown in size, sensitivity, and complexity, passing through the 30 or 40 m of arm length of the prototypes in Garching, Glasgow, and Caltech, to the 600 m of GEO600, still active as a test bench for new technologies. The last 15 years have seen the construction, commissioning, and operation of the km-scale interferometers, some of the largest research infrastructures in the world: the scientific teams have grown from local groups of a dozen scientists to international collaborations of several hundreds people; the two 4 km observatories of LIGO, in the USA and the 3 km Virgo detector, in Italy (a French–Italian joint venture, now grown to a European collaboration with The Netherlands, Poland, Hungary) have observed the cosmos, producing relevant upper limit science (an overview of these results can be found in Chap. 2) but missing the "smoking gun" of an evidence for direct detection of GW. This task is thus passed on to the next generation of devices, the advanced detectors, now under construction.

The present of the experimental search of GW is what this book is mainly about: after several years of successful data taking, the large interferometers have been decommissioned and dismantled, and the construction of the advanced detectors has begun within the same infrastructures (the end buildings and the 3 or 4 km long vacuum system), both on the two LIGO sites and at Virgo. This "second generation" is characterized by physical and technological advances that are described in detail in the Chaps. 3–11 and will provide a sensitivity ten times better: it is common belief that direct detection of GW cannot be missed by these detectors.

And finally the future of GW detectors: in the short term, the years 2014–2017, will see the commissioning and then operation of Advanced LIGO and Advanced Virgo and, most likely, will deliver a direct detection, sought after and relentlessly chased for over 50 years. In 2018, according to optimistic schedules, also KAGRA will join the detectors' network: a Japanese effort that will fold into an Advanced detector some features, like cryogenics and underground location, typically considered as further, delayed improvements, so that we can call it a "second and half generation." More detectors may appear in the coming years: LIGO-India, a joint venture of LIGO and several Indian scientific institutions (Indigo) is, at the moment of writing, seriously considered by funding agencies.

For the not-so-near-future, running over into the next decade, a third generation of detectors, again ten times more sensitive, is already being conceived and designed: the last two chapters describe this long-term effort, that should move the status of GW observations from detection to routine astronomy. The Einstein

Telescope will probably complement the wealth of physics information that we can expect from a LISA space mission.

This book springs as a by-product of a specialized course, held at the EGO-Virgo site in May 2012, that focused on the science and technology that are at the core of the so-called Advanced Interferometers. The course, that I was called to direct, attracted many enthusiastic students and a dozen of excellent lecturers: there was immediately a large consensus and interest in collecting the lecture notes in a publication, that could serve as a guide for future schools and as a handbook for any GW experimentalist. The table of contents of this book closely retraces the course syllabus, and all of the authors (with one exception) had lectured the same topic in the course. Each one of them is an experienced specialist, extremely knowledgeable, and deeply involved in the development that she or he discusses, chosen among the very best scientists in that particular field. Each of their contribution has been thoroughly revised both by the book editor and by the internal refereeing system of the author's own scientific collaboration (Virgo, GEO, or LIGO) In this respect, we are confident that we are providing contents that are rigorous, nontrivial and up-to-date.

Acknowledgments The Virgo-Ego Science Forum (VESF) has provided the framework for the course and this book. The VESF coordinator Valeria Ferrari had the correct intuition of diversifying the VESF Schools (we now have Introduction to GW, Theory, Data Analysis, Advanced Detectors…) and was instrumental in pushing forward the idea that schools are a pivotal tool to approach new generations of young researchers and to convey them toward (and integrate into) the GW community. VESF is closely dependent on its parent institutions: the EGO observatory consortium and the Virgo Scientific Collaboration. We are therefore grateful to Federico Ferrini, the EGO Director, and to the Virgo spokesperson Jean-Yves Vinet, for providing hospitality, motivation and financial support for the school and for supporting this project. Thanks to M.me Sevèrine Perus of EGO who handled the school logistic. A special thanks goes to all the authors who bore with my periodic reminders and with the pressure to deliver their manuscripts; but to some of them I am particularly grateful: Gabriele Vajente lead the way, submitting two chapters before we even considered writing this book. Andrea Viceré and Catherine Nary Man filled in for unexpected defections with very little notice, Giovanni Losurdo was the key adviser for the syllabus. Finally, thanks to all the reviewer who contributed, through accurate and critical reading, to further improve the quality of each chapter…and to the Springer editors, for their patience during gestation of this work.

An exciting time is ahead of us all, gravitational waves hunters: the Holy Grail of detection has never been so likely and so near!

Rome, October 2013 Massimo Bassan

References

1. P.R. Saulson, *Fundamentals of Interferometric Gravitational Wave Detectors* (World Scientific, Singapore, 1994)
2. M. Maggiore, *Gravitational Waves. Vol. 1: Theory and Experiments* (Oxford University Press, Oxford, 2007)
3. D.G. Blair (ed.), *The Detection of Gravitational Waves* (Cambridge University Press, Cambridge, 1991)
4. D.G. Blair, E.J. Howell, L. Ju, C. Zhao (eds.), *Advanced Gravitational Wave Detectors* [Hardcover], (Cambridge University Press, Cambridge, 2012)
5. B.S. Sathyaprakash, B.F. Schutz, Physics, astrophysics and cosmology with gravitational waves. Living Rev. Relativ. **12**(2), (2009)
6. P. Jaranowski, A. Krolak, *Analysis of Gravitational-Wave Data* (Cambridge University Press, Cambridge, 2009)
7. A. Einstein, Näherungsweise Integration der Feldgleichungen der Gravitation. In: Sitzungsberichte der Kniglich Preussischen Akademie der Wissenschaften (Berlin, 1916), p. 688
8. A. Einstein, Über Gravitationswellen. In: Sitzungsberichte der Kniglich Preussischen Akademie der Wissenschaften (Berlin, 1918), p. 154
9. R.A. Hulse, J.H. Taylor, Astrophys. J. **195**, L51 (1975)
10. J.M. Weisberg, D.J. Nice, J.H. Taylor, Astrophys. J. **722**, 1030 (2010)
11. G.E. Moss, L.R. Miller, and R.L. Forward, Appl. Opt. **10**, 2495 (1971)

Contents

Chapter 1
Towards Gravitational Wave Astronomy

Giovanni Losurdo

Abstract This chapter is meant to introduce the reader to the forthcoming network of second-generation interferometric detectors of gravitational waves, at a time when their construction is close to completion and there is the ambition to detect gravitational waves for the first time in the next few years and open the way to gravitational wave astronomy. The legacy of first-generation detectors is discussed before giving an overview of the technology challenges that have been faced to make advanced detectors possible. The various aspects outlined here are then discussed in more detail in the subsequent chapters of the book.

1.1 Introduction

This book sees light at a time when a new generation of interferometric detectors, capable in principle to expand the explorable volume of the universe by three orders of magnitude, and thus to start the era of GW astronomy, is being realized. Virgo and LIGO have stopped taking data in 2011, and started the decommissioning and then the construction of Advanced Virgo and Advanced LIGO. Thus, no new science data will be available until 2015. In the GW community, this period with no data is usually known as *the crossing of the desert*: a tough time when scientists get ready to discover a new *promised land*, working on the prototyping and construction of the new detectors, preparing their commissioning and the data analysis for the upcoming years. A good time to focus on the technology challenges that we are facing in the realization of such new detectors (and even beyond), which are discussed in this book. This first chapter aims to provide a reference frame for the following chapters: it starts with a brief summary on what has been achieved by the first-generation

G. Losurdo (✉)
Istituto Nazionale di Fisica Nucleare (INFN)–Sezione di Firenze, Via G. Sansone, 1,
I-50019 Sesto Fiorentino, FI, Italy
e-mail: losurdo@fi.infn.it

M. Bassan (ed.), *Advanced Interferometers and the Search for Gravitational Waves*,
Astrophysics and Space Science Library 404, DOI: 10.1007/978-3-319-03792-9_1,

detectors and then discusses the main technology novelties, pointing to the rest of the book for a more exhaustive description.

1.2 The Legacy of the First-Generation Interferometric Detectors

The existence of gravitational waves (GW) was predicted by Albert Einstein almost one century ago [1]. The effort to detect GW was initiated by Joe Weber, half a century ago, using resonant bars at room temperature. For several decades the resonant bars have been used, improving the technology and the sensitivity and establishing, for the first time, an observing worldwide network. In the first decade of the twenty-first century, the interferometric detectors of GW have overcome the resonant detectors demonstrating a big improvement in sensitivity and bandwidth. The operation of the kilometer-scale ground-based interferometers LIGO [2] and Virgo [3] has proven that such a technology is now suitable and reliable to realize a worldwide network of wideband detectors with duty cycle and sensitivity sufficient to aim and open the way to the GW astronomy (for a more complete description and bibliography of the history of GW research see [4] and references therein).

No discovery has been reported after some years of operation. However, interesting science results have been collected and a significant legacy has been left by the so-called *first-generation detectors* to the science community:

- *Infrastructure*: The relevant investments undertaken to create the infrastructure hosting the first-generation interferometers will continue to pay off, since the same facilities will host the advanced interferometers: the LIGO facilities (Hanford, WA and Livingston, LA), the Virgo site (Cascina, Italy), and GEO600 (Hannover, Germany) [5]. However, the expansion of the network requires the creation of new observatories. The construction of the underground 3 km detector KAGRA has been funded in Japan [6] and the possibility to move one of the three Advanced LIGOs to India is being explored.
- *Sensitivity and robustness*: LIGO and Virgo have reached or approached the design sensitivity over the whole frequency range. Some important upgrades have allowed us to enhance the detectors and go beyond the initial design sensitivity. Most of the noise contributions were understood and modeled. The reliability and robustness achieved has been quite impressive (given the complexity of these machines): all detectors could finally run with duty cycles of $\sim 80\,\%$ or larger. These achievements have eventually demonstrated the technology of the interferometric detectors of GW.
- *Commissioning experience*: Reaching the design sensitivity required about 5 years of commissioning work (alternated to data taking periods). The sensitivity (see Fig. 1.1) was improved progressively: the interferometric detectors were new instruments, never before tested in such extreme configurations. Commissioning teams had to learn how to operate and control the interferometers, cope with optical configurations of increasing complexity, cope with thermal effects and scattered

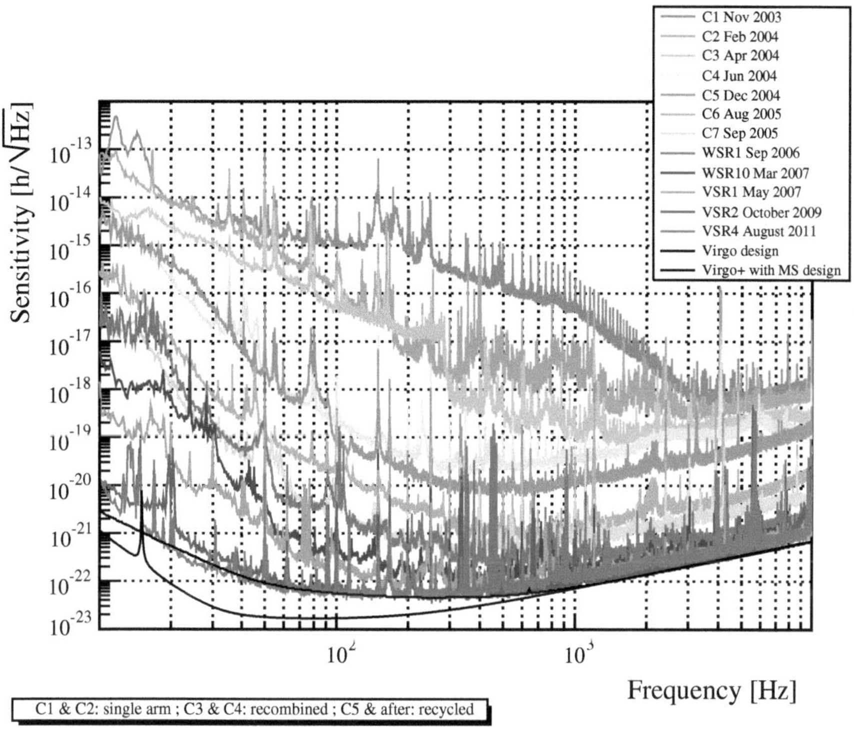

Fig. 1.1 The evolution of the Virgo sensitivity. The *black solid lines* are the design sensitivities of the initial Virgo and the enhanced Virgo+ configuration (*Credit* The Virgo Collaboration)

light, identify unknown sources of noise, and fix the detectors when unexpected problems emerged.

- *Technologies for advanced detectors*: Some technologies relevant for the second generation have been tested in most recent years on Virgo, LIGO, and GEO600, allowing debugging and further developments which will pay off in the commissioning and operation of the advanced detectors.
- *Data analysis*: An intense effort has been pursued to analyze the data of the first-generation detectors, leaving a big heritage in terms of methods, software, and expertise which will be very useful to analyze the data of the advanced detectors.
- *Network*: During the years the GW community became more and more aware that joint observation is necessary to increase the confidence in the events and the amount of science that could be extracted from them. The creation of a worldwide network of GW observatories is the cornerstone for the start of the GW astronomy.
- *Multi-messenger observations*: In addition to establishing a network, the GW community has advanced further by initiating a multi-messenger approach, signing memoranda of understanding for joint observation with optical and radio telescopes, gamma and X-ray satellites, and neutrino detectors.

- *Science*: Extensive science runs have been performed and, despite the fact that no detection has been claimed, relevant upper limits have been set.

1.3 Creating a Worldwide Network

The LIGO Scientific Collaboration (including the LIGO and GEO600 detectors) and Virgo have signed a Memorandum of Understanding for joint data taking, and publications policy joint analysis after May 2007. They have been operating as a "single machine," thus enhancing the chances of detection. Several joint science runs have been held so far:

- LIGO S5 (Nov. 2005–Sept. 2007) and Virgo VSR1 (May 2007–Oct. 2007): the overlap between the long S5 run and the VSR1 lasted 4 months.
- LIGO S6 (July 2009–Oct. 2010) and Virgo VSR2 (July 2009–Jan. 2010) / VSR3 (July 2010–Oct. 2010): LIGO and Virgo run together for 6 months at the end of 2009. Then Virgo was shut down to install monolithic suspensions. Once Virgo was back in operation (with monolithic suspensions) LIGO and Virgo run jointly for 2 more months. Then LIGO was shut down to start the upgrade to Advanced LIGO.
- Virgo VSR4 and GEO S6e (June–Aug. 2011): Virgo and GEO run together for 3 months. Then Virgo was shut down to start the upgrade to Advanced Virgo.

The network of interferometers has pushed the upper limits on gravitational wave amplitude emitted by several astrophysical sources to the lowest level ever reached (see for instance [7–11]). The most relevant results achieved so far are discussed in Chap. 2. The analysis of the data of the last runs is still ongoing and more papers are expected in the following months. Hopefully, this network approach will be undertaken also for the second generation, with a much better sensitivity. The science case for the advanced detectors and the expectations for the event rates are discussed in Chap. 2 of this book.

1.4 Advanced Detectors: The Scenario

The first decade of the century has been characterized by the operation of the first generation of the interferometric detectors and the demonstration of their technology. In the second decade a network of advanced interferometers, exploiting the results of many years of R&D, will be taking data. The construction of the Advanced LIGO detectors and of Advanced Virgo is in progress and a first science run could be possible in 2015. GEO600 is undergoing a phase of incremental upgrades (GEO-HF) aimed at reaching a competitive sensitivity in the high frequency range. Besides the upgrades of existing detectors, a 3 km underground detector (KAGRA) has also been funded in Japan and currently the tunnels are being excavated. The schedule is

Table 1.1 This table has been extracted from Aasi et al. [12]

Epoch	Estimated run duration	BNS range (Mpc) LIGO	Virgo	Number of BNS detections	% BNS localized within 5 deg^2	20 deg^2
2015	3 months	40–80	–	0.0004–3	–	–
2016–2017	6 months	80–120	20–60	0.006–20	2	5–12
2017–2018	9 months	120–170	60–85	0.04–100	1–2	10–12
2019+	(per year)	200	65–130	0.2–200	3–8	8–28
2022+ (India)	(per year)	200	130	0.4–400	17	48

It contains a summary of a plausible observing schedule, expected sensitivities, and source localization with the advanced LIGO and Virgo detectors. In the first two columns, the possible dates and durations of the science runs are listed. For each run, in the third column, the expected inspiral range (a figure of merit indicating the distance at which a coalescing binaries can be detected, averaged over the GW polarization states and the source direction in the sky) for coalescing binary neutron stars (BNS), computed on the basis of the likely sensitivity evolution of the detectors, is indicated for LIGO and Virgo. In the next columns, the corresponding number of events and network capability to localize the source in the sky are indicated. See [12] for the full explanation of the underlying assumptions

quite ambitious: the goal is to be a part of the second-generation network as soon as 2017. A further novelty might be the installation of a new detector in a location far enough from LIGO and Virgo to considerably enlarge the network baseline, and thus to improve the accuracy in the reconstruction of the source angular position: LIGO and the Indigo consortium have proposed to install a third Advanced LIGO in India.

The goal of Advanced LIGO/Virgo is to extend the distance at which a target source can be detected by a factor $\sim$10, which corresponds to increasing the event rate by $\sim$1,000 with respect to the detectors of first generation. The final sensitivity goal, as the experience with the first-generation teaches, will be reached gradually. As the sensitivity improves, the detection rate, as well as the source localization capability, will improve. The likely evolution of the network capabilities (based on the experiment construction schedules and on the former experience on the commissioning evolution) has been outlined in a recent paper by the LIGO Scientific Collaboration and the Virgo Collaboration [12] and it is summarized in Table 1.1.

The improved sensitivity of the GW detectors and the largely increased expected event rate widens the possibilities with multi-messenger science. The joint opera tion of GW detectors and high energy neutrinos (HEN) telescopes is particularly appealing. Both GW and HEN travel undisturbed over cosmological distance and escape dense media. Both carriers are likely to provide an image of the universe complementary to that given by electromagnetic information. Thus, coincident observation of GW and HEN from the same event could allow a deep investigation of the source. Moreover, current plans of advanced GW detectors and neutrino telescopes (IceCube [13] and KM3NET [14]) promise observation on the same timescale. Among the astrophysical targets for a simultaneous detection of GW and HEN there are several sources that could be also observed through electromagnetic waves: gamma ray burts, supernovae, and soft gamma repeaters.

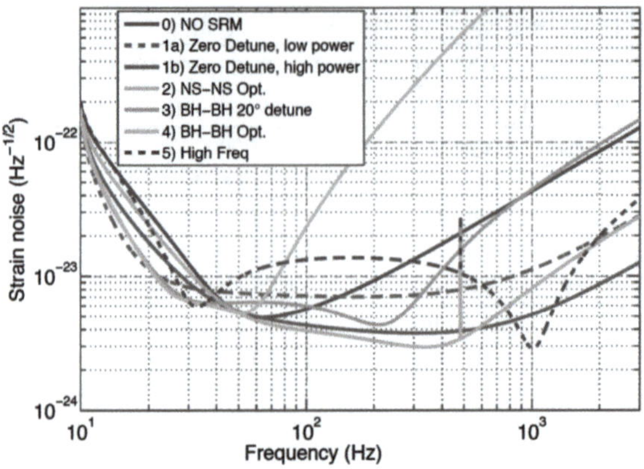

Fig. 1.2 The Advanced LIGO principal anticipated noise sources limiting the sensitivity in different configurations. The *solid red curve* (the one with the best sensitivity at 2 kHz) is the reference configuration, using tuned signal recycling, which provides a good sensitivity over a wide range. The *yellow* and *light blue curves* (respectively, the one with worst and best sensitivity at 400 Hz) correspond to a tuning which optimize the SNR for binary *black holes* and binary neutron stars respectively. The *black dashed curve* corresponds to a narrow band tuning, useful to enhance the sensitivity at the frequency of a monochromatic source. Figure from Harry [15]

In Sect. 1.5 an overview of the technologies needed to achieve such an ambitious goal is given and the following chapters in this book provide a deeper insight into each of those aspects. The crucial features allowing to increase the sensitivity are recalled here:

- the mirror technology has been improved in all the aspects: better substrates, better polishing, coatings with reduced absorption and mechanical dissipation;
- the mirrors are suspended through silica fibers, thus reducing the suspension thermal noise;
- the input laser power is increased by one order of magnitude to improve the shot noise limited high frequency sensitivity. This requires improved thermal compensation systems (to cope with thermal lensing) and heavier mirror (to reduce the effect of radiation pressure noise);
- the Advanced LIGO bandwidth is extended at lower frequencies by implementing a new vibration isolation system. This is not needed in the Virgo case thanks to the existing Superattenuator;
- signal recycling allows to tune the detector bandwidth and enhance the sensitivity in a chosen frequency range (see Fig. 1.2).

The path toward the best sensitivity will obviously take time. Both the operation of signal recycling and the increase in the input power by almost one order of magnitude with respect to the first generation will put forward relevant commissioning challenges. Therefore, Advanced LIGO/Virgo will start their operation with low input power and increase the complexity and performance in steps.

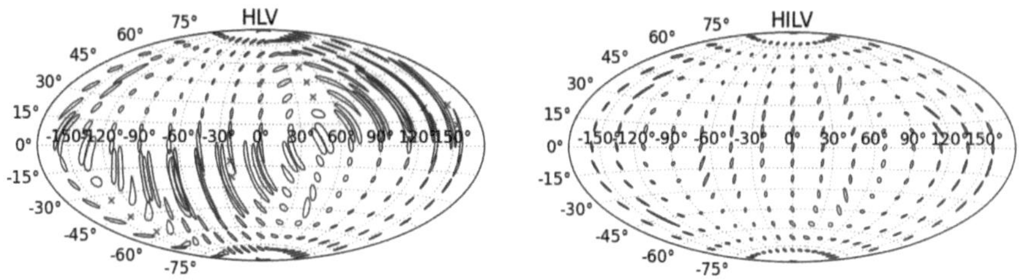

Fig. 1.3 The angular resolution of the LIGO-Virgo network without and with LIGO-India. Figure from Fairhurst [17]

1.4.1 Advanced LIGO

The Advanced LIGO project has been approved by the National Science Foundation in April 2008. It foresees the upgrade of the three LIGO interferometers to second generation. At the time of this writing, the construction works are close to completion while the commissioning of components of the detector has already started in 2012. At the design sensitivity (in the configuration optimized for BNS detection) the BNS inspiral range will be of 200 Mpc, which corresponds to detecting a BNS up to a distance of 450 Mpc if optimally oriented. With such a sight of distance, tens of BNS events per years are expected [16].

The LIGO laboratory and the Indigo Consortium are studying the possibility to move the third interferometer (Hanford-2) to India. This would require the indian funding agencies to provide the vacuum system and the infrastructure, while all the rest of the equipment will be shipped by LIGO. Such a choice would allow to considerably increase the angular resolution of the worldwide network [17] (see Fig. 1.3).

1.4.2 Advanced Virgo

Advanced Virgo [18] has been approved by INFN and CNRS at the end of 2009 (beside the Virgo funding agencies, Advanced Virgo will be contributed by the dutch laboratory NIKHEF). The Advanced Virgo design is described in the Technical Design Report [19].

In Fig. 1.4 the noise budget is shown. The sensitivity below 50 Hz is limited by a combination of suspension thermal noise and radiation pressure noise. The first is minimized by suspending the large mirrors (42 kg mass) with fused silica fibers similarly to what has been done for Virgo+. Around 100 Hz the main limitation to the sensitivity is due to the thermal noise in the mirror coatings. The sensitivity shown in the figure assumes current coating technology. So, any further improvements from this ongoing R&D will have a direct positive impact on the detector sensitivity. The

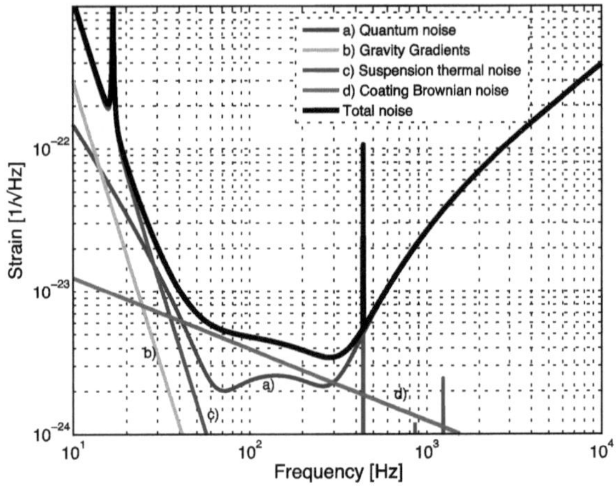

Fig. 1.4 The Advanced Virgo target sensitivity with the main noise contributions in the BNS-optimized configuration

effect of coating thermal noise is minimized running the Fabry–Perot cavities near the co-focal configuration, thus maximizing the beam size on the mirrors.

In order to reduce the laser shot noise and thus increase the interferometer sensitivity at higher frequencies the available laser input power will be around 200 W. A different choice with respect to LIGO has been made for the Advanced Virgo laser: a fiber amplifier will be used instead of a solid-state one. To cope with the larger radiation pressure effects due to the larger power injected heavier test masses (42 kg) will be used and will be suspended, as it was in Virgo+, through fused silica fibers. To this extent, the experience done with the new Virgo+ payloads will be precious to understand all the features of this technology, improve it further, and reduce the risk for the advanced detectors.

The optical scheme is modified by the introduction of a signal recycling cavity. The sensitivity shown in the figure assumes that the signal recycling is detuned in such a way as to have the best sensitivity to binary neutron stars coalescences. In this case, the coalescences of binary neutron stars will be visible up to distances of about 135 Mpc. Other tunings are possible and can be used to target other sources. Also, the finesse of the 3 km Fabry–Perot cavities will be larger (~450). A DC detection scheme will be adopted to reduce some technical noises.

The vacuum system and the infrastructures will also be upgraded to improve the sensitivity in the low and mid-frequency ranges. Large cryotraps will be installed at the extremes of the 3 km vacuum pipes to isolate them from the rest of the vacuum system. This, combined with the baking of the tubes, will allow reducing the residual gas pressure to the target of 10^{-9} mbar. The air-conditioning machines as well as several other sources of acoustic and seismic noise will be moved out of the central experimental hall to reduce the sources of vibrations of vacuum pipes, optical mounts, or any other parts that could scatter photons and generate stray light noise.

All the photodiodes to be used in science mode will be in vacuum and seismically isolated by compact vibration isolators.

Virgo+ has been shut down in November 2011 and the installation of Advanced Virgo has started. The installation of the upgrades is expected to be completed in 2015.

1.4.3 KAGRA

KAGRA (KAmioka GRAvity and gravitational waves) is a 3 km underground detector to be realized in Japan, in the mine of Kamioka. KAGRA will start its operation at room temperature and will then be upgraded to the cryogenic (baseline) configuration. In its final configuration, the KAGRA test masses will be cooled down to 20 K. Therefore, KAGRA will anticipate some of the features envisaged for the third generation: being underground and cryogenic.

The reduction of the sapphire mirror thermal noise by cooling has been recently demonstrated on CLIO, a 100 m interferometer in the Kamioka mine used as a test facility for KAGRA technologies [20].

The construction of the KAGRA detector has officially started on January 2012. KAGRA aims to joining Advanced LIGO, Advanced Virgo and GEO-HF in the world network of second generation from 2017.

1.5 Advanced Detectors: The Technologies

In this section, we discuss the relevant technologies that allowed to approach the design sensitivity of the first-generation interferometric detectors and how these technologies have been further developed to realize the advanced interferometers.

1.5.1 Optical Configurations

The optical configurations of the advanced detectors will maintain the general scheme of the Michelson interferometer with Fabry–Perot cavities but will be updated in several features:

- *Fabry–Perot cavities in the arms*: the finesse of the Fabry–Perot cavities will be increased to $\sim$450 (for Advanced LIGO/Virgo).
- *Signal recycling*: a *signal recycling mirror* suspended between the beam splitter and the output port of the interferometer can *recycle* the audio sidebands containing the signal created by the impinging GW and increase the signal-to-noise ratio in a limited bandwidth [21]. The position of the signal recycling mirror and its reflectivity determine the central frequency and the bandwidth, respectively. Signal recycling has already been used by GEO600 in the first generation, and will be adopted by Advanced LIGO/Virgo.

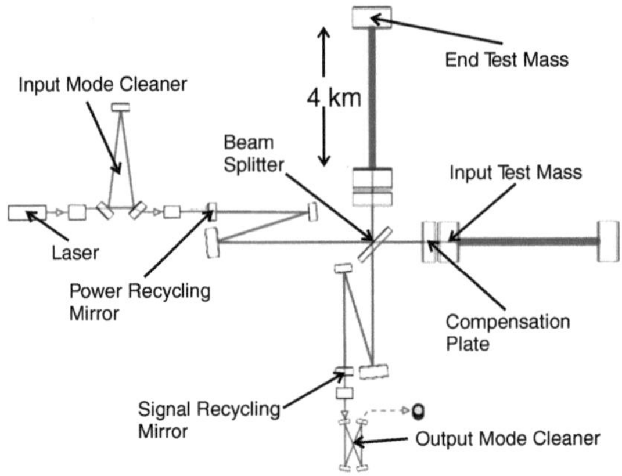

Fig. 1.5 Optical layout of Advanced LIGO

- *Stable recycling cavities*: the power recycling cavity of the initial Virgo and LIGO is *marginally stable*: the Gouy phase of the cavity is close to zero and high-order modes can resonate into it. This makes the cavity more sensitive to aberrations, induced by thermal effects, optical defects, or misalignments. To increase the robustness with respect to such effects Advanced LIGO will be provided with *stable* recycling cavities (see Fig. 1.5). The non-degeneracy is achieved by folding the beam path and having the mode matching telescope inside the cavity.

The optical schemes for Advanced LIGO/Virgo are shown in Figs. 1.5 and 1.6. Chapter 3 explores in detail the optical configuration aspects mentioned here.

1.5.2 Mirrors

The quality of the mirrors is a crucial element for the correct operation of the detectors. It may affect the final performance in several different ways since it plays a role in the thermal noise budget, scattered light, optical aberrations, and radiation pressure noise:

- the *diameter* of the mirrors must be large to allow for large beam spot size (which helps reducing the impact of thermal noise). The Advanced Virgo test masses are 350 mm in diameter as the Virgo mirrors, but twice as thick (20 cm). The beam splitter will be exceptionally large (55 cm in diameter) to reduce the risk of spurious scattering (see Fig. 1.7);
- the weight of the mirrors should be sufficient to reduce the effect of radiation pressure fluctuations below the other noise contributions. The mass of the Advanced Virgo test masses will be 42 kg (twice as heavy as the Virgo mirrors);

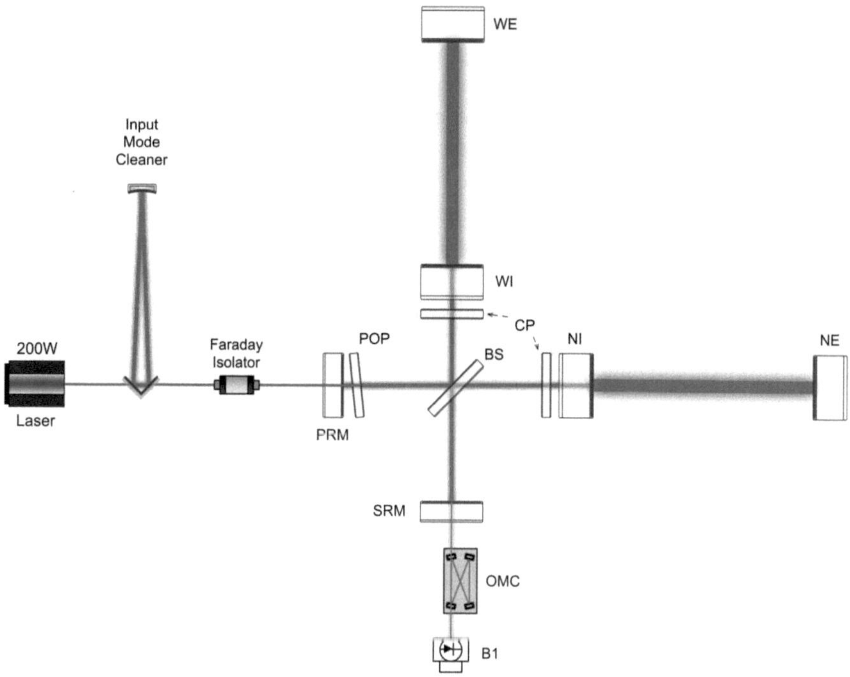

Fig. 1.6 Optical layout of Advanced Virgo

Fig. 1.7 The Advanced Virgo beam splitter

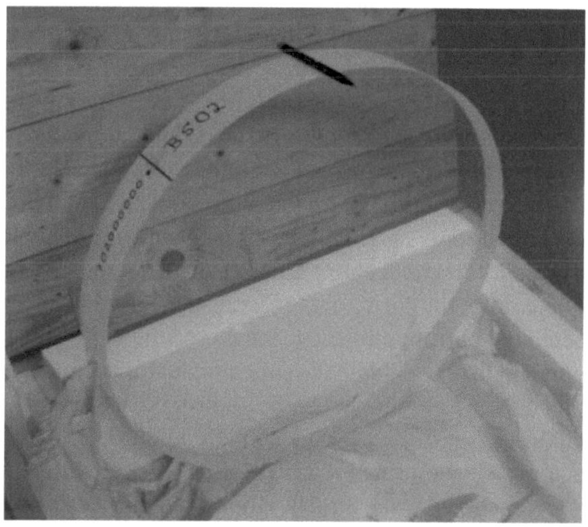

- the material used to make the mirror must be chosen in such a way as to minimize the absorption at the laser wavelength (better than 1 ppm/cm) while keeping the mechanical internal dissipation responsible for thermal noise at a minimum. Different kinds of fused silica are available today on the market that meet the stringent requirements for GW interferometers;
- the polishing of the surface must be excellent, since the deviations from perfect flatness and the residual microroughness are responsible for scattering losses (see Chap. 10). The flatness required for the Virgo test masses was 8 nm rms, while for the advanced detectors it should be better than 0.5 nm rms. Superb polishing figures (~0.2 nm rms) have been already achieved on the Advanced LIGO test masses by ion beam milling;
- the coating should guarantee low absorption and low thermal noise. In the advanced detectors, the thermal noise generated in the high reflectivity coating layers limits the sensitivity in the mid-frequency range, while its absorption is mainly responsible for the thermal lensing aberrations. Progress has been made in reducing the coating losses by doping the high index layer of Ta_2O_5 with Titanium [22, 23]. A very good uniformity in the thickness of the coating layer (at the level of 0.5 nm rms) must also be achieved in order to avoid excess optical losses (see Chap. 8 for more details).

1.5.3 Vibration Isolation

The test masses and the other sensitive optical components of interferometric detectors must be suspended and isolated from ground vibrations. The isolator must also allow to apply suitable control forces to the mirrors to bring the interferometer into operating conditions and maintain it as such. The approach to the problem of the vibration isolation has been different for the various projects. Virgo was the only one to pursue, since its beginning, the development of a vibration isolation system able to extend the detector bandwidth down to ~10 Hz: the superattenuator [24]. It was conceived to suppress seismic vibrations by about 300 dB at 10 Hz. This is achieved by an 8 m long vertical chain of mechanical filters, able to low-pass vibrations in 6 degrees of freedom above their resonance frequencies. Passive filtering is not the end of the story: the chain of filters is suspended from an inverted pendulum platform [25] actively controlled to compensate for the earth tidal strain over the interferometer arm length and to damp the superattenuator normal modes [26]. Moreover, the mirror position and alignment can be controlled at the level required by the interferometer control system. Three main actuation points are envisaged and the mirror control is achieved by splitting the actuation over them with a hierarchical approach [27]. The superattenuator was already conceived as a *second-generation device*: it has been thoroughly and successfully tested in the operation of Virgo and its isolation performance has been demonstrated to be compliant with the requirements of Advanced Virgo [30]. More details are provided in Chap. 7.

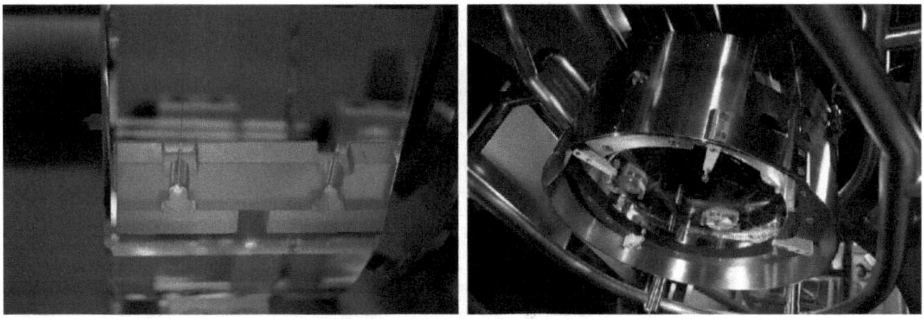

Fig. 1.8 The monolithic payload installed in Virgo in 2010

Unlike Virgo, initial LIGO was provided with less performing vibration isolation: the mirrors were suspended from a bench laying on a stack of metal and rubber layers [31]. One of the most important upgrades to Advanced LIGO is the implementation of new isolators capable of pushing the seismic wall down to 10 Hz [32]. It is made of three different parts:

- a hydraulic external pre-isolator (HEPI) for low frequency control;
- a two-stage actively controlled platform, designed to suppress the seismic noise by $\sim 10^{-3}$ at 10 Hz;
- a quadruple pendulum passive isolation system providing passive isolation above a few Hz.

1.5.4 Monolithic Suspensions

The Virgo low frequency sensitivity was limited by the thermal noise of the steel wires suspending the mirrors. To beat that limit both Advanced Virgo and Advanced LIGO will use *monolithic suspensions*, that is, to suspend the test masses by means of fused silica fibers. The intrinsic dissipation of fused silica is a factor $\sim$1,000 better than steel, so there is a large potential improvement of the corresponding thermal noise. The pioneering work on this technology was pursued at GEO600 (see [5] and references therein), which has been using *monolithic suspensions* for years. Their design was not pushed to the level required by advanced detectors (for what concerns thickness of the fibers and weight of the mirrors). However, this has triggered a long R&D work that finally allowed to engineer monolithic payloads suitable for advanced detectors.

A further step forward has been achieved when Virgo built monolithic payloads with 21 kg standard mirrors and operated them successfully on the interferometer [33] (see Fig. 1.8). For the first time the *first generation* suspension thermal noise limit was beaten. The realization of monolithic payloads has required several important technological achievements:

Fig. 1.9 The Advanced LIGO quadruple pendulum

- the production of the fibers: the fibers are produced starting from a fused silica rod. The rod is heated in the central part by a CO_2 laser and its two ends are then pulled away in a controlled way (the speed profile determines the shape of the fiber) [34].
- the handling of the fibers: monolithic payloads have proven to be very robust in the usual operation on the interferometers. However, the handling of the fibers must be very careful: any contact of the fiber with an external body, no matter how gentle and soft, will cause microcracks on the fiber surface that will lead to its breaking.
- the attachment of the fiber to the mirror: this is achieved in two steps. First an intermediate fused silica block must be attached to the mirror side. This is done using a hydroxy-catalyst bond (better know as *silicate bonding*), a chemical procedure able to joint the parts in a low loss way (unlike using ordinary glues) [28, 29]. The second step is the welding of the silica fiber to the intermediate block, which is done again using a CO_2 laser.

The technology has been further refined for the advanced detectors. Advanced LIGO will use a quadruple pendulum where the last stage (test mass) and the penultimate stage are made in silica and connected through silica fibers (Fig. 1.9).

1.5.5 Laser

New lasers are needed for the advanced detectors which plan to improve the high frequency sensitivity by reducing the laser shot noise. The nominal power of such laser is 200 W. The lasers for GW detectors must satisfy stringent requirements:

- continuous, single frequency, single mode;
- high power (10–20 W for the first generation, 200 W for the advanced detectors);
- low noise (frequency, power, geometry);
- long lifetime;
- good efficiency.

In the early phases of interferometer design Argon-ion lasers emitting at 514 nm had been considered and used on prototypes (see for example [35]). Their performance was not satisfactory, mostly for efficiency and reliability. They were phased out as the Nd:YAG amplifiers were developed. Virgo chose the 1,064 nm wavelength first and in 1989 the first 18 W injection locked Nd:YAG amplifier was developed [36–39].

Eventually, all the first-generation detectors were equipped with infrared sources at 1,064 nm (Nd:YAG or Nd:YVO) providing power of 10–20 W. The advanced detectors have been designed assuming an available power of $\sim$200 W. A solid-state laser using a three-stage injection locked oscillator scheme has been developed at the Albert Einstein Institute and Laser Zentrum Hannover and will be used in Advanced LIGO [40, 41].

A different technology has been chosen for Advanced Virgo: a fiber amplifier. Fiber lasers are more efficient (and so it is easier to manage thermal effects), more compact, tunable in power, and have very good performance in terms of noise and stability. They are cheaper than solid-state laser and show great potential for extrapolation to higher power. The development of such amplifier is still in progress. In fact, Advanced Virgo will start its operation with the old solid-state laser of Virgo+, which can provide up to 60 W, more than enough for the first phase of the operation. Implementation of the 200 W fiber laser is foreseen in 2017 or 2018 (see Chap. 4).

1.5.6 Thermal Compensation

Heat absorption in the mirror bulk and in the coatings can induce both geometrical distortions (depending on the thermal expansion coefficient) and purely optical ones (through the temperature dependence of the refraction index). Such *thermal lensing* causes aberrations of the beams and proper thermal compensation systems are needed to cope with it. When the first generation of interferometers was being designed nobody foresaw to introduce actuators to cope with thermal effects. While it was well known that the heat absorption in the mirrors could be a source of thermally induced aberrations [42], their effect was expected to be negligible. During the commissioning it was found that the impact of thermal effects on the mirrors could be large enough to degrade the performance of the interferometers already at low power. Therefore, techniques to manage such effects had to be implemented. On the other hand, it was also understood that *thermal actuators* could be also used to correct for optics imperfections or to fine-tune the detector. GEO600 installed ring heaters around the tests masses to tune the radius of curvatures and reduce the aberrations observed [43]. LIGO used a flexible thermal compensation system using a CO_2 laser beam to deposit heat on the test mass [44]. A similar system was used in Virgo and was successfully used to increase the power in the interferometer.

In the second-generation detectors, the impact of such effects will be much larger due to the higher power envisaged and the design of the thermal compensation system is rather sophisticated. First of all, it is necessary to act independently to correct the thermal effects both in the Fabry–Perot cavities and in the power/signal recycling cavities.

- The main effect in the Fabry–Perot cavities is associated to the heat absorption in the high reflective coatings which causes a change in the radius of curvature of the test masses via thermoelastic deformation. This will be corrected by using ring heaters around each test mass.
- The input test masses heat up in the center due to absorption of laser power. Since the refraction index is temperature-dependent this effect causes the optical path to be not uniform, which is equivalent to having a spurious lens in the interferometer. This can be corrected by heating up the borders of the mirror in such a way as to null the thermal lensing. In LIGO and Virgo, this has been done by projecting on the mirror a CO_2 laser beam in annulus shape, obtained by an axicon. In Advanced LIGO/Virgo a similar approach will be followed. However, given the increase of power to be managed and the better sensitivity, it will not be possible to shine the CO_2 laser directly on the test masses: the amplitude noise of the CO_2 laser would spoil the detector sensitivity. To solve this problem a further optics, named "compensation plate" will be suspended and the CO_2 laser will be shone on it. Such plate will stay in the recycling cavity, thus its movement is not directly coupled to the dark fringe.

The thermal compensation system described so far can only cope with deformations with cylindrical symmetry. In some cases, non- axisymmetric deformations can play a relevant role.[1] Thus, a thermal compensation system for the non-axisymmetric part should as well be designed. The technology chosen for Advanced Virgo is a scanning laser, first developed by Ryan Lawrence in his PhD thesis [45].

The thermal compensation system needs suitable sensing: Hartmann sensors [46] and phase cameras will be used. The Hartmann sensors are able to measure the thermal deformations, while the phase cameras are suitable to measure the intrinsic optical defects (such as the inhomogeneity of the substrates crossed by the laser

[1] This is expected in Advanced Virgo, which is designed with so-called *marginally stable recycling cavities*. With such a design the detector is more sensitive to aberrations induced by thermal effects and optical defects. For instance, an inhomogeneous pattern of the refraction index of the mirrors, which can be non-axisymmetric, can cause important aberrations and must be compensated for.

beam). The combination of the information by the sensors and a proper combination of the different actuator should be sufficient to approach an *aberration free interferometer* (in practice, to reduce the aberrations to a tolerable level). This topic is widely discussed in Chap. 9.

1.5.7 Squeezing

Shot noise limits the sensitivity at high frequency. In principle, the possibility to improve it is associated to the progress of the laser technology and the available laser power. In fact, already in the first generation, thermally induced aberrations have been limiting the possibility to increase the power inside the interferometer and thermal compensation systems capable of correcting such aberrations have been implemented. The advanced detectors are designed for an available laser power ten times larger and more sophisticated thermal compensation.

However, another possibility to improve high frequency sensitivity without increasing the thermal load has been put forward by recent experiments: the injection of squeezed light [47] in the interferometer (better known as *squeezing*). Sources of squeezed light have been realized and progressively improved and a gain of 13 dB below the shot noise level has recently been achieved [48].

An experiment of squeezing on a real interferometric detector has been done first by GEO600, achieving promising results. A broadband noise reduction of up to 3.5 dB has been achieved in the shot noise-limited frequency range (Fig. 1.10). This results opens the way to implementing squeezing in the advanced detectors. Chapter 11 is dedicated to this topic.

1.6 Third Generation: The Einstein Telescope

Advanced detectors at their best sensitivity promise to detect tens of events per year. However, the expected signal-to-noise ratios might be insufficient for accurate astronomical and astrophysical studies. Therefore, a natural question arised in the GW community: could we conceive a third-generation detector, able to enhance the sensitivity of the advanced detectors by an order of magnitude and further extend the low frequency bandwidth toward 1 Hz? A large part of the European GW community participated in a design study, funded by the EU within the Framework Program 7, and came up with the proposal of the Einstein Telescope, a large underground and cryogenic interferometer aiming to reach that sensitivity goal.

The sensitivity goals of ET cannot be met within the current infrastructure. On ground, the gravity gradient noise, generated by the fluctuations of the local gravity fields due to seismic oscillations of the ground, couples directly to the test masses and short-circuits the vibration isolators. The only way to suppress such noise is to

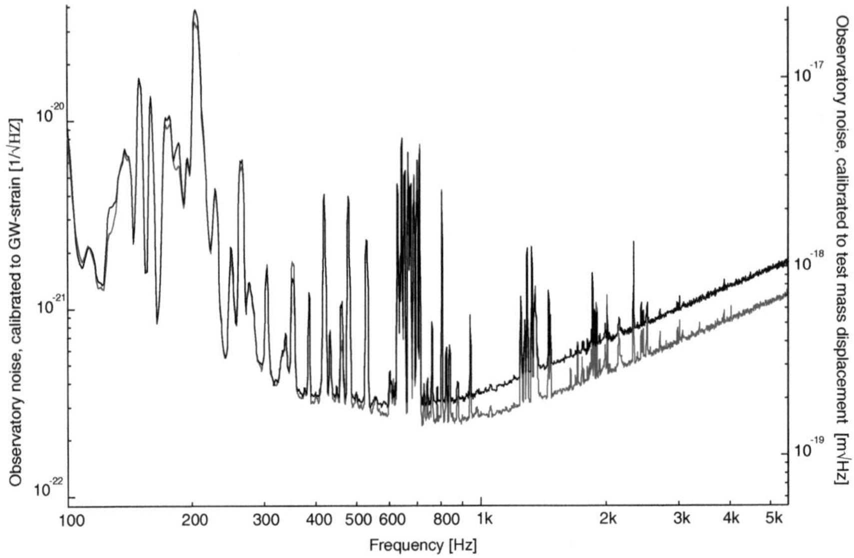

Fig. 1.10 Nonclassical reduction of the GEO600 instrumental noise using squeezed vacuum states of light. Figure from LIGO Scientific Collaboration [49]

go underground, where seismic activity is orders of magnitude lower than on the surface. The following features are envisaged [50]:

- 10 km armlength, underground detector;
- very long vibration isolators, able to push the *seismic wall*[2] down to 1 Hz;
- cryogenic test masses for thermal noise suppression;
- non-gaussian beam profiles, for thermal noise and thermal lensing suppression;
- high power: ~1 kW laser, high finesse Fabry–Perot arm cavities, high power recycling factor;
- signal recycling and squeezed states injection;
- heavy test masses for suppression of radiation pressure noise.

The observatory will have the shape of an equilateral triangle with 10 km long side (thus, 30 km of tunnel should be excavated). Such infrastructure would host three co-located interferometers.

The science reach of such an instrument would be formidable [51]:

- the high SNR would allow to investigate with great details the physics of compact stars (e.g., constraining the equation of state of neutron stars);
- the details of the waveforms would allow to constrain alternative metric theories of gravity;

[2] Vibration isolators act as very steep low-pass filters. Therefore, the residual seismic noise transmitted to the mirrors has a spectrum that behaves as $\sim 1/f^N$ with $N \gg 1$, looking like a "wall" limiting the sensitivity at very low frequencies.

- cosmological parameters could be measured using standard sirens[3];
- detection of GW and neutrino from a supernova event would allow to constrain the graviton and neutrino masses.

Chapter 13 is dedicated to the ET project while Chap. 14 describes the cryogenics aspects.

References

1. A. Einstein, Ann. der Phys. **49**, 769 (1916)
2. B. Abbot et al., The LIGO Scientific Collaboration. Reports on Progress in Physics **72**, 076901 (2009)
3. F. Acernese et al., The Virgo Collaboration. Class. Quantum Gravity **25**, 114045 (2008)
4. M. Cerdonio, G. Losurdo, La, Rivista del Nuovo Cimento **8**, 389–480 (2012)
5. H. Grote et al., The GEO600 Collaboration. Class. Quantum Gravity **25**, 114043 (2008)
6. Y. Aso et al., The KAGRA Collaboration. Phys. Rev **D88**, 043007 (2011)
7. J. Abadie et al., The LIGO Scientific Collaboration and The Virgo Collaboration. Phys. Rev. **D81**, 102001 (2010)
8. The LIGO Scientific Collaboration and The Virgo Collaboration, Nature **460**, 990 (2009)
9. B.P. Abbott et al., The LIGO Scientific Collaboration and The Virgo Collaboration. Ap. J. **713**, 671–685 (2010)
10. B.P. Abbott et al., The LIGO Scientific Collaboration and The Virgo Collaboration. Ap. J. **715**, 1438–1452 (2010)
11. J. Abadie et al., The LIGO Scientific Collaboration and The Virgo Collaboration. Ap. J. **715**, 1453–1461 (2010)
12. Aasi J, et al. The LIGO Collaboration and The Virgo Collaboration, arXiv:1304.0670v1 (2013), to be published on Living reviews of relativity.
13. F. Halzen, S.R. Klein, Rev. Sci. Instrum. **81**, 081101 (2010)
14. U.F. Katz, The KM3NET consortium, Nucl. Instrum. Meth. A, S57, 626–627
15. G.M. Harry, The LIGO Scientific Collaboration. Class. Quantum Gravity **27**, 084006 (2010)
16. J. Abadie et al., The LIGO Collaboration and The Virgo Collaboration. Class. Quantum Gravity **27**, 173001 (2010)
17. S. Fairhurst, Class. Quantum Gravity **28**, 105021 (2011)
18. T. Accadia et al., The Virgo Collaboration, *Proceedings of the MG12 World Scientific Conference*, (2012)
19. The Virgo Collaboration Virgo Internal Report VIR-0128A-12 (2012) https://tds.ego-gw.it/itf/tds/file.php?callFile=VIR-0128A-12.pdf
20. T. Uchiyama et al., Phys. Rev. Lett. **108**, 141101 (2012)
21. B.J. Meers, Phys. Rev. **D38**, 2317 (1998)
22. B. Cimma et al., Appl. Opt. **45**, 1436 (2006)
23. I.W. Martin et al., Class. Quantum Gravity **25**, 055005 (2008)
24. S. Braccini et al., The Virgo Collaboration. Astrop. Phys. **23**, 557 (2005)
25. G. Losurdo et al., Rev. Sci. Instrum. **70**, 2508 (1999)
26. G. Losurdo et al., Rev. Sci. Instrum. **72**, 3653 (2001)
27. F. Acernese et al., The Virgo Collaboration. Astrop. Phys. **20**, 629 (2004)
28. S. Rowan et al., Phys. Lett. A **233**, 303 (1997)
29. L. Cunningham et al., Phys. Lett. A **374**, 3993 (2010)
30. F. Acernese et al., The Virgo Collaboration. Astrop. Phys. **33**, 182 (2010)

[3] GW analog of the standard candles.

31. J. Giaime et al., Rev. Sci. Instrum. **67**, 208 (1999)
32. B.P. Abbott et al., Class. Quantum Gravity **19**, 1591 (2002)
33. M. Lorenzini, The Virgo Collaboration. Class. Quantum Gravity **27**, 084021 (2010)
34. A. Heptonstall et al., LIGO Internal Report, (2005) http://www.ligo.caltech.edu/docs/T/T050206-00.pdf
35. D.H. Shoemaker et al., Phys. Rev. **D38**, 423 (1988)
36. D.H. Shoemaker, A. Brillet, C.A. Man, O. Cregut, Opt. Lett. **14**, 609 (1989)
37. O. Cregut et al., Phys. Lett. A **140**, 284 (1989)
38. F. Bondu, P. Fritschel, C.A. Man, A. Brillet, Opt. Lett., **21**, 582 (1996)
39. R. Barillet et al., Meas. Sci. Tech. **7**, 162 (1996)
40. B. Willke et al., Class. Quantum Gravity **25**, 114040 (2008)
41. L. Winkellman et al., Appl. Phys. B **102**, 529 (2011)
42. P. Hello, J.Y. Vinet, Phys. Lett. A **178**, 351 (1993)
43. H. Lück et al., Class. Quantum Gravity **21**, S985 (2004)
44. R. Lawrence et al., Class. Quantum Gravity **19**, 1803 (2002)
45. R. Lawrence, Active wavefront correction in laser interferometric gravitational wave detectors, Ph.D thesis, MIT (2003) http://hdl.handle.net/1721.1/29308
46. T.L. Kelly et al., Appl. Opt. **46**, 861 (2007)
47. D.F. Walls, Nature **306**, 141 (1983)
48. T. Eberle et al., Phys. Rev. Lett. **104**, 251102 (2010)
49. LIGO Scientific Collaboration, Nature Phys. (2011) doi:10.1038/nphys2083
50. M. Punturo et al., Class. Quantum Gravity **27**, 084007 (2010)
51. M. Abernathy et al. The ET Science Team, Einstein gravitational wave telescope–conceptual design study, (2011) https://tds.ego-gw.it/itf/tds/file.php?callFile=ET-0106C-10.pdf

Chapter 2
The Science Case for Advanced Gravitational Wave Detectors

Andrea Viceré

Abstract We provide an introduction to the science case for advanced interferometric gravitational wave detectors, which are currently being realized in the world and in the second half of the present decade will achieve a sensitivity ten times larger than first generation instruments. Starting from the results obtained with first generation instruments, this paper will motivate the experimental effort and will provide an introduction to the science that will be harvested by second generation detectors.

2.1 Introduction

The first-generation, km-scale interferometric gravitational wave detectors completed their science runs in 2011, having achieved the design sensitivity and having collected years of observation: no signal was detected in data published so far, and the LIGO Scientific Collaboration (LSC) [1] and the Virgo Collaboration [2] could set upper limits on several of the possible Gravitational Wave (GW) sources. On this solid basis, we now look forward to the second-generation instruments and to the science that will be harvested thanks to their increased sensitivity, starting in 2015 when the detectors will progressively come back online [3].

Before looking forward, it is important to underline the success of first-generation interferometers: when the construction of LIGO [4], Virgo [5] and other large interferometers began in the 1990s, reaching the declared sensitivity goals seemed a formidable challenge in face of the many noises which would plague the instruments. Although each of these noises had been estimated and appropriate countermeasures had been devised, it still seemed incredible that a mirror could be located in space with an accuracy better than 10^{-18} m, and over a wide frequency range.

A. Viceré (✉)
Dipartimento di Scienze di Base e Fondamenti, Via S.Chiara 27, I-61029 Urbino, Italy
e-mail: andrea.vicere@uniurb.it

M. Bassan (ed.), *Advanced Interferometers and the Search for Gravitational Waves*, Astrophysics and Space Science Library 404, DOI: 10.1007/978-3-319-03792-9_2,

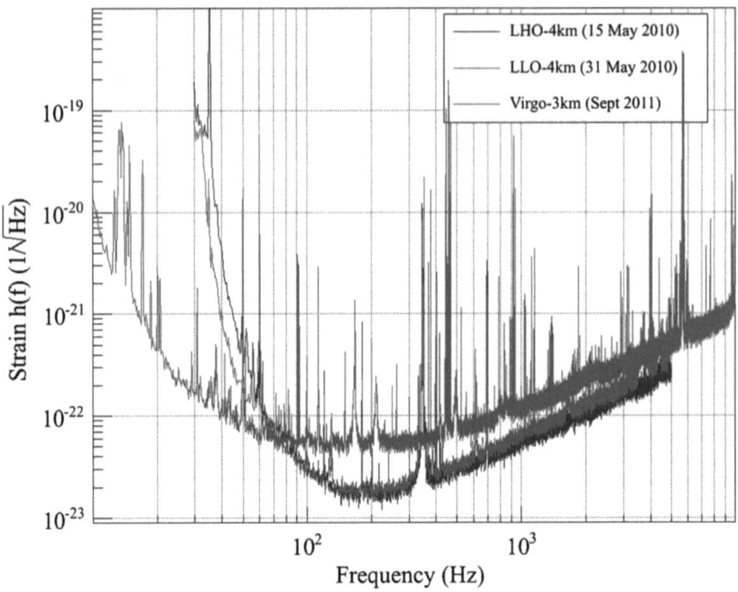

Fig. 2.1 Results from the first generation large interferometric detectors. The two almost superposing curves are representative of the sensitivities achieved by the LIGO detectors at Hanford (LHO) and at Livingston (LLO), USA; the third curve is instead representative of the sensitivity of the Virgo detector at Cascina, Italy. Note that LIGO detectors are more sensitive above 70Hz, whereas Virgo is better at lower frequencies, thanks to its advanced seismic isolation. See Eq. 2.9 for the definition of the linear spectral density of h, displayed in this figure

Despite legitimate doubts, the first-generation goals have been reached and even surpassed: for instance we show in Fig. 2.1 representative sensitivities achieved during the latest runs of the LIGO and Virgo instruments, which could be compared with design sensitivities, and found consistent [6]. The curves show the linear spectral density of the metric deformation $h = g - \eta$ that would be needed to produce in the detector the same level of signal as the one observed.

Such sensitivities allowed to achieve an "inspiral range" of about 20 Mpc in first-generation LIGO detectors: this is the distance at which a binary coalescence of two neutron stars could be seen by the detectors with a signal-to-noise ratio (SNR) of 8, after averaging over the directions of the source and polarizations.[1]

An instrument devoted to astrophysical observations has to achieve not just its design sensitivity, but has also to deliver a good duty cycle: as an example, during its second science run (VSR2), between July 2009 and January 2010, the Virgo interferometer was in a locked state for more than 85 % of time, and could record science data over 80 % of the run [7].

[1] At the same distance, an optimally located and oriented source, namely at the zenith or nadir of the detector, could be seen with an SNR as high as 18.

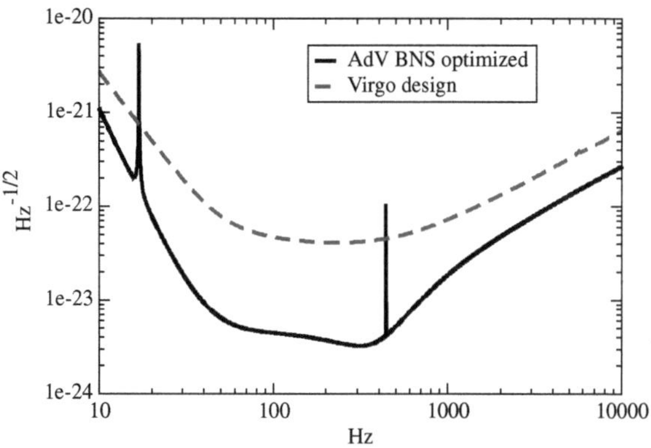

Fig. 2.2 Comparison of AdV sensitivity with the design sensitivity of Virgo. The "BNS optimized" curve is just a representative of a range of possible sensitivity tunings

The advanced gravitational detectors, like Advanced LIGO (aLIGO in the following) [8] and Advanced Virgo (AdV in the following) [9], are designed to improve the amplitude sensitivity by a factor of about 10 over a large fraction of the bandwidth: as an example, we show in Fig. 2.2 a comparison of AdV and Virgo design sensitivities [9]. Note that the AdV sensitivity will be tunable, allowing to optimize the detection probability for different sources: in this plot we show the so-called BNS optimized curve, tuned for Binary Neutron Star coalescences, which is a convenient benchmark. The sensitivity improvement will allow detecting transient sources about 10 times farther, thus increasing the chances of observation by a factor of about 10^3, due to the enlargement of the observed space volume. The same improvement will raise by a factor 100 the detector sensitivity to a continuous signal like the cosmological stochastic background. It is therefore one of the goals of this chapter to convince the reader that the advanced detector era (ADE) will be an interesting time!

In the rest of this chapter we will try to provide the scientific motivations for the realization of aLIGO and AdV, discussing the most interesting GW sources and what could be learned by detecting their signals or by placing even more stringent upper limits. For each of these sources we will also recapitulate the main results obtained, in terms of upper limits, thanks to first-generation instruments; this will help us understand how a sensitivity translates into an astrophysical statement. We will show that aLIGO and AdV will harvest their scientific potential thanks to a close collaboration: their instruments constitute a long baseline network of observatories, capable to coherently exploit the time, amplitude, and phase information coming from the different sites, thus reconstructing the source position, the polarization of the impinging waves, and ultimately giving access to the physics of the sources. We will also see that further information, coming from electromagnetic and neutrino observatories, will be important and even in some cases critical in order to fully harvest the scientific potential of a GW observation: to this purpose, a wide program of collaboration with the astronomical community has been pioneered with first-generation detectors, and will be fully developed with second-generation ones.

Several very good and extensive reviews on possible astrophysical sources for gravitational waves already exist [10–15]: we could refer interested readers particularly to [13] for sources for ground-based interferometers.

2.2 Generalities

Before discussing the individual sources and their interest for advanced detectors, it is convenient to introduce some conventions and notations which will be used in the following.

We recall that an interferometric detector produces a single time series $s(t)$, composed of a signal $h(t)$, proportional to the metric deformation induced by the GW, and of a background noise $n(t)$. We assume that the instrument transfer function has already been compensated for [7], so that $h(t)$ is linearly related to the metric change $h_{\text{ab}}(t)$, as follows: assuming that the interferometric detector arms are oriented along the unit vectors $\hat{X}$, $\hat{Y}$, the output is normalized to have

$$h(t) \equiv h_{\text{ab}}(t)\frac{1}{2}\left(\hat{X}^a\hat{X}^b - \hat{Y}^a\hat{Y}^b\right) = \frac{1}{2}Tr\left[\mathbf{h}\cdot(\mathbf{X}\otimes\mathbf{X} - \mathbf{Y}\otimes\mathbf{Y})\right]\,; \tag{2.1}$$

the detector moves in space, and the vectors $\hat{X}$, $\hat{Y}$ rotate with time, modulating the response to the metric h_{ab}; this effect can be disregarded if the timescale of the signal is much shorter than the rate of variation of the detector orientation, a condition true for ground-based detectors when considering transient signals, but false when dealing with permanent signals like those emitted by distorted neutron stars.

Expanding the signal in series of plane waves, one can write

$$h_{\text{ab}}(t, \mathbf{x}) = \sum_{A=+,\times} \int_{-\infty}^{+\infty} \mathrm{d}f \int_{S^2} \mathrm{d}\hat{\Omega}\, h_A(f, \hat{\Omega}) e^{i2\pi f(t-\hat{\Omega}\cdot\mathbf{x}/c)} \varepsilon^A_{\text{ab}}(\hat{\Omega}) \tag{2.2}$$

as a sum over frequencies f, directions $\hat{\Omega}$ and polarizations $\boldsymbol{\varepsilon}^A$. For instance, considering a plane wave traveling along $\hat{\Omega}$ and taking a reference frame in which axes $\hat{x}$, $\hat{y}$ are orthogonal to $\hat{\Omega} = \hat{z}$, one has[2]

$$\mathbf{h} = \begin{pmatrix} 0 & 0 & 0 & 0 \\ 0 & h_+ & h_\times & 0 \\ 0 & h_\times & -h_+ & 0 \\ 0 & 0 & 0 & 0 \end{pmatrix} = h_+\boldsymbol{\varepsilon}^+ + h_\times\boldsymbol{\varepsilon}^\times\,. \tag{2.3}$$

More generally, the response of the detector, located at $\mathbf{d}(t)$, will be a sum

[2] We use 4-vectors specified as (ct, x, y, z).

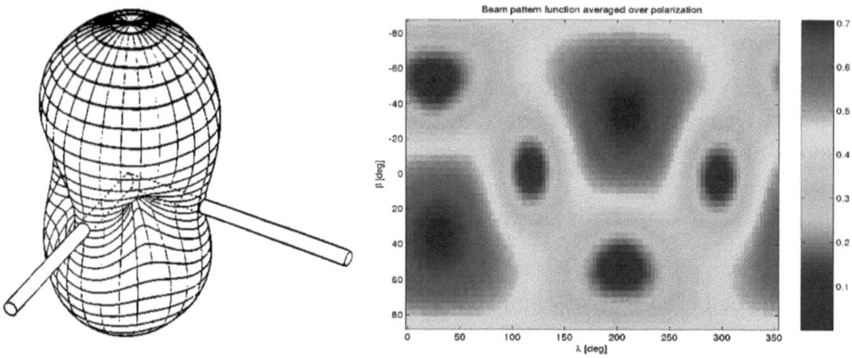

Fig. 2.3 The antenna pattern of a interferometric detector, averaged over the possible signal polarizations. *Left* the two orthogonal beams represent the arms of the detector, and the distance between a point on the surface and the intersection of the arms represents the quantity in Eq. 2.6. *Right* the same quantity is represented as a pattern of colors on the sky

$$h(t) = \sum_{A=+,\times} \int_{-\infty}^{+\infty} \mathrm{d}f \int_{S^2} \mathrm{d}\hat{\Omega}\, h_A(f, \hat{\Omega}) e^{i2\pi f (t - \hat{\Omega}\cdot\mathbf{d}/c)} F^A(\hat{\Omega}) \tag{2.4}$$

in which the antenna patterns $F^{+,\times}(\hat{\Omega})$ are introduced:

$$F^A(\hat{\Omega}) \equiv \sum_{\mathrm{ab}} \varepsilon^A_{\mathrm{ab}}(\hat{\Omega}) \frac{1}{2} \left(\hat{X}^a \hat{X}^b - \hat{Y}^a \hat{Y}^b \right) = \frac{1}{2} Tr \left[\boldsymbol{\varepsilon}^A \cdot (\mathbf{X} \otimes \mathbf{X} - \mathbf{Y} \otimes \mathbf{Y}) \right]; \tag{2.5}$$

they depend on the sky location $\hat{\Omega}$ and specify how a GW signal of given polarization translates into the output of the detector.

The average response of the detector to different polarizations, as a function of the sky location $\hat{\Omega}$, can be defined as

$$\bar{F}(\hat{\Omega}) = \sqrt{\left[F^+(\hat{\Omega})\right]^2 + \left[F^\times(\hat{\Omega})\right]^2} \tag{2.6}$$

and its dependence on $\hat{\Omega}$ is shown in Fig. 2.3: we note the limited directionality, with favored locations and a few blind spots.

Note that if the output of the detector is normalized (a procedure called h-reconstruction) so as to have the relation in Eq. 2.1 between the metric and the output, then different detectors can compare their data directly, without knowing instrument details, apart from position and orientation of the arms.

We have already said that the signal $s(t)$ output by a detector is the sum of a gravitational signal $h(t)$ and a noise $n(t)$: the detector sensitivity $\tilde{h}(f)$ is actually defined in terms of the spectral density of this noise. Assuming that the noise is Gaussian and stationary, its autocorrelation function

$$C_n(t_a - t_b) = \langle n(t_a)n(t_b) \rangle, \tag{2.7}$$

depends only on the time difference of the noise samples, and determines the noise statistical characteristics completely: its Fourier transform allows to define the (one-sided) noise spectral density $S_n(f)$ as

$$\frac{1}{2}S_n(f) = \int e^{i2\pi f t} C_n(t) \mathrm{d}t \tag{2.8}$$

and the detector sensitivity is simply defined as

$$\tilde{h}(f) \equiv \sqrt{S_n(f)}; \tag{2.9}$$

This is the quantity which has been displayed, for instance, in Fig. 2.1, for first-generation detectors.

2.3 Transients

We define transients as those events which have a duration significantly shorter than the observation time: for ground-based detectors, such events include a variety of phenomena, of which the coalescence of compact objects is perhaps the best understood, and will deserve a specific treatment. Many other phenomena are also relevant, most notably the core-collapse of supernovae, and the emission of transient signals by excited compact objects.

2.3.1 Compact Binary Coalescence

We define Compact Binary Coalescence (CBC) as the coalescence of binary systems formed by neutron stars (NS–NS), by black holes (BH–BH), or mixed systems formed by neutron stars and black holes (NS–BH); although only the existence of NS–NS systems is observationally verified (see for instance [16] for a review), there are good arguments to anticipate the existence also of systems in other classes.

2.3.1.1 The Signal from a Coalescence

The signal emitted during a coalescence is conventionally divided into three phases, better illustrated pictorially by the signal shown in Fig. 2.4, emitted by a pair of BH: we have an initial "inspiral" phase, during which the orbit shrinks adiabatically and the signal is a chirp with amplitude and frequency increasing with time; a "merger" phase, during which the objects collide, generally the most difficult phase to model

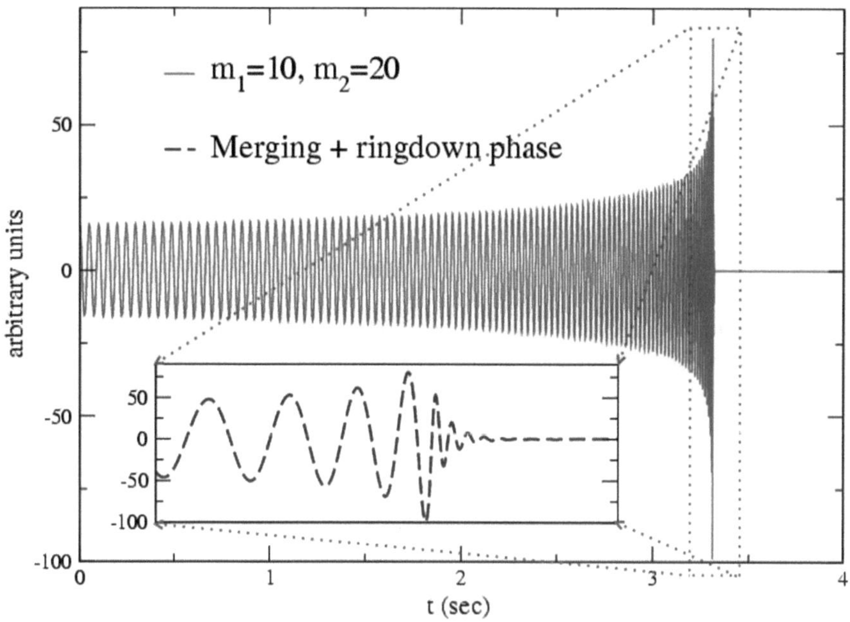

Fig. 2.4 The final seconds of the signal emitted by a pair of (non-spinning) black-holes (10+20 $M_\odot$) in coalescence

(but also perhaps the most interesting); finally a "ring-down" phase during which the resulting object relaxes toward a stable state, emitting damped sinusoids. The relative importance of the three phases, as seen by a ground-based detector, depends on the system's mass: focusing first on binary neutron stars, whose existence is attested by the observation of pulsars in binary systems and for which mass parameters have been estimated to be close to 1.4 $M_\odot$, the inspiral phase is by far the dominant one, and the emission of energy in GW's is given by

$$\Delta E_{\rm rad} \simeq 4.2\times 10^{-2} M_\odot c^2 \left(\frac{\mathscr{M}}{1.22\,M_\odot}\right)^{5/3} \left(\frac{f_{\rm max}}{1\,{\rm kHz}}\right)^{2/3}, \qquad (2.10)$$

where we have introduced the chirp mass $\mathscr{M} \equiv \mu^{3/5} M^{2/5}$, in terms of the total mass $M \equiv m_1 + m_2$ and the reduced mass $\mu \equiv m_1 m_2/M$ of the binary system. We introduce as well $f_{\rm max}$, the GW frequency at the end of the inspiral phase.

It is therefore a rather efficient mechanism, which makes it possible to see such NS–NS signals at large distances, as far as about 200 Mpc in aLIGO [17]. For larger mass systems the SNR grows approximately as $M^{5/6}$, thus allowing to detect events further out, as far as 1 Gpc for BH–BH systems of moderate mass: however the $f_{\rm max}$ frequency becomes smaller as the mass grows, and eventually the inspiral signal falls below the low-frequency cutoff due to the detector's seismic and thermal noise, so the SNR becomes dominated by the merger and the ring-down phases.

The waveform at the inspiral phase is fairly well understood theoretically, and its dominant features can be grasped even using just Newtonian mechanics, complemented with the knowledge that the dominant contribution to the GW emission by a distribution of mass $\rho(\mathbf{x})$, at a distance R by the observer, is given by the time variation of its quadrupole component

$$h_{ij} = -\frac{2G}{c^4 R}\frac{\mathrm{d}^2}{\mathrm{d}t^2}\int \rho(\mathbf{x})x_i x_j \mathrm{d}V = -\frac{2c}{R}\frac{\mathrm{d}^2}{\mathrm{d}t^2}\int \rho(\mathbf{x})x_i x_j \mathrm{d}V \Big/ \left(\frac{c^5}{G}\right) \tag{2.11}$$

where the factor $c^5/G = 3.63 \times 10^{52}$ W $= 2.03 \times 10^5\ M_\odot c^2/s \equiv L_o$ is a luminosity of 10^5 solar masses per second! Stellar size distributions of masses are required to generate sizable signals.

Assuming for simplicity a pair of objects with equal mass m, in a circular orbit characterized by an angular frequency ω and a semi-axis a, the equivalence of gravitational and centripetal forces translates into Kepler's law $\omega^2 = Gm/4a^3$, which yields

$$h_{ij} \propto \frac{2\,(Gm)^2}{c^4 a\ R} = \frac{(2Gm)^{5/3}}{c^4 R}\omega^{2/3} \tag{2.12}$$

hence the signal increases if the objects' separation a decreases; and this in fact happens, because the mechanical energy

$$E = -Gm^2/4a \tag{2.13}$$

of the system decreases by radiation in gravitational waves. In linearized general relativity, an energy momentum $t_{\mu\nu}$ can be associated to the GW, and the source luminosity is

$$L_{\mathrm{GW}} \equiv -\frac{\mathrm{d}E}{\mathrm{d}t} = \frac{G}{5c^5}\left\langle \dddot{Q}_{ij}\dddot{Q}^{ij}\right\rangle \text{with } Q_{ij} = \int \rho\left(x_i x_j - \frac{1}{3}\delta_{ij}x^k x^k\right)\mathrm{d}V \tag{2.14}$$

which, for a binary system in circular orbit, gives

$$L_{\mathrm{GW}} = \frac{128}{5}\frac{G}{c^5}\left(ma^2\omega^3\right)^2 = \frac{32G^{7/3}}{5\sqrt[3]{4}c^5}(m\omega)^{10/3}\,. \tag{2.15}$$

Comparing with Eq. 2.13 and using again Kepler's law, it is straightforward to compute the time evolution of ω

$$\omega(t) = \omega_0\left[1 - \frac{256\,G^{5/3}\mu\,M^{2/3}}{5\,c^5\omega_0^{-8/3}}t\right]^{-3/8}, \tag{2.16}$$

which shows the chirping behavior of the signal. Note that after a time τ

$$\tau = \frac{5c^5}{256 G^{5/3} \mu\, M^{2/3} \omega_0^{8/3}} \tag{2.17}$$

the angular frequency goes to infinity, signaling a breakdown of the approximation, close to the collision. Integrating the expression for $\omega(t)$ we find the phase and instantaneous frequency of the signal

$$\phi(t) = \frac{16\pi \nu_0 \tau(\nu_0)}{11}\left[1 - \left(1 - \frac{t}{\tau(\nu_0)}\right)^{5/8}\right]; \ \nu(t) = \nu_0 \left(1 - \frac{t}{\tau(\nu_0)}\right)^{-3/8}, \tag{2.18}$$

in which $\nu_0 = 2\omega_0$ is the GW frequency (twice the orbital frequency) at $t = 0$ and $\tau(\nu_0)$ is the time to coalescence:

$$\tau(\nu_0) \equiv \frac{5}{256} \frac{c^5}{G^{5/3}} \frac{(\pi \nu_0)^{-8/3}}{\mathcal{M}^{5/3}}. \tag{2.19}$$

This simplified treatment tells nothing about the merger and ring-down phase: a more complex analysis is required to compute accurately enough the signal, even in the inspiral phase (see [18] for a review). Still, theorists are able to model the emitted GW signal in an accurate way: as long as the compact objects, orbiting each other at a distance d, are sufficiently separated, say up to $(dc^2)/(MG) > 10^3$, post-Newtonian expansions allow to precisely predict the waveform. The analytical methods fail when the post-Newtonian expansion (in powers of v/c), fails, and this is an issue particularly for large mass systems, as well as for neutron stars when they start merging, but significant progress with numerical relativity [19] is enabling to model the signal through the last stages of the collision and down to the relaxation processes which take place in the compact resulting system (generally a black-hole).

This said, many characteristics of the signal can be deduced on the basis of these simple formulas: in particular, it is immediate to estimate the inspiral's phase duration in a certain frequency band $[\nu_l, \nu_u]$ as

$$t_u - t_l = \tau(\nu_0)\left[\left(\frac{\nu_l}{\nu_0}\right)^{-8/3} - \left(\frac{\nu_u}{\nu_0}\right)^{-8/3}\right]: \tag{2.20}$$

Extending the sensitivity toward lower frequencies ($\nu_l \rightarrow 0$) increases the signal's observable duration faster than $1/\nu_l^2$, which motivates the quest for sensitive detectors at low frequencies.

Coalescences therefore appear on ground-based detectors as transients just because of the restricted detectors' bandwidth: we start observing only above a few tens of Hz in first-generation detectors, and above 10 Hz in advanced detectors. For more massive sources (larger $\mathcal{M}$), the observation time becomes shorter, to the point that large mass binaries are just not accessible to observation by ground-based laboratories.

2.3.1.2 The Rate of NS–NS Signals

There are basically two ways by which astrophysicists can predict the number of NS–NS events: empirical, based on a few observations of compact binary systems which will coalesce on a timescale comparable with the age of the Universe; and theoretical, based on models of binary star formation and evolution.

For an account of the results of the empirical method, the reader could refer to [20]; on the basis of the observations, corrected for selection factors linked to the fact that we only see binary systems in which at least one component is a visible pulsar, the authors quote a rate range $\mathscr{R}_{\text{emp}} \in 3\text{–}190\,\text{Myr}^{-1}\ \text{MWEG}^{-1}$, where MWEG stands for Milky-Way Equivalent Galaxy.

The theoretical models display different cases, depending on assumptions about the dominant channel by which a pair of massive stars evolves into a pair of close compact objects, surviving two supernova explosions without being disrupted, and whether during a common envelope phase[3] the binary system undergoes instead a premature merger as a result of friction.

The reader is referred to [21] for an example of the comparison of different models, and the resulting recent re-estimate of the rates; the authors introduce several models, two of which are more pessimistic, since they assume that binaries in a common envelope phase are more likely to undergo a premature merger; the third model instead allows for survival of stars through the common envelope phase and leads to larger estimates.

The range of variability of the results within each theoretical model depends also on the way they are calibrated, either by means of assumptions about the star formation rate, or by means of the rate of supernova events in a Milky Way type galaxy; low rates are obtained with star formation calibration, high rates with supernova calibration.

The wide range of models available in the literature has been taken into account into a summary document published by LSC and Virgo [17]: the realistic NS–NS rate value is expected to be $\mathscr{R}_{\text{re}} \simeq 100\,\text{Myr}^{-1}\text{MWEG}^{-1}$, but models allow this figure to vary by orders of magnitudes above and below.

The other ingredient for estimating the rate of detectable events is the density of MWEG, which can be assumed to be

$$\rho_{\text{gal}} \simeq 0.01\,\text{Mpc}^{-3}\,; \tag{2.21}$$

for NS–NS events, the detection range is sufficiently close not to require correcting this density to take into account cosmological effects.

To which distance can advanced detectors see NS–NS signals? Conservatively, we rely only on the accurate knowledge of the waveform in the inspiral phase, which yields an SNR for an optimally oriented source

[3] A common envelope phase takes place when the larger of the two stars fills its Roche lobe and a mass transfer is initiated toward the companion.

$$\mathrm{SNR} = \sqrt{4\int_{\nu_l}^{\nu_u} \frac{|\tilde{s}(f)|^2}{S_n(f)} \mathrm{d}f} = \frac{(G\mathscr{M})^{5/6}}{r\,c^{3/2}\,\pi^{2/3}} \left(\frac{5}{6}\right)^{1/2} \sqrt{\int_{\nu_l}^{\nu_u} \frac{f^{-\frac{7}{3}}}{S_n(f)} \mathrm{d}f}$$
$$= 1.56 \times 10^{-19} \left[\frac{\mathscr{M}}{M_\odot}\right]^{5/6} \left[\frac{\mathrm{Mpc}}{r}\right] \sqrt{\int_{\nu_l}^{\nu_u} \frac{f^{-\frac{7}{3}}}{S_n(f)} \mathrm{d}f}\,, \tag{2.22}$$

where we recall that $S_n(f)$ is the detector one-sided spectral density, defined in Eq. 2.8, namely the square of the sensitivity as shown in Fig. 2.1. The actual value of the ν_u beyond which the merger begins is somewhat model dependent; a conservative estimate is given by the frequency at the "innermost stable circular orbit", ISCO. To lowest order in the PN expansion, this frequency can be written as

$$f_{\mathrm{ISCO}} \simeq \frac{4.4\,\mathrm{kHz}}{(M/M_\odot)}\,. \tag{2.23}$$

Given the SNR expression, we define the "inspiral range" d_{range} as the distance at which an advanced detector would see an event with SNR = 8, a value conventionally assumed as safe for claiming a detection.[4] This distance results to be

$$\frac{d_{\mathrm{range}}}{1\,\mathrm{Mpc}} = \frac{1}{2.26} \times 1.95 \times 10^{-20} \left[\frac{\mathscr{M}}{M_\odot}\right]^{5/6} \sqrt{\int_{\nu_l}^{f_{\mathrm{ISCO}}} \frac{f^{-\frac{7}{3}}}{S_n(f)} \mathrm{d}f}\,, \tag{2.24}$$

where the factor 1/2.26 accounts for an average over source location and polarization, leading to a reduction of the SNR as compared with a optimally oriented source.

As an example, if we assume an NS–NS event, with $m_1 = m_2 \simeq 1.4\,M_\odot$, one has $\mathscr{M} \simeq 1.22\,M_\odot$, and $f_{\mathrm{ISCO}} \simeq 1.4\,\mathrm{kHz}$; in AdV, the resulting sight distance is about 155 Mpc; the resulting volume observable by AdV alone is therefore $V_{\mathrm{BNS}} \simeq 1.5 \times 10^7\,\mathrm{Mpc}^3$, which allows to compute a rate through the simple formula

$$\rho = V_{\mathrm{BNS}} \times \mathscr{R}_{\mathrm{MW}} \times \rho_{\mathrm{gal}} \tag{2.25}$$

where $\mathscr{R}_{\mathrm{MW}}$ is the merger rate in a MWEG.

Actually, the operation in network of aLIGO and AdV further boosts the detection rate, by allowing a better coverage of the sky: again referring to [17] we simply quote that a realistic rate of 40 events/year is anticipated, to be compared with two events/century in first-generation instruments. Again, these numbers are to be taken with great care, and could be affected by very large errors.

---

[4] Actually, experience has taught that actual detectors may yield spuria well above this limit, depending on the quality of the detector noise; this is also the reason why a network of detectors in coincidence is required.

2.3.1.3 The Rate of BH–NS and BH–BH Coalescences

Again, there are essentially two ways by which a population of BH–BH or BH–NS systems can be inferred: it is possible to use the same population synthesis models applied to the NS–NS systems, and partially calibrated by the observation of binary pulsars, to infer the population of more massive stars; or one can exploit the few observations of binary systems including a massive star and a BH, to infer the rate of BH–BH production.

The population synthesis models, for instance the one discussed in [21], display for BH–BH the same variation of the results, internal within each model and across models that we have discussed for the NS–NS sources. The most pessimistic model predicts for BH–NS pairs galactic merger rates of 0.07–0.11 Myr^{-1}, whereas for BH–BH systems it predicts chirp masses in the range 4–8 $M_\odot$ and galactic merger rates in 0.02–0.03 Myr^{-1}. Optimistic models assume a more favorable evolution of the binary system through the common-envelope phase, and predict galactic merger rates in 3.2–4.8 Myr^{-1} for BH–NS sytems, and in 7.7–11 Myr^{-1} for BH–BH systems.

The large variability of the models calls for caution by itself, and population synthesis models are not extended easily to higher masses, because very large mass stars have a complex evolution, marked by explosive events and large mass losses over short periods. However, the few observations of very massive binary systems lead to believe that high mass BH pairs are actually formed and are detectable by advanced GW detectors.

In particular, the IC10 X-1 and NGC300 X-1 X-ray sources are tight binary systems comprising a 20–30 $M_\odot$ BH accreting from a massive $\sim$20 $M_\odot$ Wolf-Rayet star,[5] whose future evolution has been studied in-depth in [22]. The authors predict that the systems will evolve in BH pairs, with a chirp mass $\mathscr{M} \sim 15\,M_\odot$; systems with such a mass will be detectable by advanced detectors at distances beyond 1 Gpc. The other important prediction in [22] is that the systems will remain tight enough to coalesce in about 3 Gyr: the authors are able to estimate the abundance of such systems and predict a rate of detection, by the network of advanced detectors, of several thousand events per year.

The variability of these predictions is summarized in [17]; for first-generation detectors plausible estimates of BH–NS and BH–BH events are of the order of a few per thousand year, but could be as high as one every few years. The estimates for aLIGO and AdV are much more favorable, respectively 10 BH–NS and 20 BH–BH per year, and could be as high as 300 and 1,000.

---

[5] A Wolf-Rayet is a very hot star ($T > 25000\,\mathrm{K}$) surrounded by a large envelope of rapidly expanding gas carried by stellar winds up to a billion times stronger than our Sun's. WR stars are believed to be a stage in the evolution of massive stars, with $M > 20\,M_\odot$.

2.3.2 Core Collapse Supernovae

The collapse of the core of a massive star, the so-called Supernova Type II events, was in the past considered a very promising source of GW emission, with optimistic values for the energy emitted in GWs, of the order of $10^{-2}\, M_{\odot}c^2$, comparable to a NS–NS coalescence.

More recent estimates of the energy emitted in GWs have shown, though, that the GW energy emitted is likely to be $\leq 10^{-8}\, M_{\odot}c^2$, and this fact, together with the limited knowledge of the waveform emitted, reduced the prospects of detection by first-generation interferometric detectors to galactic events, at best.

Galactic events are known to be rare: the latest galactic supernova observed has been the explosion of the blue star Sanduleak $-69°$ 202a, which led to the famous SN1987a event. The estimates of less than one event per two decades in the Milky Way and in the local group of galaxies is supported by indirect evidence, provided by the Integral experiment. Integral has detected γ rays emitted in the radioactive decay of the ^{26}Al isotope, copiously produced in Supernova explosions; the galactic abundance of this Al isotope, about $3\, M_{\odot}$, is consistent with the supernova rate quoted above [23].

A recent review [24] recalls, however, that there may be 1 SN occurring about every other year within 3–5 Mpc from Earth. The recent SN 2008bk, which exploded roughly 3.9 Mpc away, is an example SN from this region of space. Initial LIGO and Virgo class detectors had no chances, though, of detecting GWs from SN 2008bk; however, within a few years of observation, the advanced detectors will possibly detect a few such events in our local universe up to a few Mpc [25].

The shape of the signals emitted during supernova events is not as well modeled as for coalescences: although models exist, based on complex simulations of the core collapse [26–28], the large uncertainty in the models, and in the model parameters, give rise to very different signal shapes; some examples are shown in Fig. 2.5. This uncertainty makes it impractical to classify the signals in terms of SNR, which would imply a good knowledge of the signal template: instead, it is preferred to quantify the strength of the signals with a "root sum square" (rss) amplitude, defined for a signal $h(t)$ as

$$h_{\rm rss} \equiv \sqrt{\int |h(t)|^2\, {\rm d}t}\,, \tag{2.26}$$

where $h(t) = F_+ h_+(t) + F_\times h_\times(t)$ in terms of the two polarizations of the GW waveform and the detector's antenna patterns $F_{+,\times}$: hence $h_{\rm rss}$ at the detector is different from the signal's energy.

Let us introduce the GW power crossing an area element ${\rm d}A$

$$\rho_{\rm GW} \equiv \frac{{\rm d}^2 E_{\rm GW}}{{\rm d}A\, {\rm d}t} = \frac{c^3}{32\pi G}\left\langle \frac{{\rm d}h_{ij}(t)}{{\rm d}t}\frac{{\rm d}h^{ij}(t)}{{\rm d}t}\right\rangle; \tag{2.27}$$

using Eq. 2.3 it can be rewritten in terms of polarization components

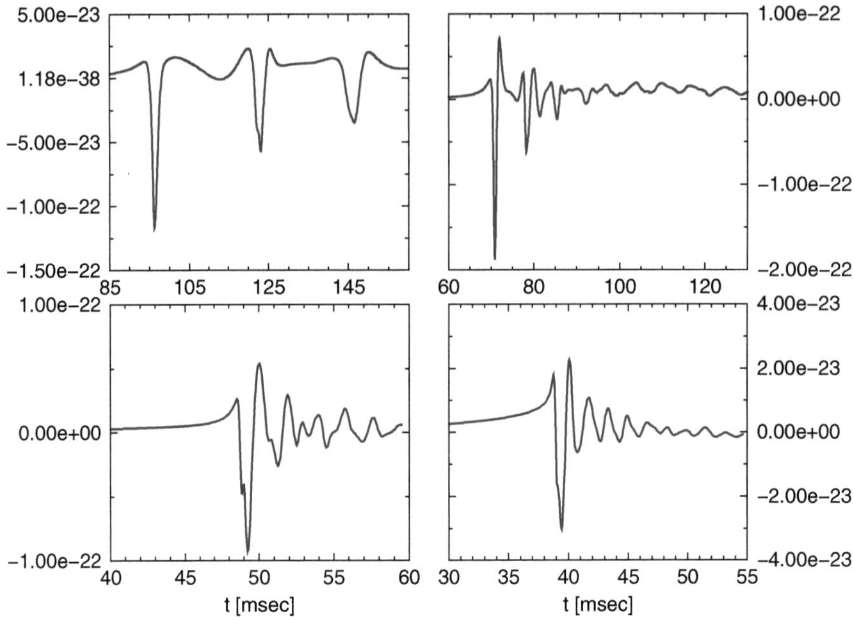

Fig. 2.5 Examples of burst GW signals emitted by a simulated core collapse supernova, taken from [26]

$$\frac{\mathrm{d}^2 E_{\mathrm{GW}}}{\mathrm{d}A\,\mathrm{d}t} = \frac{c^3}{16\pi G}\left\langle \left(\frac{\mathrm{d}h_+(t)}{\mathrm{d}t}\right)^2 + \left(\frac{\mathrm{d}h_\times(t)}{\mathrm{d}t}\right)^2 \right\rangle \tag{2.28}$$

or equivalently as a power per unit area and frequency, in terms of the Fourier transform of the signal $\hat{h}_\pm$

$$\frac{\mathrm{d}^2 E_{\mathrm{GW}}}{\mathrm{d}A\,\mathrm{d}f} = \frac{\pi c^3}{2G} f^2 \left[\left|\hat{h}_+(f)\right|^2 + \left|\hat{h}_\times(f)\right|^2\right] . \tag{2.29}$$

This quantity can be related to h_{rss} in a signal-dependent way: for instance for a sine-Gaussian signal with central frequency f_0 and "quality factor" Q

$$h_{\mathrm{sg}}(t) = h_c \sin(2\pi f_0 t) \exp\left[-(2\pi f_0 t)^2 / \left(2Q^2\right)\right] , \tag{2.30}$$

emitted by a source at (non-cosmological) distance r, one obtains

$$E_{\mathrm{GW}} = \frac{r^2 c^3}{4\pi} (2\pi f_0)^2 h_{\mathrm{rss}}^2 . \tag{2.31}$$

Formulas like Eq. 2.31 allow one to translate the received signal into an energy emitted by the source, in a model-dependent way, and h_{rss} is used as a measure of the signal strength generally for any transient event of unknown shape.

2.3.3 Scientific Results on Transients

Having accumulated years' worth of observation, first-generation instruments data have been used to establish upper limits on two types of quantities: on the *rates* of transient signals, such as the coalescence of binary compact objects, and on the *amplitude* of events possibly emitted in coincidence with some external trigger.

2.3.3.1 Untriggered Searches

The analysis of LIGO S6 run, which lasted from July 2009 to October 2010, and Virgo VSR2 (July 2009–Jan 2010), VSR3 (July 2010–Oct 2010) data provides the best results obtained so far on the *low-mass* binaries [6], namely systems with M_{tot} in the 2–25 $M_\odot$ range, including NS–NS coalescences, which could be searched as far away as 40 Mpc, and heavier systems which would have been visible further out, obtaining limits on the rate $R_{NS-NS} < 1.3 \times 10^{-4}\text{yr}^{-1}\text{Mpc}^{-3}$, $R_{NS-BH} < 3.1 \times 10^{-5}\text{yr}^{-1}\text{Mpc}^{-3}$ and $R_{BH-BH} < 6.4 \times 10^{-6}\text{yr}^{-1}\text{Mpc}^{-3}$.

The same data allowed to set the best limits so far on heavier BH–BH events: the larger masses required to search in the data not only for the inspiral phase, but also the merger and ringdown phase of the signal [29]. For M_{tot} in the 25–100 $M_\odot$ interval, the search achieved a range of 300 Mpc for (20, 20) $M_\odot$ BH–BH systems, and set a limit $R < 3.3 \times 10^{-7}\ \text{yr}^{-1}\text{Mpc}^{-3}$ for systems with non-spinning components in the 19–28 $M_\odot$ range.

When searching instead for unmodeled signals, the best results obtained by interferometers so far are due again to LIGO S6 and Virgo VSR2,3 science runs: in an untriggered search [30], over a total observation time of 207 days, signals shorter than 1 s were targeted over a 64–5000 Hz frequency band, setting an upper limit on the rate of 1.3 events/year, for signals having h_{rss} in the range 5×10^{-22}–$10^{-20}\ \text{Hz}^{-1/2}$, depending on the kind of waveform searched. These sensitivities would have allowed to detect, with a 50 % efficiency, sources emitting an energy of about $2.2 \times 10^{-8}\ M_\odot c^2$ at 10 kpc, a plausible scenario for a galactic supernova.

2.3.3.2 Triggered Searches

The use of a network of gravitational wave detectors brings at least two advantages: first, in a simultaneous analysis of data from different instruments it is possible to introduce artificial relative time shifts that allow to measure the background of false coincidences. Second, it is possible to increase the sensitivity to specific directions in the sky by specifying appropriate time delays in the analysis and by imposing conditions on the relative signal amplitudes, as resulting from the antenna pattern of each instrument.

These procedures become crucial when searching for events coincident with some external trigger, for which we generally know precisely both the sky localization and the timing.

As an example, considerable interest was raised by GRB070201, a short hard γ-ray burst electromagnetically localized in the direction of M31 (Andromeda), about 770 kpc away, comfortably within the range of first-generation instruments. The LIGO Hanford detectors were taking data, which were analyzed under the assumption of a coalescing binary progenitor. The search could exclude at 99 % confidence level the occurrence of a coalescence in M31: if the event had really been associated with a coalescence, the GRB should have been emitted further away along the line of sight [31]. Or it could have been due to a soft γ-ray repeater (SGR) [32], a source emitting periodically short ($\sim$0.1 s) flares of soft γ rays, with peak luminosity up to 10^{35} J/s. These sources are believed to be magnetars, that is, neutron stars with very large magnetic fields, in excess of 10^{15} gauss, in which crust rearrangements may occur, thus exciting non-radial modes which decay emitting also GW [33]. Also a search for generic bursts of GWs was carried out, which set an upper limit of less than $4.4 \times 10^{-4}\, M_{\odot}c^2 = 7.9 \times 10^{43}$ J on the emitted energy; the loose energy constraint, significantly larger than the energy expected by an SGR, implies that the non-observation of a GW counterpart is compatible with the SGR hypothesis.

Gravitational waves emitted by SGR sources were searched for with high sensitivity in data acquired by GEO600 [34], LIGO and Virgo detectors between 2006 and 2009, looking for GW bursts in coincidence with soft γ-ray bursts from six galactic magnetars [35]: in a sample of 1279 γ events, no evidence of coincident GW burst was found, and upper limits could be obtained, particularly interesting for SGR 0501+4516, a source likely to be just 1 kpc away, setting limits on GW energy between 3×10^{37} J and 1.4×10^{40} J, depending on the signal model.

A different kind of triggered search was carried out in correspondence of a timing glitch of the Vela pulsar, a sudden change in the pulsar's e.m. signal, of a few parts per million, observed in the radio spectrum on August 12, 2006: such a glitch is indicative of a sudden jump of the pulsar's rotational frequency and could be due to a "star-quake", exciting NS's normal modes. A search for oscillatory, damped waveforms was carried out in [36], under the assumption of GW signals in the 1–3 kHz range and damping times in the 50–500 ms range, placing an upper limit on the signal amplitude ranging from $6.3 \times 10^{-21}\,\mathrm{Hz}^{-1/2}$ to $1.4 \times 10^{-20}\,\mathrm{Hz}^{-1/2}$, as a function of the excited mode, corresponding to energy limits from 5×10^{37} to 1.3×10^{38} J.

Recent LIGO and Virgo data, from runs S6, VSR2, and VSR3, were searched for GW events in coincidence with a sample of 154 GRBs, as usual searching both for coalescences of compact object pairs, or for generic bursts [37]: an overall median lower limit on the distance of the events $D \sim 17$ Mpc could be deduced, whereas for NS–NS and NS–BH systems the median limit was 16 and 28 Mpc respectively.

It should be underlined that these non-detections were expected: GRBs distances, when measured, are significantly larger, for instance GRB030329 had a redshift $z = 0.1685$ corresponding to a distance $d \simeq 800\,\mathrm{Mpc}$, out of reach for first-generation detectors, but possibly within the range of advanced instruments.

2.3.4 Transients in the ADE: Science and Prospects

We have seen that the advanced detectors, aLIGO and AdV, promise a sensitivity 10 times better, in amplitude, than first-generation instruments: when achieved, this improvement will allow searching for coalescing binaries events in a space volume 1,000 times larger, and we have already discussed in Sects. 2.3.1.2 and 2.3.1.3 what is the expected rate of events. As discussed also in [38], under a realistic scenario advanced detectors should not just perform the first detection, but collect a sizable sample of events, and really start gravitational astronomy: for instance, it has been argued that early detections of GWs with reasonable localization would be possible even before the binary merger [39–43], thus opening interesting perspectives in collaboration with e.m. observatories.

We should admit that pessimistic scenarios exist, under which a non-detection of NS–NS events would not violate strongly the astrophysical expectations; however, a non-detection by the LSC-Virgo network over several years of observation would push the models of sources and of source populations to their limits.

For unmodeled sources, signals with $h_{\mathrm{rss}} \sim 10^{-23}\,\mathrm{Hz}^{-1/2}$ will be accessible, making it possible, for instance, to detect galactic supernova events with energies released in GWs as small as a few $10^{-9}\, M_{\odot}c^2$. Such a sensitivity should grant the detection of a galactic supernova, or place extremely tight limits on the models of the core collapse.

2.3.4.1 Science from NS–NS Detections

It should be underlined that, indirectly, we already believe that binary NS emit gravitational waves: there is no mechanism, other than the emission of GW predicted in general relativity (GR), which could explain the very good agreement of the period evolution of the Hulse-Taylor binary pulsar [44, 45].

However, this is an *indirect* evidence, whereas the detection of GW waves from the *inspiral* phase would be a very clean, *direct* confirmation of GR, and the detection of a number of such events would also allow to test alternative theories of gravity in a model independent way [46].

The detection of a good number of NS–NS events would also constitute a new way to calibrate the distances: in fact, a network of detectors allows to reconstruct the waveform amplitude, which is a known function, as in Eq. 2.12, of the source distance and binary masses. On the other hand, masses can also be deduced by the time evolution of the waveform, as in Eq. 2.16. Hence the distance of a NS–NS event can be reconstructed, and if the event can be associated with an astrophysical object, it will serve as distance calibration, and potentially compete with other methods for determining the Hubble constant [47]: such a program could be implemented for instance by associating GW events with GRBs [48]. In the absence of electromagnetic counterparts, it has been proposed to measure the Hubble constant by relying on the narrow observed distribution of NS masses: then the information in the GW waveform allows to directly deduce the event redshift [49]; an alternative approach exploits catalogs of galaxies with measured redshifts to associate the GW events with a host [50].

Perhaps even more interesting is what could be learned by studying the *merger* phase of the NS–NS coalescence: when the two NS are close enough, and start to be disrupted by the tidal forces, they cannot be considered point particles anymore and the emitted GWs become dependent on the details of the NS structure, and on the Equation of State (EoS) of the nuclear matter [51]. The effects are most evident during the merger phase, which is however very complicated to model, and requires sophisticated simulations which take into account general relativity and nuclear physics in order to predict the emitted GW [52] and possibly a simultaneous γ-ray burst [53]. Since the merger signal in NS–NS will be relatively weak, and the SNR carried will thus be limited, the prospects of extracting information which could strongly constrain the complex physics of the NS will be reduced, even in the ADE. It may be more promising to rely on the last stages of the inspiral phase, during which an NS becomes distorted in the tidal field of the other; this distortion induces a phase change in the GW signal which could be detected, and in [54] it is shown that a few tens of such observations could allow to discriminate between hard, moderate, and soft EoS.

The *ringdown* phase of the resulting BH, characterized by the emission of waves corresponding to the quasi-normal-modes of the system, will be another clean laboratory: the modes are entirely determined by the mass and the spin parameters of the final objects, and the measurement of their frequencies and damping times will allow to completely determine the final BH.

2.3.4.2 Science from NS–BH and BH–BH Detections

Differently from NS–NS systems, no BH–BH or BH–NS system has been observed yet; consequently, the assumptions about the coalescence rate of such systems is somewhat less firm, as already discussed. However, since the SNR scales roughly as $M^{5/6}$, larger mass systems are visible to greater distances and their interest is quite substantial.

The BH–BH collisions will be a unique laboratory; each BH is solely characterized by its mass and spin, which completely determine the dynamic of the event horizon, and the theoretical problem is *conceptually* simple because the only relevant equations are those of general relativity. Hence a direct BH–BH observation would be a clean test of GR in a regime of strong fields, which could allow, for instance, to discriminate between alternative formulations of the theory. However, the deviations from the simple Newtonian formulas become relevant not just for the merger, but also for the inspiral phase, and an accurate calculation of the effects has called for a substantial theoretical effort, contributing to the development of the field of numerical relativity (NR), which was born in the 1980s mainly to study the core-collapse events [55]. One important milestone was the simulation of a head-on BH–BH collision [56, 57], whereas later the progress in the simulation methodologies made it possible to study realistic collisions in three dimensions, including eventually spin effects, which are expected to be important in binary black-holes: it is not possible to account here for the rich literature on the subject, and the reader could consider for instance the reviews [58, 59].

From the point of view of gravitational wave searches, the results of NR have been exploited to assess the sensitivity of the search waveforms [60], and a significant effort is still ongoing to further translate the knowledge gained by several NR groups into waveforms usable for the data analysis: in the ADE, this effort is likely to lead to a close collaboration of theorists and experimentalists.

Some of the considerations made for NS–NS events apply also to BH–BH and BH–NS events, in particular the possibility to measure the event distance, and actually to much larger distances: however, no mechanism is known by which a BH–BH event could result also in an e.m. event, and therefore the contribution to determining the Hubble constant is uncertain.

The BH–NS collisions have in principle a good potential to shed light on the EoS of the NS; for a wide range of mass parameters, the latter should be disrupted by the tidal forces due to the BH, and the resulting GWs will be within the frequency range of GW detectors, thus allowing a direct probe of the NS structure [61]. It has been recently argued though, considering NS–BH pairs with a non-spinning BH and a range of BH masses considered most likely, that the resulting GW signal would have only a weak dependence on the NS characteristics, to the point that a NS–BH event may be indistinguishable, in simulation, from a BH–BH event with the same masses [62]: this happens for instance in systems comprising a NS and an $8M_{\odot}$ BH, because the NS is "swallowed whole" by the BH, before being tidally disrupted.

Together with NS–NS events, it should also be noticed that BH–NS coalescences are one of the proposed mechanisms for producing the violent shocks that would generate Gamma Ray Burst events; the importance of collaborating with GRB detectors is therefore evident.

2.3.4.3 Association with Gamma and X-ray Bursts

The gamma-ray bursts, detected by instrumentation on board of satellites, are believed to be originated by shocks in matter rapidly accreting onto newly formed black holes, as a consequence either of a coalescence of compact objects, or of the collapse of particularly massive supernovae. Observationally, the events are divided into two broad classes, a minority of *short* bursts (lasting less than about 2 s) and a majority of *long* ones, lasting anywhere between 2 s to several minutes, with an average of 30s duration.

There is mounting evidence (see for instance [63]) that long GRBs, or at least those with a softer γ spectrum, are associated with *collapsars*, that is massive, rapidly spinning Wolf-Rayet stars that collapse in Kerr black-holes with accretion disks. Instead, good arguments [64] exist to assert that short bursts are related with coalescing NS–NS or NS–BH pairs. In fact, it was argued in the past that NS–BH coalescences could also be an alternative mechanism for long γ-ray bursts, but now the collapsar model is favored.

The γ rays would be powered by the accretion disk orbiting around the newly formed Kerr black hole; a disk, for instance, formed by material resulting by the tidal disruption of part of the two colliding objects, in the case of short GRBs: it is worth recalling that the infall of matter on a rotating BH is the most efficient known mechanism for turning matter into energy [65], far more efficient than thermonuclear reactions.

If the pictures for *short* and *long* gamma ray bursts are correct, GW observatories may be able to detect a signal from the inspiral and merger event, or by the supernova respectively. As we have seen, such events have already been searched for in the data of the first-generation detectors (see for instance [66] for a search triggered by a long γ-ray burst), finding no evidence of a correlation, but this was not a surprise since for most of the short GRB there is observational evidence, by the observation of faint X-ray afterglows, that they occur at cosmological distances ($z \geq 0.1$), beyond the range of first-generation GW detectors.

Advanced detectors, with their 10-fold larger range, will have greater chances, and a confirmed association of a CBC with a GRB would have tremendous consequences.

First of all, the GRB trigger could give a greater confidence in the GW event, thanks to the tight time association. For short GRBs, the models show that it is necessary to allow only for about 2 minutes before and after the GRB maximum in order to over-cover the time windows where a GW event is expected. Considering a GRB rate $\sim$ 1/day, false alarms can be abated by factors $O(10^3)$. At fixed false-alarm rate, the coincidence with GRBs allows thus lowering the thresholds applied to GW events, thus gaining in sensitivity.

From the physical point of view, as we have already discussed, the observation of a NS–NS (or NS–BH) GW event would allow reconstructing the mass parameters of the progenitors, through observation of the inspiral phase, and potentially also the mass and spin parameters of the formed Kerr BH, by observing its relaxation through GW emission at the frequencies of quasi-normal modes; this information would be an input for constraining the models of the GRB process and could help in understanding the complex dynamics in the accretion disk. For instance, the GW event would provide information about the event distance, thus allowing a precise, absolute calibration of the energy emission.

When the advanced detectors era comes, the NASA's SWIFT [67] satellite, which is a veritable multi-messenger observatory, carrying also optical and X-ray telescopes, will no longer be operational, but will be replaced by the international Fermi Gamma-ray Space Telescope (formerly known as GLAST) [68], which retains the capability of "swift" pointing towards GRB sources, while providing greater sky coverage and sensitivity over a much wider range of energies, up to $\sim$300 GeV.

Although γ-ray events have captured most of the interest, also X-ray prompt bursts or longer tails have been observed in conjunction with core-collapse supernovae and GRB events, by a range of X-ray satellites, including Chandra [69] by NASA and X-MM Newton [70] by ESA. While both Chandra and X-MM Newton missions can be expected to last until the beginning of the ADE, we can hope that the joint ESA-NASA X-Ray Evolving Universe Spectrometer (XEUS) [71] will be operating as a successor.

2.3.4.4 Soft γ-Ray Repeaters

Besides the very powerful GRBs, soft gamma repeaters ("SGRs") are X-ray stars that emit bright, repeating flashes of soft (i.e., low-energy) gamma rays. While four SGRs have been identified in our galaxy so far, many other millions almost certainly exist, and a similar number probably exists in every other galaxy. According to the "magnetar" model, SGR are associated to neutron stars (NS) with very intense magnetic fields, such that the star crust may break under accumulated magnetic stress [33]. A hot fireball forms which cools down through the emission of bursts of electromagnetic (EM) radiation. During the crust quake, the star's non-radial seismic modes could be excited, thus leading to the emission of a GW burst as well, in fact theoretical estimates predict comparable EM and GW energy release, within a few orders of magnitude. We have seen in Sect. 2.3.3.2 that the best upper limit on the strength of the GW emission placed with current LIGO data is 3×10^{37} J, which is within the range 10^{37}–10^{39} J for the observed EM emission: the advanced detectors, thanks to a sensitivity improvement (in amplitude) by a factor of 10, will be able to detect GW energies 100 times smaller, hence the SGRs will be very promising sources, either for a detection or for placing stringent constraints on the models.

At times (though rarely), SGRs undergo "giant flares," also called *hyperflares*, during which the emission is dramatically increased with a total energy about a thousand times higher than in a typical burst; this has been the case on 27 December 2004 for SGR 1806–1820 [72]. Quasi-periodic oscillations (QPO) in the tails of the giant flares have been observed, thought to be driven by the star's seismic modes, the same that are expected to emit GWs. This information could be used to narrow down the search in the GW spectrum to the observed QPO frequencies, leading to an improvement in sensitivity. For hyperflare events occurring at the time of advanced detector operation, the observable GW energetics at 100 Hz would be close to 10^{36} J, leading to stringent constraints on the magnetar model [73].

2.3.4.5 Association with Neutrino Observations

We have seen that advanced GW detectors should be able to detect supernova events emitting as little as $10^{-9}\, M_{\odot}c^2$ in GWs: this sensitivity should allow to confirm a detection with a high confidence level, but probably will not allow a detailed characterization of the GW signal.

This said, the significant uncertainty about the mechanism of the SN collapse, in which neutrino, magneto-hydrodynamic and acoustic effects are expected to be relevant [74], implies that even a mere detection, or a non-detection, of GWs from a nearby SN could constrain significantly the core-collapse explosion mechanism.

Actually, we expect that about 99 % of the gravitational energy released during the collapse of the SN is carried out of the core by the so-called thermal neutrinos, having energies of a few tens of MeV, released in a short burst (a few seconds) through the inverse β-decay process of neutronization. The existing neutrino detectors, like Super-Kamiokande [75], LVD [76] and IceCube [77] can detect these abundant neutrinos from sources in the Milky Way and beyond the Magellanic clouds,[6] and operate under an agreement [78], exchanging time information about events, and regularly searching for coincidences which could correspond to a real SNe.[7] Both Super-Kamiokande and IceCube are expected to be in operation in the advanced detectors era, and a correlated detection of an event by neutrino and GW detectors would first of all strengthen its statistical significance. Perhaps more interestingly, the neutrino emission is expected to be separated by a very small delay, of a few ms, from the emission of GWs: this tight coincidence at the source level could be partially spoiled in presence of a nonzero neutrino mass, which will induce a delay $t_{\rm prop}$ at detection level, expressed as [81]

$$t_{\rm prop} = 5.2\,{\rm ms}\frac{d}{10\,{\rm kpc}}\left(\frac{m_\nu}{1\,{\rm eV}}\right)^2\left(\frac{10\,{\rm MeV}}{E_\nu}\right)^2 \tag{2.32}$$

which means that a good accuracy in determining this delay could allow constraining neutrino masses tightly.

In addition to the detection of thermal, or low-energy, neutrinos from SNe, there is considerable interest for detecting High Energy ν, which is actually the main goal of IceCube and of the future KM3NeT detector [82]: their respective locations at the south pole and in the Mediterranean sea allow them to cover the entire sky. These neutrino detectors use an array of photomultipliers, regularly spaced in a large volume of ice or water to collect the Cerenkov light emitted by charged particles, which result from the collision of the high-energy ν's with the nuclei in water. The quasi-simultaneous excitation of several photomultipliers allows to reconstruct the direction of the charged particles and to infer the one of the incoming ν. An attempt at identifying coincidences between high-energy neutrinos detected by Antares (the predecessor of the KM3NeT) and GW candidate events detected by LIGO and Virgo has been carried out in [83], starting the development of a methodology.

There exist potential sources of joint high energy ν and GW emission; for instance, it has been speculated [64] that HE ν's could be emitted during the binary mergers,

---

[6] Even though IceCube is designed for high energy ν, it can detect SN neutrinos in the Large Magellanic Cloud with a 5σ significance.

[7] The Sudbury Neutrino Observatory (SNO) [79] was also part of the agreement SNEWS (Super-Nova Early Warning System), and has been recently decommissioned. Plans are underway, though, to upgrade it to SNO+ [80].

and thus be associated with short GRBs as well. More generally, it is expected that HE ν's are produced in the shocks by accelerated protons, for instance in collapse events associated with long GRBs: the resulting fluxes could be sufficient to be detected at cosmological distances by km^2 scale ν detectors like IceCube and KM3NeT. Also, during the giant flares from Soft Gamma Repeaters, the violent crustal reorganization of the magnetar is believed to emit both GW and ν's, with fluxes that might be detectable or at least allow to constrain the models.

As with other joint observations, the simultaneous detection of GWs and HE ν's would thus give enormous insight into the generation processes.

2.3.4.6 The Electromagnetic Follow-up of Transient GW Events

We have seen that GW experiments have carried out searches of signals triggered by external events. The converse starts to be true: GW candidate events are becoming targets for detectors of electromagnetic radiation, and optical, X-ray or γ observations can be exploited to look for transients that could be associated with the same source of a GW signal candidate.

There are some difficulties: first, a network of three sites like the one formed by LIGO and Virgo detectors is capable of reconstructing the candidate's source position with rather limited accuracy, not better than tens of degrees, depending on the candidate signal's strength. The lack of accurate localization calls for observing wide areas of the sky, using instruments generally designed with smaller fields of view, thus requiring the development of specific methods for the analysis of images. Second, one needs a low-latency GW search, capable of delivering candidate events a few minutes after data acquisition: this requires that a number of data processing steps are performed on-time, including calibration, data quality and vetoes, trigger generation, and event reconstruction.

In [84] this program has been implemented and tested, focusing on signals from coalescing pairs of neutron-stars, and using terrestrial telescopes, whereas in [85] both neutron-star binaries and un-modeled signals were targeted, using the Swift satellite observatory. During the advanced detector era this approach will be further expanded, by covering more extensively the spectrum of GW signals, and by involving several partner observatories, both on Earth and in space.

2.4 Continuous Signals

Besides transient signals, the ground-based advanced GW detectors will target also permanent signals, which are expected to be essentially of two kinds: the periodic ones emitted by rotating, distorted neutron stars, and the stochastic ones emitted either by astrophysical sources, or immediately after the Big Bang.

A permanent source is clearly particularly appealing, but the price to pay is that signals are expected to be extremely weak, and first-generation detectors have already set stringent limits: still, the science potential is immense, as we will try to discuss.

2.4.1 *Rotating Neutron Stars*

Spinning compact objects emit gravitational waves when their mass distribution displays a time-varying quadrupolar component: for example, a freely precessing rotor could exhibit a momentum of inertia along one axis $\hat{z}$ different from the ones along the other two axes. The emitted signal would be particularly intense in compact objects like the neutron stars: considering the extremely high stability of the electromagnetic signal emitted by a particular type of neutron stars, the pulsars, we anticipate that the emitted GW radiation would be essentially monochromatic, with a frequency stability comparable to that of the rate of pulsar electromagnetic pulses.

What could cause the deviation from spherical shape in a compact object like an NS? It could be due to internal magnetic stresses (e.g., [86, 87]) or to accretion onto neutron stars residing within low mass X-ray binaries [88].

2.4.1.1 The Signal from a Rotating NS

A freely precessing NS emits a periodic signal with a characteristic amplitude, as seen at distance R

$$\begin{aligned} h_c &= \frac{2G}{c^4}\frac{\varepsilon I_{xx}}{R}\Omega^2, \qquad \text{with} \quad \varepsilon \equiv \frac{I_{zz} - I_{xx}}{I_{xx}} \\ &\simeq 10^{-26}\left[\frac{I_{xx}}{10^{38}\,\mathrm{kg\,m^2}}\right]\left[\frac{10\,\mathrm{kpc}}{R}\right]\left[\frac{f}{100\,\mathrm{Hz}}\right]^2\frac{\varepsilon}{10^{-6}} \end{aligned} \tag{2.33}$$

where Ω is the angular frequency of the spinning object; in the second line, we have exposed typical values for the momentum of inertia I of the star, and plausible values for its "ellipticity" parameter ε, which expresses its degree of asymmetry.[8] In order to compare this amplitude with a detector's sensitivity, like the linear spectral densities shown in Fig. 2.1, one should multiply h_c by the square root of the observation time T, which determines the accessible frequency resolution. One year of observation corresponds to $\sqrt{T} \simeq 5600\ s^{1/2}$, therefore a perfectly monochromatic signal with an amplitude as in Eq. 2.33 would yield peaks at about $10^{-23}\,\mathrm{Hz}^{-1/2}$ in the linear

[8] A similar formula holds for a tri-axial rotator, in which case one has

$$h_c = \frac{2G}{c^4}\frac{\varepsilon I_{zz}}{R}\Omega^2 \qquad \text{with} \quad \varepsilon \equiv \frac{I_{xx} - I_{yy}}{I_{zz}}.$$

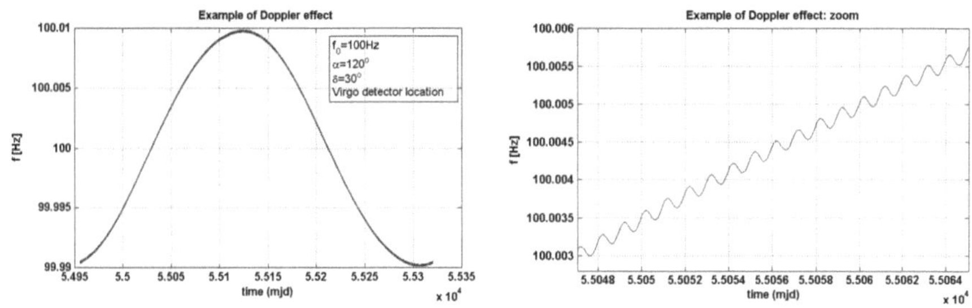

Fig. 2.6 The evolution of the frequency of the signal emitted by a pulsar, as received by a detector located on Earth. *Left* the yearly evolution of the frequency, dominated by the Earth revolution. *Right* a zoom over 18 days which shows the daily modulation due to Earth rotation, superposed to the frequency drift due to the revolution

spectra of the data; this shows that searches for such signals are limited to galactic sources. However, the number of galactic sources is easily estimated to be very large: $10^8 - 10^9$ neutron stars in the Milky Way alone, on the basis of the supernova rate (about 1 every 30 years), the Milky Way age (about 13 billion years) and the assumption that most supernovae evolve in NS and not BH.

Besides, the signal emitted is simply a periodic wave, in general with pulsation Ω and 2Ω, but the detection is made difficult by the Doppler effect due to Earth rotation and revolution, which modulates the signal phase and leads to an evolution of the instantaneous frequency with time, as shown for example in Fig. 2.6.

To appreciate the size of the effect, let us consider just the Earth's revolution motion, modeled as a circular orbit with angular frequency ω_m: the phase of a periodic signal with pulsation ω received from direction $\hat{n}$ is altered as

$$\phi(t) = \omega\left(t + \frac{\hat{n}\cdot\mathbf{r}_{\oplus}(t)}{c}\right) = \omega\,(t + \beta\cos\omega_m t) \tag{2.34}$$

where $\mathbf{r}_{\oplus}(t)$ is Earth position in a reference frame centered on the Sun, hence $\beta = O(1\mathrm{AU}/c)$; the signal itself can be expanded in a series of sidebands

$$\begin{aligned}\cos(\omega(t+\beta\cos\omega_m t)) = &\sum_{k=-\infty}^{+\infty}(-1)^k J_{2k}(\omega\beta)\cos(\omega+2k\omega_m)t \\ &- \sum_{k=-\infty}^{+\infty}(-1)^k J_{2k+1}(\omega\beta)\sin(\omega+(2k+1)\omega_m)t\,.\end{aligned} \tag{2.35}$$

The Bessel functions $J_l(\omega\beta)$ determine the amplitude of each sideband: if $\omega\beta \ll \pi/2$, only the lowest two l values matter. In the case of Earth revolution though, $\beta \sim 5\times10^2$

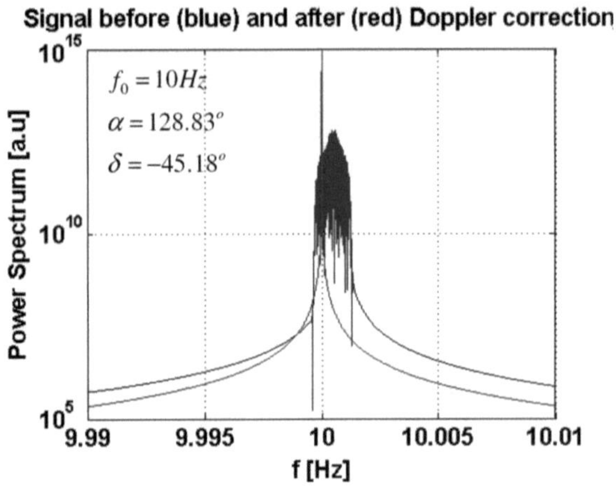

Fig. 2.7 A simulated example of the correction of the Doppler effect due to Earth rotation and revolution: the signal received by the detector, which is spread over many sidebands, is transformed back into a monochromatic signal

s, hence for $\omega = 2\pi \cdot 100$ Hz we obtain $\omega\beta \sim 3 \times 10^5$, and a similar number of small sidebands contributes,[9] each correspondingly weaker.

The Doppler effect can be corrected for, as shown in Fig. 2.7, but only if the source location is known with high accuracy, hence signals associated with known pulsars can be searched for with a higher sensitivity.

Besides, the fundamental period of the source will evolve with time: a rotating neutron star will lose energy as

$$\frac{\mathrm{d}E}{\mathrm{d}t} = -\frac{G}{5c^5}\left\langle \dddot{Q}_{ij}\dddot{Q}_{ij}\right\rangle = -\frac{32G}{5c^5} I_{zz}^2 \varepsilon^2 \Omega^6 \; ; \tag{2.36}$$

given the rotational kinetic energy $E = \frac{1}{2} I_{zz}\Omega^2$, it follows

$$\dot{\Omega} = -\frac{32G}{5c^5} I_{zz}\varepsilon^2\Omega^5 \tag{2.37}$$

hence the following relation holds for the rotation period P:

$$\dot{P} \simeq 10^{-14} \frac{I_{zz}}{10^{38}\,\mathrm{kg\,m^2}} \left(\frac{P}{1\,\mathrm{ms}}\right)^{-3} \left(\frac{\varepsilon}{10^{-6}}\right)^2 . \tag{2.38}$$

A measured braking $\dot{P}$ provides therefore an upper limit on ε:

---

[9] It is useful to recall the formula

$$\frac{\sum_l l^2 J_l(x)}{\sum_l J_l(x)} = \frac{x^2}{2}$$

which relates the variance of the index l in the distribution of $J_l(x)$ to the square of the argument x: for large x, large values of l are important.

$$\varepsilon \leq \varepsilon_{\max} \equiv 3 \times 10^{-9} \left(\frac{P}{1\,\text{ms}}\right)^{3/2} \left(\frac{\dot{P}}{10^{-19}}\right)^{1/2} . \tag{2.39}$$

As an example, for the Crab $\dot{P} = 38\,\text{ns/day} = 4.4 \times 10^{-13}$ hence

$$\varepsilon_{\max}(\text{Crab}) \simeq 1.2 \times 10^{-4} . \tag{2.40}$$

In deriving Eq. 2.39 we have assumed a braking due solely to GW emission; using Eq. 2.33 the value $\varepsilon_{\max}$ could be translated into a maximum h_c value, also called the *spin-down* limit. This is known to be an upper limit, since the EM emission is a competing, and likely dominant, mechanism for the loss of rotational energy, which actually alters also the dependence of the braking on the period P [89].

2.4.1.2 Existing Limits on NS Signals

The LIGO and Virgo runs provide the most stringent limits so far on periodic signals emitted by compact objects; for instance a targeted search was carried out using S5 data (Nov 2005–Nov 2007) [90], considering 116 known millisecond and young pulsars, and exploiting also radio and X ray observations to deduce ephemerides for the phase of the signal. This search allowed to set the best limit so far on the Crab's GW emission, a factor of 7 below the spin-down limit; this being a limit on the amplitude, it implies that less than $\sim 2\,\%$ of the energy loss could be due, in principle, to GW emission. For several other young pulsars, the limits are only slightly above the spin-down limits: the best results were obtained for pulsar J1603-1720, with a limit $h_c < 2.3 \times 10^{-26}$, and for J2124-3358, with a limit $\varepsilon < 7 \times 10^{-8}$, in the ballpark of plausible values.

Despite their lesser sensitivity, for lack of precise knowledge of the source location and phase, all-sky searches are important because of the expected large number of neutron stars which are not detected as pulsar: the LIGO S5 data [91] were exploited to this end, searching a frequency band 50–800 Hz, and assuming $-6 \times 10^{-9}\,\text{Hz/s} \leq d\nu/dt < 0\,\text{Hz/s}$. At the frequency of 150 Hz, where the detector displays its best sensitivity, an upper limit $h_c < 10^{-24}$ was obtained for linearly polarized signals, whereas at the upper end of the frequency band a limit $h_c < 3.8 \times 10^{-24}$ was obtained for all polarizations and sky locations.

Using Virgo VSR2 data, thanks to the good low-frequency sensitivity, an upper limit on the emission by the Vela pulsar (expected to emit at $\nu \sim 22.47\,\text{Hz}$) could be placed [92] at $h_c <$1.9–2.2 $\times 10^{-24}$, the interval being determined by the assumptions on the signal, thus obtaining a limit more stringent than the spin down limit $h_{c,\max} = 3.3 \times 10^{-24}$ derived from the pulsar's period evolution.

2.4.1.3 Science from NS Continuous Signals

From the numbers discussed above, it should be clear that even in the advanced detectors era, the chances to detect continuous signals emitted by compact objects will remain limited to sources in our own galaxy; however, the sensitivity improvement will allow detecting signals with h_c ten times fainter, and consequently to probe neutron-star ellipticities as small as $\varepsilon \sim 10^{-9}$, therefore significantly improving the prospects of a detection or the relevance of the resulting upper limits.

Even limiting to AdV, and considering only the known pulsars, it is possible to show that the improved sensitivity will allow to beat the spin-down limit by orders of magnitude, for about 40 of the pulsars presently known [93].

The increased sensitivity will open new prospects also for electromagnetically quiet NS: on the basis of the neutron star birthrate [94] it is possible to estimate that there exist $O(10^5\text{–}10^6)$ such objects in a radius of just 1 kpc; within this volume the sensitivity of advanced detectors may allow discovering a close enough NS.

Within the theoretical framework of GR, the detection of continuous GW signals from galactic pulsars would provide good insight into the source physics, thanks to the simple form of the expected signal, and its direct relation to the source parameter, in particular to the ellipticity parameter ε, and this in turn to the rigidity of the NS crust.

For pulsars, the relationship between the GW emission frequency and the rate of EM pulses would help discriminating among competing mechanisms generating this asymmetry. Moreover, it is well known that pulsar rotation may undergo sudden phase discontinuities, believed to be related to "starquakes", or rearrangements of the NS crust stresses. If such events could be related with analogous losses of phase coherence in the GW signals, this observation could shed light on the nature of the deformation and ultimately on the rich physics of these stars, which possess huge magnetic fields and are believed to contain a core with the characteristics of a superfluid [95].

Besides pulsars, we should mention another potentially interesting class of spinning neutron stars; the Low Mass X-ray Binaries, constituted by a NS which is spun-up by the infall of material through an accretion disk. The evidence is that the spin of these systems results from the equilibrium between the torque due to the accretion mechanism and some braking mechanism, the most likely being the GW emission [96]. The resulting GW frequencies could be stabilized by this feedback mechanism and are expected in the range 600–1200 Hz; these sources could be targeted especially well by advanced detectors operating in narrowband mode, when their rotational frequency is measured; this is not the case for instance for Sco X-1, for which however interesting limits would be placed also operating in wide-band mode [97].

As discussed already, the targeted search of GW emitted by known objects relies on the assumption that the phase evolution is coherent over long timescales. If a "star quake" occurs, it can introduce a "glitch" both in the series of EM pulses and in the GW signal. Knowing whether a glitch has occurred is possible, if the radio telescopes regularly check the count of pulses for slight variations in the expected arrival of the event. A confirmed correlation between GW and EM glitches would tell us a lot about the structure of the NS crust and its dynamics.

2.4.2 Stochastic Background

Besides periodic signals, we anticipate the existence of stochastic signals, either of cosmological or of astrophysical origin. The same inflationary mechanism that is believed to have amplified the vacuum fluctuations of the early universe should have led to the emission of GWs shortly after the Big Bang [98], whereas a stochastic signal can result also from the incoherent superposition of transient signals, emitted by a large population of distant sources [99]. It has been shown [100] that advanced detectors are likely to detect the stochastic signals caused by binary coalescences, and that these could form a "foreground" potentially masking the GW background due to cosmological sources.

2.4.2.1 The Stochastic Signal as Seen by a Detector

A detector l will receive a superposition of signals from different directions

$$h_l(t) = \sum_p \int \mathrm{d}\Omega\ F_l^p(\hat{\Omega}) \int_{-\infty}^{+\infty} \mathrm{d}f\ e^{i2\pi f\left(t-\frac{\mathbf{d}_l\cdot\hat{\Omega}}{c}\right)} h_p\left(f,\hat{\Omega}\right)\ ; \tag{2.41}$$

here $\mathbf{d}_l$ specifies the location of the detector, $\hat{\Omega}$ is a unit vector pointing to a source, and F_l^p is the detector's l antenna pattern, for polarization p; therefore $h_p(f,\hat{\Omega})$ represents the p component of the stochastic signal from direction $\hat{\Omega}$ at frequency f.

Signals of cosmological origin are expected to be stationary, isotropic and Gaussian, hence the statistical properties of h are specified entirely by the second-order moment

$$\langle h_p(f,\hat{\Omega})h_q^*(f',\hat{\Omega}')\rangle = \frac{1}{4\pi}\delta(f-f')\delta^2(\hat{\Omega}-\hat{\Omega}')\delta_{pq}\frac{1}{2}H^2(f) \tag{2.42}$$

in which $H(f)$ is a one-sided linear spectral density: by computing

$$\langle h_l(t)h_l(t')\rangle = \sum_p \frac{1}{4\pi}\int \mathrm{d}\Omega\,[F_l^p(\hat{\Omega})]^2 \cdot \int_{-\infty}^{+\infty} \mathrm{d}f\ e^{i2\pi f(t-t')}\frac{1}{2}H^2(f) \tag{2.43}$$

one sees that, apart factors of the order of unity, $H^2(f)$ directly contributes to the detector's spectral noise density. Using the energy density of GW

$$\rho_{gw} = \frac{c^3}{32\pi G}\left\langle \frac{\mathrm{d}h_{ij}(t)}{\mathrm{d}t}\frac{\mathrm{d}h^{ij}(t)}{\mathrm{d}t}\right\rangle \tag{2.44}$$

already defined in Eq. 2.27, it can be directly shown that

$$\frac{\mathrm{d}\rho_{gw}}{\mathrm{d}\log f} = \frac{\pi c^3}{2G} f^3 H^2(f)\,. \tag{2.45}$$

It is customary to use a normalized logarithmic energy density

$$\Omega_{\mathrm{GW}}(f) \equiv \frac{1}{\rho_c}\frac{\mathrm{d}\rho_{gw}}{\mathrm{d}\log f} \tag{2.46}$$

in terms of the critical energy density ρ_c required to close the Universe, defined as

$$\rho_c = \frac{3H_0^2 c^3}{8\pi G} \tag{2.47}$$

where the Hubble constant $H_0 = h_0 \times 100 \times \mathrm{km}/(\mathrm{s}\cdot\mathrm{Mpc})$. One obtains

$$h_0^2\Omega_{\mathrm{GW}}(f) = \frac{4\pi^2 h_0^2}{3H_0^2} f^3 H^2(f) \tag{2.48}$$

as a dimensionless measure of the spectral strength of the GW background, independent on the experimental uncertainty on h_0 (recall the most recent estimates $h_0 \simeq 0.714 \pm 0.016$ [101]).

The models for the cosmological stochastic background in general foresee a spectral density well approximated by a power law, at least in the frequency band accessible to ground-based interferometric detectors [98]:

$$\Omega_{\mathrm{GW}}(f) = \Omega_n\left(\frac{f}{f_0}\right)^n \tag{2.49}$$

for instance, the standard inflationary theory foresees a flat spectrum $n = 0$, while other models, including string models, foresee $n > 1$, and therefore spectra increasing with the frequency [93].

Taking into account existing bounds, one should aim at $h_0^2\Omega_{\mathrm{GW}}(f) \leq 10^{-6}$, or

$$H(f) \leq \sqrt{\frac{3}{4}\frac{H_0}{\pi h_0}} f^{-3/2} \times 10^{-6} \simeq 10^{-21}\frac{\mathrm{Hz}}{f^{3/2}} \tag{2.50}$$

a quantity which should be compared the detector's sensitivity: at 100 Hz one should detect $H(f) \simeq 3\times 10^{-24}\,\mathrm{Hz}^{-1/2}$, about a factor 10 below the best sensitivity attained so far.

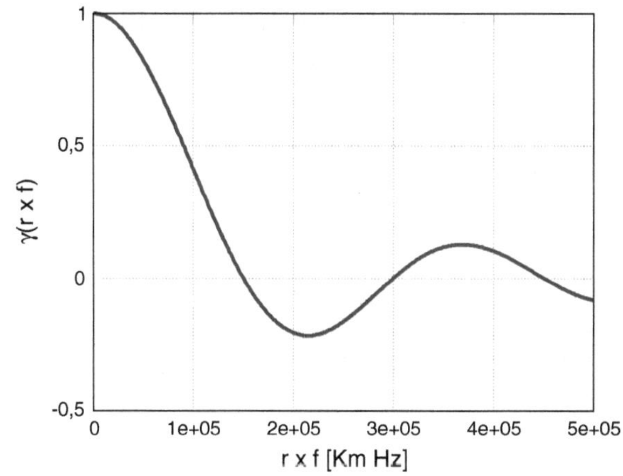

Fig. 2.8 An example of overlap reduction function $\gamma(r \times f)$. Setting a detector distance, say $r = 150$ km, the function displays maxima and minima, with the first lobe ending at about 150 Hz

In fact, correlating the output of several instruments, it is possible to do much better, as we will see later.

2.4.2.2 Existing Limits on the Stochastic Background, and Prospects

A stochastic background appears in a single detector as an additional noise, and Eq. 2.50 shows that this would be below the instruments's noise floor.

However, the noise is coherent across different instruments, and can be exposed by a cross-correlation of their outputs: this correlation can be written as

$$< h_l(t) h_m(t') > = \int e^{i2\pi f(t-t')} \gamma(r_{lm}, f) H^2(f) \, \mathrm{d}f \tag{2.51}$$

where γ is a "overlap reduction function" depending on the product $r_{lm} \times f$, with r_{lm} the distance between the two detectors l, m.

An example of γ is shown in Fig. 2.8: for distant detectors, it is similar to a sinc function of the argument, displaying an oscillating pattern, and tends to 1 for parallel, co-located detectors.

The best published limits have been obtained by correlating LIGO data from the two distant detectors at Hanford and Livingston, acquired during the S5-VSR1 runs [102]:

$$h_0^2 \Omega_{\mathrm{GW}} < 6.9 \times 10^{-6} \qquad @ \, 100\mathrm{Hz}\,, \tag{2.52}$$

a result which improved over direct limits from Big Bang nucleosynthesis [98] and from CMB [103], and which constrained also more exotic cosmological models, like those involving cosmic strings [104]. During the same run, a limit at high frequencies could be set using the LIGO and Virgo detectors [105]: for instance, assuming the power law in Eq. 2.49 with $n = 3$, one obtains

$$h_0^2 \Omega_{\rm GW} = h_0^2 \Omega_3 \left(\frac{f}{900\text{Hz}}\right)^3 < 0.16 \times \left(\frac{f}{900\text{Hz}}\right)^3 \quad \text{in } 600\text{–}1000\,\text{Hz}\,. \tag{2.53}$$

Relaxing the assumption of complete isotropy of the stochastic background, it was possible to investigate potential anisotropies, either due to point sources or to extended sources, again using S5 data [97], obtaining:

$$\begin{aligned} H^2(f) &< 2\text{–}20 \times 10^{-50}\text{Hz}^{-1}(\text{point sources}) \\ H^2(f) &< 5\text{–}35 \times 10^{-49}\text{Hz}^{-1}\text{sr}^{-1}(\text{extended sources}) \end{aligned} \tag{2.54}$$

and limits as low as 7×10^{-25} could be set on the rms GW strain emitted from selected interesting directions, like the galactic center or Sco X-1, at the best sensitivity of 160 Hz.

It is immediate, on the basis of Eq. 2.51, to see that a 10-fold improvement in the detectors sensitivity provides a limit on $H(f)$ 10 times better as well, and therefore 100 times better on $\Omega_{\rm GW}$, which will further constrain the theoretical models, restricting the range of the allowed parameters for the equation-of-state [106]. Most of the space of parameters for the cosmic (super) string models will become accessible [107], and even very exotic theories like the so-called pre-Big Bang models will be constrained [108].

2.5 Conclusions

A large experimental effort is underway in order to start, with the second generation of interferometric gravitational wave detectors, the era of gravitational astronomy. The successes of the first generation make us confident that the promised sensitivity improvements will be achieved, and the mature state of the analysis methods guarantees that all the accessible science will be harvested from the data. Science that, we have tried to argue, is rich: not only will we test general relativity in its most elusive prediction, the gravitational waves, but we will also shed light on some of the most energetic events in the Universe, we will acquire information about exotic states of the matter, and further extend our knowledge of the Big Bang; sufficient value, we hope, in return for taxpayer's money.

Acknowledgments The author thanks the LIGO Scientific Collaboration and the Virgo Collaboration for permission to use the sensitivity plots in Figs. 2.1 and 2.2; warm thanks are due to Chris Van Den Broek for a careful reading and for very useful comments.

References

1. http://www.ligo.org
2. http://www.virgo.infn.it
3. LIGO Scientific Collaboration, Virgo Collaboration, Prospects for Localization of Gravitational Wave Transients by the Advanced LIGO and Advanced Virgo Observatories, arXiv:1304.0670 [gr-qc] (2013)
4. B. Abbott et al. (The LIGO Scientific Collaboration), Rep. Prog. Phys. **72**, 076901 (2009)
5. T. Accadia et al. (The Virgo Collaboration), J. Instrum. **7**, P03012 (2012)
6. J. Abadie et al. (LIGO Scientific Collaboration and Virgo Collaboration), Phys. Rev. D **85**, 082002 (2012)
7. T. Accadia et al. (The Virgo Collaboration), Class. Quant. Grav. **28**, 025005 (2011)
8. G.M. Harry (for the LIGO Scientific Collaboration), Class. Quant. Grav. **27**, 084006 (2010)
9. The Virgo Collaboration, Advanced Virgo Technical Design Report, note VIR-0128A-12, https://tds.ego-gw.it/ql/?c=8940 (2012)
10. W.H. Press, K.S. Thorne, Annu. Rev. Astron. Astrophys. **10**, 335 (1972)
11. K.S. Thorne, *Gravitational Radiation* (1989), p. 330
12. D.R. Lorimer, Living Rev. Relat. **11**, (2008)
13. B. Sathyaprakash, B.F. Schutz, Living Rev. Relat. **12** (2009)
14. D.G. Blair, E.J. Howell, L. Ju, C. Zhao (Eds), *Advanced Gravitational Wave Detectors* (Cambridge University Press, Cambridge, 2012)
15. K. Riles, Prog. Part. Nucl. Phys. **68**, 1 (2013)
16. J.A. Faber, F.A. Rasio, Living Rev. Relat. **15** (2012)
17. J. Abadie et al. (LIGO Scientific Collaboration and Virgo Collaboration), Class. Quant. Grav. **27**, 173001 (2010)
18. L. Blanchet, Living Rev. Relat. **9** (2006)
19. G.B. Cook, Living Rev. Relat. **3**, (2000)
20. C. Kim, V. Kalogera, D. Lorimer, Effect of PSRJ0737-3039 on the DNS Merger Rate and Implications for GW Detection. arXiv:0608280 [astro-ph] (2006)
21. K. Belczynski, R.E. Taam, V. Kalogera, V. Rasio, T. Bulik, Astrophys. J. **662**, 504–511 (2007)
22. T. Bulik, K. Belczynski, A. Prestwich, Astrophys. J. **730**, 140 (2011)
23. R. Diehl et al. (INTEGRAL), Radioactive ^{26}Al from massive stars in the Galaxy. Nature **439**, 45–47 (2006)
24. C.D. Ott, Class. Quant. Grav. **26**, 063001 (2009)
25. C.D. Ott, Class. Quant. Grav. **26**, 204015 (2009)
26. H. Dimmelmeier, J.A. Font, E. Müller, Astrophys. J. **560**, L163–L166 (2001)
27. T. Zwerger, E. Müller, Astron. Astrophys. **267**, 623 (1993)
28. H. Dimmelmeier, C.D. Ott, A. Marek, H.-T. Janka, Phys. Rev. D **78**, 064056 (2008)
29. J. Aasi et al. (LIGO Scientific Collaboration and Virgo Collaboration), Phys. Rev. D **87**, 022002 (2013)
30. J. Abadie et al. (LIGO Scientific Collaboration and Virgo Collaboration), Phys. Rev. D **85**, 122007 (2012)
31. B. Abbott, et al., Astrophys. J. **681**, 1419 (2008)
32. K. Hurley, Adv. Space Res. **47**, 1326–1331 (2011)
33. R.C. Duncan, C. Thompson, Astrophys. J. Lett. **392**, L9–L13 (1992)
34. http://www.geo600.org
35. J. Abadie et al. (LIGO Scientific Collaboration and Virgo Collaboration), Astrophys. J. Lett. **734**, L35 (2011)
36. J. Abadie, et al. (The LIGO Scientific Collaboration), Phys. Rev. D **83**, 042001 (2011)
37. J. Abadie, et al. (LIGO Scientific Collaboration and Virgo Collaboration), Astrophys. J. **760**, 12 (2012)
38. J. Abadie, et al. (LIGO Scientific Collaboration and Virgo Collaboration), Class. Quant. Grav. **27**, 173001 (2010)

39. K. Cannon, C. Hanna, D. Keppel, Phys. Rev. D **84**, 084003 (2011)
40. J. Luan, S. Hooper, L. Wen, Y. Chen, Phys. Rev. D **85**, 102002 (2012)
41. S. Hooper, S.K. Chung, J. Luan, D. Blair, Y. Chen, L. Wen, Phys. Rev. D **86**, 024012 (2012)
42. K. Cannon et al., Astrophys. J. **748**, 136 (2012)
43. Q. Chu, L. Wen, D. Blair, JPCS **363**, 012023 (2012)
44. R.A. Hulse, J.H. Taylor, Astrophys. J. **195**, L51 (1975)
45. J.M. Weisberg, D.J. Nice, J.H. Taylor, Astrophys. J. **722**, 1030–1034 (2010)
46. T.G.F. Li et al., Phys. Rev. D **85**, 082003 (2012)
47. B.F. Schutz, Nature **323**, 310 (1986)
48. S. Nissanke, Astrophys. J. **725**, 496 (2010)
49. S.R. Taylor, J.R. Gair, I. Mandel, Phys. Rev. D **85**, 023535 (2012)
50. W. Del Pozzo, Phys. Rev. D **86**, 043011 (2012)
51. V. Ferrari et al., Phys. Rev. D **81**, 064026 (2010)
52. J.A. Faber, F.A. Rasio, Binary neutron star mergers. Living Rev. Relat. **15** (2012)
53. S. Rosswog, Compact binary mergers: an astrophysical perspective Invited review "Nuclei in the Cosmos 2010". *Proceedings of Science (NIC XI)*, 032
54. W. Del Pozzo et al., Phys. Rev. Lett. **111**, 071101 (2013)
55. R.F. Stark, T. Piran, Phys. Rev. Lett. **55**, 891 (1985)
56. R.A. Matzner et al., Science **270**, 941–947 (1995)
57. U. Sperhake et al., Phys. Rev. D **71**, 124042 (2005)
58. T.W. Baumgarte, S.L. Shapiro, Numerical Relativity: Solving Einstein's Equations on the Computer (Cambridge University Press, New York, 2010)
59. M. Shibata, K. Taniguchi, Coalescence of black hole-neutron star binaries. Living Rev. Relat. **14** (2011)
60. B. Aylott et al., Class. Quant. Grav. **26**, 165008 (2009)
61. M. Vallisneri, Phys. Rev. Lett. **84**, 3519 (2000)
62. F. Foucart et al., arXiv:1307.7685 [gr-qc] (2013)
63. S.E. Woosley, J.S. Bloom, The supernova—gamma-ray burst connection. Ann. Rev. Astron. Astrophys. **44**, 507–556 (2006)
64. W.H. Lee, E. Ramirez-Ruiz, New J. Phys. **9**, 17 (2007)
65. K.S. Thorne, Astrophys. J. **191**, 507–519 (1974)
66. F. Acernese et al. (The Virgo Collaboration), Class. Quant. Grav. **25**, 225001 (2008)
67. SWIFT, http://www.nasa.gov/mission_pages/swift/bursts/index.html
68. GLAST, http://www.nasa.gov/mission_pages/GLAST/team/index.html
69. Chandra, http://chandra.harvard.edu
70. X-MM Newton, http://xmm.esac.esa.int
71. XEUS, http://sci.esa.int/xeus
72. J. Borkowski D. Götz, S. Mereghetti et al., GCN Circ. 2920 (2004)
73. B. Abbott et al. (LIGO Scientific Collaboration), Phys. Rev. D **76**, 062003 (2007)
74. H.T. Janka, Explosion mechanisms of core-collapse supernovae. arXiv:1206.2503 [astro-ph.SR] (2012)
75. S. Fukuda et al., The super-kamiokande detector. Nucl. Instr. Meth. Phys. Res. A **501**, 418–462 (2003)
76. The LVD Collaboration, The 1 kton LVD neutrino observatory. *Proceedings of 27th ICRC*, vol. 1 (Hamburg, 2001), p. 1093
77. S.R. Klein, for the IceCube Collaboration IceCube, A cubic kilometer radiation detector. http://www.icecube.wisc.edu/science/publications/update2008/update.pdf
78. SuperNova Early Warning System, http://snews.bnl.gov
79. Sudbury Neutrino Observatory, http://www.sno.phy.queensu.ca
80. SNO+, http://snoplus.phy.queensu.ca
81. N. Arnaud et al., Phys. Rev. D **65**, 033010 (2002)
82. The KM3NeT Consortium, KM3NeT: conceptual design for a deep sea research infrastructure incorporating a very large volume neutrino telescope in the mediterranean sea, http://www.km3net.org/CDR/CDR-KM3NeT.pdf

83. S.A. Martínez et al. (ANTARES Collaboration, LIGO Scientific Collaboration and Virgo Collaboration), A first search for coincident gravitational waves and high energy neutrinos using LIGO, Virgo and ANTARES data from 2007, arXiv:1205.3018 [gr-qc] (2012)
84. J. Abadie et al. (LIGO Scientific Collaboration and Virgo Collaboration), Astron. Astrophys. **539**, A124 (2012)
85. P.A. Evans, J. Aasi, et al. (LIGO Scientific Collaboration and Virgo Collaboration), Astrophys. J. S **203**, 28 (2012)
86. C. Cutler, Phys. Rev. D **66**, 084025 (2002)
87. A. Mastrano, A. Melatos, A. Reisenegger, T. Akgün, Mon. Not. R. Astron. Soc. **417**, 2288 (2011)
88. L. Bildsten, Astrophys. J. Lett. **501**, L89 (1998)
89. C. Palomba, A&A **354**, 163–168 (2000)
90. B.P. Abbott et al. (LIGO Scientific Collaboration and Virgo Collaboration), Astrophys. J. **713**, 671 (2010)
91. J. Abadie et al. (LIGO Scientific Collaboration and Virgo Collaboration), Phys. Rev. D **85**, 022001 (2012)
92. J. Abadie et al. (LIGO Scientific Collaboration and Virgo Collaboration), Astrophys. J. **737**, 93 (2011)
93. The Virgo Collaboration, Advanced virgo baseline design, note VIR-027A-09, https://tds.ego-gw.it/ql/?c=6589 (2009)
94. R. Narayanayan, Astrophys. J. **319**, 162 (1987)
95. D. Page, M. Prakash, J.M. Lattimer, A.W. Steiner, Superfluid neutrons in the core of the neutron star in Cassiopeia A, arXiv:1110.5115 [astro-ph.HE] (2011)
96. J.S. Heyl, Astrophys. J. Lett. **574**, L57 (2002)
97. J. Abadie et al. (LIGO Scientific Collaboration and Virgo Collaboration), Phys. Rev. Lett. **107**, 271102 (2011)
98. M. Maggiore, Phys. Rep. **331**, 283 (2000)
99. T. Regimbau, Res. Astron. Astrophys. **11**, 369 (2011)
100. C. Wu, V. Mandic, T. Regimbau, Phys. Rev. D **85**, 104024 (2012)
101. E. Calabrese et al., Cosmological parameters from pre-planck CMB measurements. arXiv:1302.1841 [astro-ph] (2013)
102. B.P. Abbott et al. (LIGO Scientific Collaboration and Virgo Collaboration), Nature **460**, 990 (2009)
103. T.L. Smith, E. Pierpaoli, M. Kamionkowski, Phys. Rev. Lett. **97**, 021301 (2006)
104. S. Sarangi, S.H.H. Tye, Phys. Lett. B **536**, 185–192 (2002)
105. J. Abadie et al. (LIGO Scientific Collaboration and Virgo Collaboration), Phys. Rev. D **85**, 122001 (2012)
106. L. Boyle, A. Buonanno, Phys. Rev. D **78**, 043531 (2008)
107. X. Siemens, V. Mandic, J. Creighton, Phys. Rev. Lett. **98**, 111101 (2007)
108. V. Mandic, A. Buonanno, Phys. Rev. D **73**, 063008 (2006)

Chapter 3
Interferometer Configurations

Gabriele Vajente

Abstract Gravitational waves induce a differential strain between free-falling test masses. The most sensitive instruments to measure this kind of effect are laser interferometers. This chapter introduces the working principles of the different optical configuration that were and will be used in gravitational wave detectors: Michelson interferometer, Fabry-Perot resonant cavity, power and signal recycling techniques. Advanced detectors will feature high power levels, therefore the important issue of radiation pressure effects is addressed. Finally, a brief introduction to the topic of diffraction limited beams and high order transverse electromagnetic modes is included.

3.1 The Interferometer as a Transducer

The effect of a gravitational wave (GW) is a change in the local space-time metric, which can be considered equivalent to a change in the distance measured between free-falling test masses. The GW polarization nature tells us that the distance variation is intrinsically differential: two measurements performed along orthogonal directions provide opposite results. Given a baseline length L, the measured variation δL is linked to the gravitational wave amplitude (strain) h by the simple relation [1–3]

$$\delta L = \pm \frac{1}{2} h \, L \tag{3.1}$$

where the sign depends on the orientation of the length being measured with respect to the wave polarization. To all practical effects, the detection of a gravitational wave

G. Vajente (✉)
INFN Sezione di Pisa, Largo B. Pontecorvo 3, 56124 Pisa, Italy

California Institute of Technology, MC 100-36, 1200 E. Colorado Blvd,
Pasadena, CA 91125, USA
e-mail: vajente@caltech.edu

M. Bassan (ed.), *Advanced Interferometers and the Search for Gravitational Waves*,
Astrophysics and Space Science Library 404, DOI: 10.1007/978-3-319-03792-9_3,

signal is equivalent to the measurement of a small differential variation of the distance between test masses.

Interferometers are today among the most sensitive instruments to measure such relative distance changes. Any optical configuration chosen for our interferometer can be considered as a transducer, which converts a length variation (caused hopefully by a gravitational wave signal) to a variation of the phase of a laser field. One of the key parameters of any detector is its *optical gain* G which gives the phase of the output field as a function of the length variation. Since the expected effect of a gravitational wave is really small, any instrument can be considered as linear:

$$\delta\phi = G\,\delta L \tag{3.2}$$

The optical gain G is in principle a function of the signal frequency. The first goal in designing a gravitational wave detector is to make its gain as large as possible with respect to the intrinsic noise limitation of the measurement. In other words, we want to optimize our system in order to have the largest possible signal-to-noise ratio (SNR).

In this chapter, we are not taking into account how to measure the phase variation of the output laser field, or in other words how to build a phase-meter. This will be one of the topics covered in Chap. 6. We simply assume to have a perfect way to produce a signal proportional to the field phase variation with respect to a reference value.

3.1.1 Plane Wave Description of Laser Fields

The electromagnetic field at any point inside the interferometer is in principle described by the vector amplitudes of both magnetic and electric fields. In most of the applications of our interest however the input laser beam is polarized. Therefore only the amplitude and the phase of the field is relevant, meaning that we can use a scalar description. It is also customary to use complex number representations:

$$\Psi = Ae^{i\phi} \tag{3.3}$$

Propagating solutions of the vacuum Maxwell equations can be described in terms of plane waves [4] of well-defined frequency, corresponding to the main laser wavelength λ:

$$\Psi(t,z) = \Psi_0 e^{i\omega t - ikz} \tag{3.4}$$

$$\omega = \frac{2\pi c}{\lambda} \tag{3.5}$$

$$k = \frac{2\pi}{\lambda} = \frac{\omega}{c} \tag{3.6}$$

In the so-called *plane-wave approximation*, we can describe the laser field at any point inside the interferometer as a coherent plane wave, thus neglecting the transverse shape of the laser beam generated by diffraction effects. This approximation is good enough to describe most of the dynamics of the laser field inside a GW interferometer. Later on in this chapter, we will give a brief introduction on diffraction effects.

In summary, we will consider only the longitudinal evolution of the laser field amplitude and phase, assigning to each point inside our optical system a complex number which gives the amplitude of the field at that point. We can also choose our units in such a way that the power at each point is simply given by the squared modulus of the field:

$$P(t, z) = |\Psi(t, z)|^2 \tag{3.7}$$

Our optical system is mainly composed of mirrors, therefore it is important to properly describe what happens when the laser impinges on a reflecting surface. The amplitude of the reflected and transmitted fields are linked to the incoming one by the (real valued) *reflectivity* r and *transmissivity* t of the mirror:

$$|\Psi_r| = r|\Psi_i| \tag{3.8}$$

$$|\Psi_t| = t|\Psi_i| \tag{3.9}$$

The conservation of energy requires that

$$r^2 + t^2 = 1 - L \tag{3.10}$$

where L are the total power losses in the mirror, due for example to internal absorption of light. Equations 3.8 and 3.9 are still missing the phase information to completely determine the output fields. There is some freedom in choosing the phases. Here we follow the convention of [5]:

$$\Psi_r = ir\Psi_i \tag{3.11}$$

$$\Psi_t = t\Psi_i \tag{3.12}$$

The need of the additional i comes again from the conservation of energy, which requires a phase jump of $\pi/2$ between the transmitted and reflected fields, as explained in [5, Sect. 1.1.2]. Different conventions are used in other texts, but the physical results are clearly equivalent.

3.1.2 Effect of a Gravitational Wave

Let us consider a laser beam propagating along a distance L. The effect of a gravitational wave passing by is a variation of the distance, as seen by the laser, of an amount proportional to the GW amplitude. If Ψ_0 is the laser field at the beginning of the space, the propagated one will be given by

$$\Psi_L = \Psi_0 \cdot e^{-ik(L+\frac{h(t)}{2}L)} = \Psi_L^{(0)} e^{-ikL\frac{h(t)}{2}} \tag{3.13}$$

The effect is therefore an additional dephasing, with respect to the propagation in the absence of GW, proportional to the GW amplitude. In the previous equation, we have assumed that the GW amplitude varies on a timescale which is much longer than the typical period of variation of the source laser field. If the GW signal is monochromatic $h(t) = h_0 \cos(\omega_{gw} t)$, then the previous equation can be rewritten at first order:

$$\Psi_L = \Psi_0 e^{-ikL} \left(1 + \frac{ikLh_0}{4} e^{i\omega_{gw}t} + \frac{ikLh_0}{4} e^{-i\omega_{gw}t}\right) \tag{3.14}$$

In other words, the effect of a GW passing through our detector is the creation of two *signal sidebands* around the main *carrier* field. The laser is no more purely monochromatic, but it contains two new components spaced by the GW frequency from the main laser one. Since the frequency of the GW we expect to be able to detect is comprised between 10 Hz and 10 kHz, these sidebands are also commonly called *audio sidebands*.

3.2 The Michelson Interferometer

Since GW are intrinsically differential, the best configuration to convert them to an optical signal is a Michelson interferometer, see Fig. 3.1. The input laser beam is entering the system from the *symmetric port* and it is divided into two equal parts by the *beam-splitter*. The two beams created in this way travel along the two km-long arms, are reflected back to the central vertex and recombine at the beam-splitter. The interference condition of the returning fields, exiting through the *antisymmetric port* can be tuned by changing the microscopic length difference between the two arms.

Mathematically, using the conventions shown in the figure, the following relations between the fields at different positions are easily found:

$$\begin{aligned}
\Psi_2 &= \frac{i}{\sqrt{2}}\Psi_1 + \frac{1}{\sqrt{2}}\Psi_{11} \\
\Psi_5 &= e^{-2ikL_Y} i r_Y \Psi_2 \\
\Psi_6 &= \frac{1}{\sqrt{2}}\Psi_1 + \frac{i}{\sqrt{2}}\Psi_{11} \\
\Psi_9 &= e^{-2ikL_X} i r_X \Psi_6 \\
\Psi_{10} &= \frac{i}{\sqrt{2}}\Psi_9 + \frac{1}{\sqrt{2}}\Psi_5 \\
\Psi_{12} &= \frac{1}{\sqrt{2}}\Psi_9 + \frac{i}{\sqrt{2}}\Psi_5
\end{aligned}$$

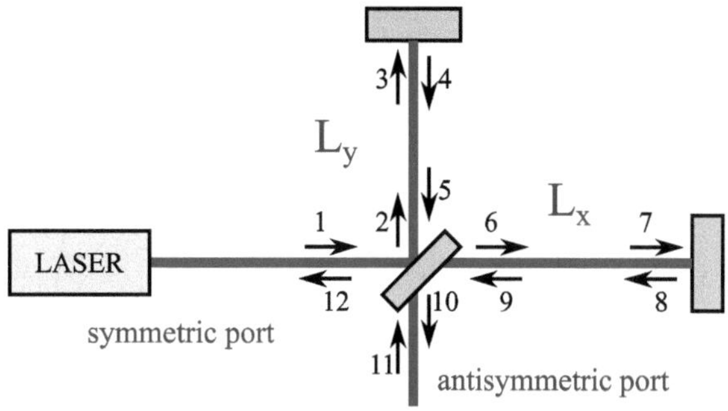

Fig. 3.1 Scheme of a simple Michelson interferometer

Here the reflectivity of the two end mirrors are indicated by r_X and r_Y and for simplicity we assumed that the beam splitter is perfectly balanced, i.e., $|r_{BS}| = |t_{bs}| = 1/\sqrt{2}$. For the moment being, no input from the antisymmetric port is considered. For our purposes, the only interesting fields are those exiting from the symmetric and antisymmetric ports, which can be found solving the above equations:

$$\Psi_{ASY} = -\frac{1}{2}\left(r_X e^{-2ikL_X} + r_Y e^{-2ikL_Y}\right)\Psi_1 \quad (3.15)$$

$$\Psi_{SYM} = \frac{i}{2}\left(r_X e^{-2ikL_X} - r_Y e^{-2ikL_Y}\right)\Psi_1 \quad (3.16)$$

In the ideal world, the two mirrors will be perfectly equal. However, in reality we might consider that the reflectivity of the two mirrors can be slightly unbalanced

$$r_X = r + \frac{\delta r}{2} \quad (3.17)$$

$$r_Y = r - \frac{\delta r}{2} \quad (3.18)$$

Moreover we assume the two arms to have possible different lengths:

$$L_X = L + \frac{\delta L}{2} \quad (3.19)$$

$$L_Y = L - \frac{\delta L}{2} \quad (3.20)$$

Using these new conventions we can rewrite the field at the antisymmetric port as follows:

$$\Psi_{ASY} = -re^{-2ikL}\left(\cos k\delta L - \frac{i}{2}\frac{\delta r}{r}\sin k\delta L\right)\Psi_1 \quad (3.21)$$

The power exiting through the antisymmetric port can be compute with the squared modulus of Eq. 3.21:

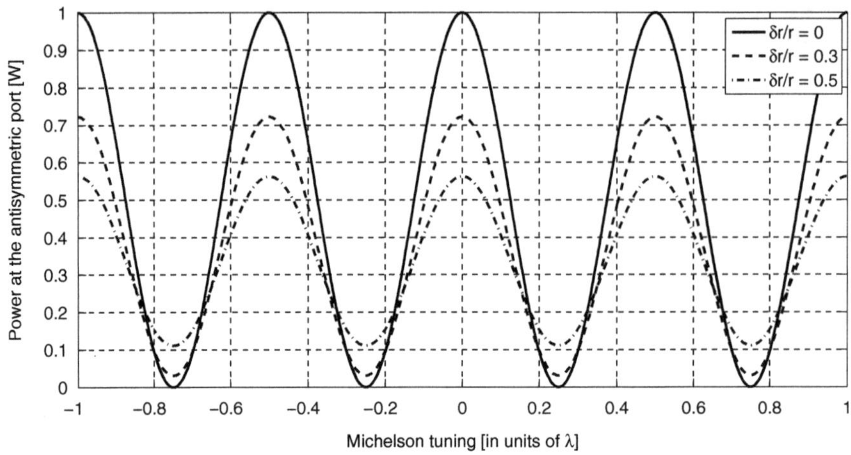

Fig. 3.2 Interference fringes in a Michelson interferometer, visible in the power exiting at the antisymmetric port as a function of the microscopic length difference of the two arms

$$P_{\mathrm{ASY}} = \left[\left(r^2 - \frac{\delta r^2}{4r^2} \right) \cos^2 k\delta L + \frac{\delta r^2}{4r^2} \right] P_{\mathrm{input}} \tag{3.22}$$

It is easy to show that the minimum possible power at the antisymmetric port is obtained (regardless of the reflectivity asymmetry) when the length difference is chosen such to have $\cos k\delta L = 0$. This is a microscopic condition, in the sense that it tells us that the length difference must be an odd multiple of half wavelength

$$\delta L = (2n + 1)\frac{\lambda}{2} \tag{3.23}$$

This condition can clearly be satisfied by an infinite number of lengths. Once the macroscopic length difference is set, one still has lots of possible values all differing by multiples of one wavelength (the so-called *interference fringes*), see Fig. 3.2. If we want to use the power at the antisymmetric port to measure the relative length change of the two arms, the best position would be the one that maximizes the slope, which is obtained at the gray fringe condition, which is the one corresponding to half of the maximum power.

The presence of asymmetries, such as the reflectivity δr makes it impossible to reach perfectly destructive interference, as visible in Fig. 3.2 where different values of $\delta r/r$ are considered. The minimum power attainable at the antisymmetric port defines the contrast defect:

$$1 - C = \frac{2P_{\mathrm{min}}}{P_{\mathrm{min}} + P_{\mathrm{max}}} \tag{3.24}$$

3.2.1 Common and Differential Signals

As we said we are using the interferometer as a transducer to convert GW signals (in the form of a differential variation $\delta L = \delta L_0 + d$ of the two arm lengths) to an optical signal. We will now compute the response of a Michelson interferometer to such a signal. In addition, we will consider how the system responds to a *common mode signal*, in this case a variation of the mean length of the two arms $L = L_0 + c$. If we consider only small signals, i.e., d and c are much smaller than one wavelength, the response of our Michelson interferometer can be computed simply taking the derivative of Eq. 3.21 with respect to L and δL:

$$\frac{d\Psi_{\mathrm{ASY}}}{dL} = 2ikre^{-2ikL_0}\left[\cos k\delta L_0 - \frac{i}{2}\frac{\delta r}{r}\sin k\delta L_0\right]\Psi_1 \tag{3.25}$$

$$\frac{d\Psi_{\mathrm{ASY}}}{d\delta L} = kre^{-2ikL_0}\left[\sin k\delta L_0 + \frac{i}{2}\frac{\delta r}{r}\cos k\delta L_0\right]\Psi_1 \tag{3.26}$$

We want to maximize the response of our detector to differential motions and, if possible, minimize the response to common motions. This happens for exactly the same condition that makes the power minimum as shown in Eq. 3.23. This is the best working point for our system. It is often called the *dark fringe condition*. The field at the antisymmetric port simplifies to:

$$\Psi_{\mathrm{ASY}} = re^{-2ikL_0}\frac{2\pi}{\lambda}\left(d - \frac{\delta r}{r}c\right)\Psi_1 \tag{3.27}$$

This equation shows one of the peculiar characteristic of Michelson interferometers: if the system is symmetric, common mode signals are not reaching the antisymmetric port. In the real world, any asymmetry will introduce a coupling of common modes to the output signal, as shown in the previous equation for the case of a mirror reflectivity asymmetry. This property of Michelson interferometers is measured by the *common mode rejection factor*, which is given in this case by $\delta r/r$.

3.2.2 Shot Noise

Equation 3.27 shows that one can improve the sensitivity of a Michelson interferometer to a GW signal by increasing the reflectivity of the end mirrors (that cannot be larger than 1 in any case) or by increasing the input laser power. Clearly, also an increase of the arm length will result in a larger differential displacement with the same GW amplitude.

The amplitude of the output signal of the Michelson interferometer must be compared to the fundamental noise limits on the read-out of such signal. We'll see more about this in a following chapter. For the moment we can say that we are using optical sensors to convert the laser field fluctuations to electric signals. The fundamental limit comes from the shot-noise of the laser itself.

The poor man explanation is that every time you have a fixed amount of power reaching your photo-sensor, the average number of photons impinging on it is fixed, but the statistics of arrivals is Poissonian. The number of photons actually hitting the diode in the unit of time is fluctuating, resulting in a limiting noise source. The time-independent nature of the fluctuations give rise to a white noise, or in other words to a noise which is independent of the frequency.

A better explanation of shot noise is to consider the quantum fluctuations of the laser field, which are converted in classical fluctuation by the sensor. You will learn more about this topic reading Chap. 11.

Without entering into too many details (that can be found in [5, Sect. 1.1] and in [6]), the shot-noise level for a certain amount of power impinging on a diode is given by the following equation

$$n_S = \sqrt{2h_P \nu P} \tag{3.28}$$

where ν is the laser optical frequency, h_P the Planck constant and P the power reaching the diode. One can argue that if a Michelson interferometer is tuned at dark fringe this value would be zero. However, we will see in Chap. 6 that this is not possible, since we need to send some additional power at the dark port in order to extract the signal. Indeed, if we tune the interferometer to have perfect destructive interference, Eq. 3.27 shows that the field will be proportional to the GW signal. Photo-diode however will detect the power, which is the squared value of the field, as given by Eq. 3.22. Therefore we will have a signal proportional only to the square of the GW signal. Since the GW is very small, its squared value is essentially zero and therefore our system will be perfectly insensitive to GWs. If we want to use the power at the antisymmetric port of a Michelson interferometer to measure the differential length change, we need to move a little bit further away from the dark fringe condition, increasing the slope but also increasing the shot noise. One can show that there is an optimum offset from dark fringe, see [1, Sect. 1.1.3] for more details.

3.3 Improving the Sensitivity: Resonant Fabry-Perot Cavities

Michelson interferometers are very simple, but unfortunately their sensitivity is not good enough. The first additional technique that is implemented in GW detectors is the replacement of the two arms with long resonant cavities.

3.3.1 Static Response of a Fabry-Perot Cavity

A Fabry-Perot cavity is made of two mirrors aligned to the optical axis: the first one (*input mirror*) is partially reflecting (with reflectivity r_i) and the second one (*end mirror*) is almost completely reflective (typically t_e^2 is few parts per million). Referring to Fig. 3.3 for the definition of the fields, one can find simple relations among them, as done in the Michelson case:

$$\begin{aligned}\Psi_2 &= t_i\Psi_1 - r_i r_e e^{-2ikL}\,\Psi_2\\ \Psi_5 &= i r_e e^{-2ikL}\Psi_2\\ \Psi_6 &= t_i\Psi_5 + i r_i\Psi_2\end{aligned}$$

We are particularly interested in the reflected field (Ψ_6) and in thc intra-cavity field (Ψ_2):

$$\Psi_{\text{cav}} = \frac{t_i}{1 + r_i r_e e^{-2ikL}}\,\Psi_1 \tag{3.29}$$

$$\Psi_{\text{ref}} = i\,\frac{r_i + r_e(t_i^2 + r_i^2)e^{-2ikL}}{1 + r_i r_e e^{-2ikL}}\,\Psi_1 \tag{3.30}$$

From Eq. 3.29 we infer that for a proper tuning of the microscopic length that gives

$$e^{-2ikL} = -1 \tag{3.31}$$

the power stored inside the cavity is maximum and given by

$$P_{\text{cav}} = \frac{t_i^2}{(1 - r_i r_e)^2}\,P_{\text{in}} = G_{\text{cav}}\,P_{\text{in}} \tag{3.32}$$

where we have defined the cavity gain G_{cav}. The condition of Eq. 3.31 shows also that there are multiple resonant peaks, spaced in length by half wave-length and in frequency by the cavity *free spectral range* of

$$\delta f_{\text{FSR}} = \frac{c}{2L} \tag{3.33}$$

If we consider a detuning δL around one of the resonant positions, the power stored inside the cavity is given by

$$P_{\text{cav}} = \frac{t_i^2}{(1 - r_i r_e)^2 + 4 r_i r_e \sin^2 k\delta L}\,P_{\text{in}} = G_{\text{cav}}\,\frac{1}{1 + \left[\frac{2\sqrt{r_i r_e}}{1 - r_i r_e}\sin\frac{2\pi\delta L}{\lambda}\right]^2}\,P_{\text{in}} \tag{3.34}$$

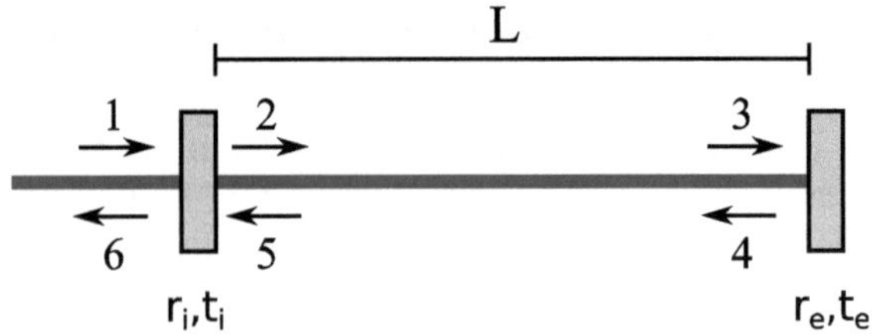

Fig. 3.3 Scheme of a resonant Fabry-Perot cavity

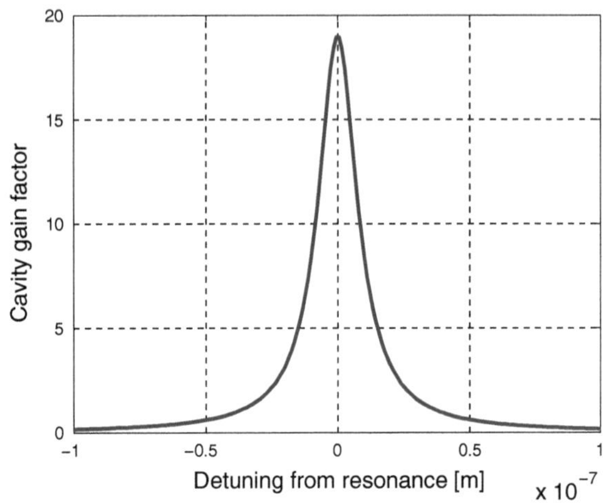

Fig. 3.4 Example of how the power stored inside a resonant Fabry-Perot cavity depends on the detuning from the resonance position. In this example the input mirror reflectivity is 0.9, corresponding to a relatively low finesse of about 30

See Fig. 3.4 for an example of the dependence of the stored power on the cavity microscopic length.

The detuning from resonance that gives half of the maximum power (*half width at half maximum*) is determined by

$$\delta L_{\mathrm{HWHM}} = \frac{\lambda}{4\mathscr{F}} \tag{3.35}$$

where we defined the *finesse* of the cavity

$$\mathscr{F} = \frac{\pi\sqrt{r_i r_e}}{1 - r_i r_e} \tag{3.36}$$

Equation 3.34 can be re-written in terms of the finesse as

$$P_{\mathrm{cav}} = \frac{G_{\mathrm{cav}}}{1 + \left(\frac{2\mathscr{F}}{\pi}\right)^2 \sin^2 \frac{2\pi\,\delta L}{\lambda}} P_{\mathrm{in}} \tag{3.37}$$

In gravitational wave interferometers the reflectivities are relatively high, and therefore the cavity finesse is much larger than unity (>50). This allows us to use

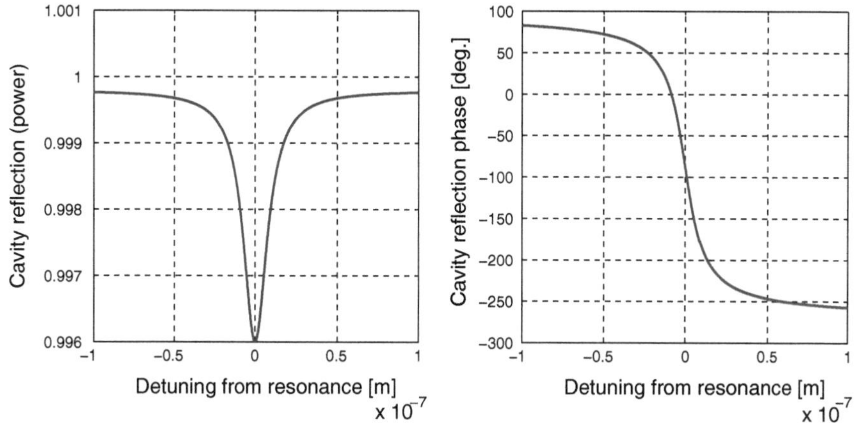

Fig. 3.5 Example of the reflection of a Fabry-Perot cavity as a function of the detuning from resonance. The cavity considered here has the same parameters used for Fig. 3.4, plus additional (large) round-trip losses of 200 ppm

the following approximate relation between the finesse and the product of the two reflectivities

$$r_i r_e = 1 - \frac{\pi}{\mathscr{F}}\sqrt{1 + \frac{\pi^2}{4\mathscr{F}^2}} - \frac{\pi^2}{2\mathscr{F}^2} \sim 1 - \frac{\pi}{\mathscr{F}} \tag{3.38}$$

Let's now turn our attention to the expression for the cavity reflectivity, given in Eq. 3.30. The term $r_i^2 + t_i^2$ is equal to one minus all the power losses in the input mirror $1 - \mathscr{L}_i$. This term can be used to account for all the sources of power loss inside the cavity, since in GW detectors these are typically small (required to be below 75 ppm in Advanced Detectors). When the cavity crosses the resonance, the amplitude of the reflected field is indeed slightly reduced (see Fig. 3.5): this is due to the fact that the circulating power is larger and the losses are more relevant. However, the most interesting feature of the reflected field is its phase variation, see again Fig. 3.5. As often happens with classical oscillators, the phase undergoes a π shift when passing through resonance. One moreover notices that when the cavity is close to resonance, the slope of the phase variation in reflection with respect to length change can get pretty large. Here is why Fabry-Perot cavities are used in GW detectors: they can optically amplify the dephasing due to a GW.

It can be easily shown with some computations, taking into account that the finesse is typically much larger than unity and total losses are small, that the slope around resonance is given by

$$\frac{d\phi}{dL} = \frac{8\mathscr{F}}{\lambda} \tag{3.39}$$

This must be compared with the case of propagation in free space, without any resonant cavity, having only the end mirror to bounce back the laser beam. The dephasing would be in this case simply:

$$\frac{d\phi}{dL} = \frac{4}{\lambda} \tag{3.40}$$

In conclusion, using a Fabry-Perot cavity we can amplify the dephasing in reflection by a factor proportional to the finesse. If two cavities like these are used in the two arms of a Michelson interferometer, the gain in sensitivity is also proportional to the finesse of the cavities. It would therefore seem a good idea to use cavities with very high finesse. However, there are many drawbacks to this solution.

We have already seen the first one: if we increase the cavity finesse, maintaining similar losses inside the cavity, the reflectivity at resonance will be smaller, since the power build-up will be larger. Therefore at the end of the day, if the finesse is too large we start losing our precious laser power inside the resonant cavity.

The second drawback is linked to the cavity line-width, which becomes smaller and smaller as the finesse increases. To maintain proper resonance conditions the length must be controlled with an accuracy much better than the line-width. This will be one of the topics of Chap. 6: however, it is obvious that the more stringent the accuracy requirement is, the more difficult it is to control the cavity length.

3.3.2 Frequency Response of a Fabry-Perot Cavity

The last and more important drawback in using very high finesse for the arm cavity is the dependence of their response on the frequency, which we neglected so far. Referring to Fig. 3.6, we can model the effect of a change in the distance between the two test masses as a motion of the end mirror with a sinusoidal time dependence at a signal frequency f_s:

$$x(t) = x_0 \cos(2\pi f_s t) = x_0 \cos \omega_s t \tag{3.41}$$

The field which is reflected from the end mirror gets an additional dephasing due to the motion. Assuming $x_0 \ll \lambda$, this can be modeled, as explained in Sect. 3.1.2, with the creation of two signal sidebands at the signal frequency:

$$\begin{aligned}\Psi_4 &= i r_e e^{-2ikx(t)} \Psi_3 = i r_e \Psi_3 + 2k r_e x(t) \Psi_3 &(3.42)\\ &= i r_e \Psi_3 + 2\pi \frac{x_0}{\lambda} \left(e^{-i\omega_s t} + e^{i\omega_s t} \right) \Psi_3 &(3.43)\end{aligned}$$

The effect of the moving mirror is therefore to inject two new signal sidebands at frequencies $\pm f_s$ inside the cavity, with an amplitude proportional to the displacement and to the static field inside the cavity. These additional sidebands propagate inside the cavity and undergo interference, being partially recycled. If we consider only the field at the signal frequency $+f_s$ we get the following equation

Fig. 3.6 Scheme of a resonant Fabry-Perot cavity with a moving end mirror

$$\Psi_4(\omega_s) = ir_e ir_i e^{-2i(k+\frac{\omega_s}{c})L}\Psi_4(\omega_s) + 2r_e\pi\frac{x_0}{\lambda}\Psi_3(0) \tag{3.44}$$

$$\Psi_4(\omega_s) = \frac{2\pi\Psi_3(0)}{1 - r_i r_e e^{-2i\frac{\omega_s}{c}L}}\frac{x_0}{\lambda} \tag{3.45}$$

and in the last equation we used the fact that the carrier is resonant in the cavity to set $e^{-2ikL} = 1$. We are again interested in the field exiting from the reflection port of the cavity, which is easily computed:

$$\Psi_R(\omega_s) = \left[\frac{2\pi}{\lambda}r_e\frac{it_i e^{-i\frac{\omega_s}{c}L}}{1 - r_i r_e e^{-2i\frac{\omega_s}{c}L}}\frac{it_i}{1 - r_i r_e}\Psi_1\right]x_0 \tag{3.46}$$

If we consider signal frequencies small enough (such that $\frac{\omega_s L}{c} \ll 1$) and we remember the expression in Eq. 3.38 for the reflectivities of a high finesse cavity, we can simplify Eq. 3.46 to the final result:

$$\Psi_R(\omega_s) = -\frac{2\pi}{\lambda}r_e\frac{1}{1 + i\frac{f_s}{\frac{c}{4L\mathcal{F}}}}\frac{t_i^2}{(1 - r_i r_e)^2}\Psi_1\, x_0 \tag{3.47}$$

The third fraction in the previous equation is telling us what we already discovered before: the response of a Fabry-Perot cavity is enhanced by the increased stored power inside it. However now we have also a frequency dependent term, given by the second fraction. This is a *simple pole* at a frequency given by the free spectral range of the cavity divided by twice the finesse. This is the *cavity pole*: the frequency response of the cavity is almost flat below it, and decreases like $1/f$ above, see Fig. 3.7.

In conclusion, if we use a Fabry-Perot resonant cavity in a GW interferometric detector, we can increase the sensitivity of a factor proportional to the finesse, but the bandwidth of our detector is limited by the cavity pole. It is important to note that the origin of the gain in sensitivity and the limitation of bandwidth is slightly different. We gain in sensitivity because of the increased circulating power inside the cavity: any variation of the length will create signal sidebands with an amplitude proportional to the static circulating field. These sidebands are then filtered, since they are forced to circulate inside the cavity: for low frequencies there is an additional gain since the sidebands are inside the cavity resonant line-width; for higher frequency the sidebands get an additional dephasing (due to their not zero frequency) which starts

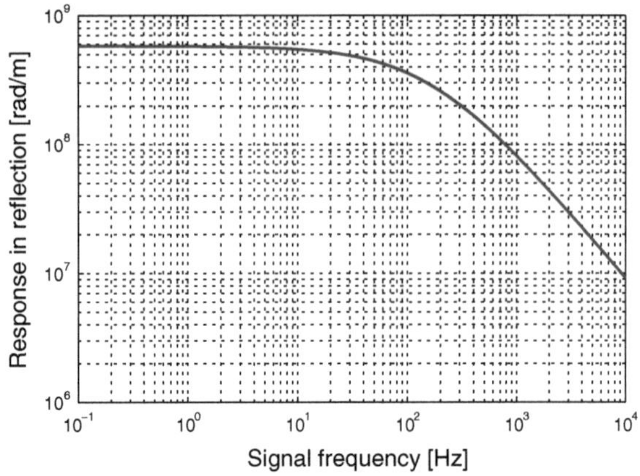

Fig. 3.7 Absolute value of the response in reflection of a Fabry-Perot cavity to a length variation, as a function of the frequency. The cavity considered here have a length of 3 km and a finesse of about 150, resulting in a pole frequency of about 160 Hz

to destroy the constructive resonance and therefore reduce the gain. The distinction between these two effects might seem a little bit academic here, but its relevance will become more clear when considering the signal recycling technique.

3.4 Increasing the Circulating Power: Power Recycling Technique

So far we improved the sensitivity by replacing the Michelson arms with resonant Fabry-Perot cavities. In this way, we could largely improve the detector sensitivity at low frequencies (below few hundreds Hz). Another simple way to improve the signal-to-noise ratio is to increase the laser power: as already stated in Eq. 3.28 the shot noise increases proportionally to the square root of the laser power, while the signal typically increases with the power. Therefore we can gain in SNR proportionally to the square root of the power which is stored inside the arm cavities.

When the Michelson interferometer is tuned close to the dark fringe condition, no power exits through the anti-symmetric port. Moreover, we already saw that the arm cavities have reflectivities very close to unity, therefore almost no power is transmitted through the end mirrors and a very small amount is lost inside the Fabry-Perot cavities. It's a simple matter of energy conservation to understand that in this configuration, almost all the input power is reflected back to the symmetric port. In other words, the Michelson interferometer is seen, from the symmetric port, as an equivalent mirror with very high reflectivity, given by the mean of the two arm resonant cavity reflectivities. Here comes into play the *power recycling technique*: an additional semi-transparent mirror is added at the symmetric port, see Fig. 3.8. This mirror forms, together with the rest of the Michelson interferometer, an additional resonant cavity (the *power recycling cavity* PRC). If the length of this new cavity is properly controlled to maintain constructive resonance, we can increase the power circulating

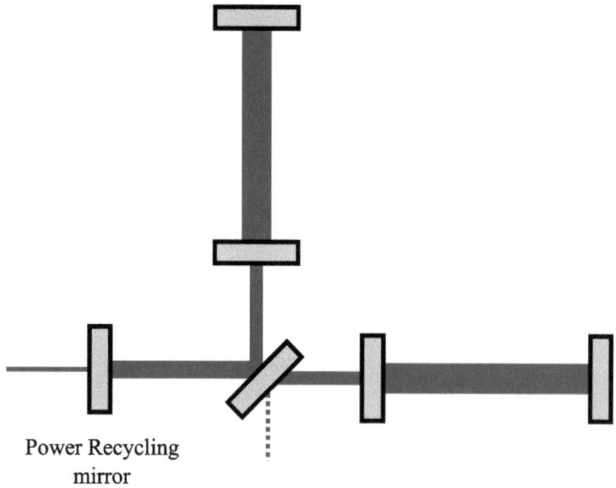

Fig. 3.8 Scheme of a power recycled Michelson interferometer with resonant Fabry-Perot cavities in the arms

inside it as in the case of the arm Fabry-Perot. In this way, the effective laser power that enters into the arms is increased by the *recycling gain* which can be computed with a variation of Eq. 3.32 where the end mirror reflectivity is replaced by the Fabry-Perot one at resonance. In the hypothesis of large finesse $\mathscr{F}$ and small round-trip losses $\mathscr{L}$, the arm cavity reflectivity (Eq. 3.30) can be quite well approximated by

$$R = -i(1 - \frac{\mathscr{F}}{\pi}\mathscr{L}) \tag{3.48}$$

This means that the reflectivity of the Michelson interferometer is completely determined by the losses inside the arms and by their finesse. It is obvious that one should make the losses as small as possible, to increase the effective reflectivity. The recycling gain is then given by

$$G_{\text{rec}} = \frac{t_R^2}{\left[1 - r_R\left(1 - \frac{\mathscr{F}}{\pi}\mathscr{L}\right)\right]^2} \tag{3.49}$$

where r_R and t_R are the reflectivity and transmissivity of the power recycling mirror. Figure 3.9 shows the dependence of the recycling gain on the PR mirror reflectivity. The maximum possible gain is obtained for a reflectivity of the PR mirror which is equal to the effective reflectivity of the Michelson interferometer. In this case, the PR cavity is *optimally coupled* and the reflected power is zero. However, it is clear from Fig. 3.9 that the gain is not much decreased for small changes of the PR reflectivity away from the optimal one. It is finally easy to compute the maximum possible recycling gain. Again in the hypothesis of large finesse and small losses, it is simply given by

$$G_{\text{max}} = \frac{\pi}{2\mathscr{F}\mathscr{L}} \tag{3.50}$$

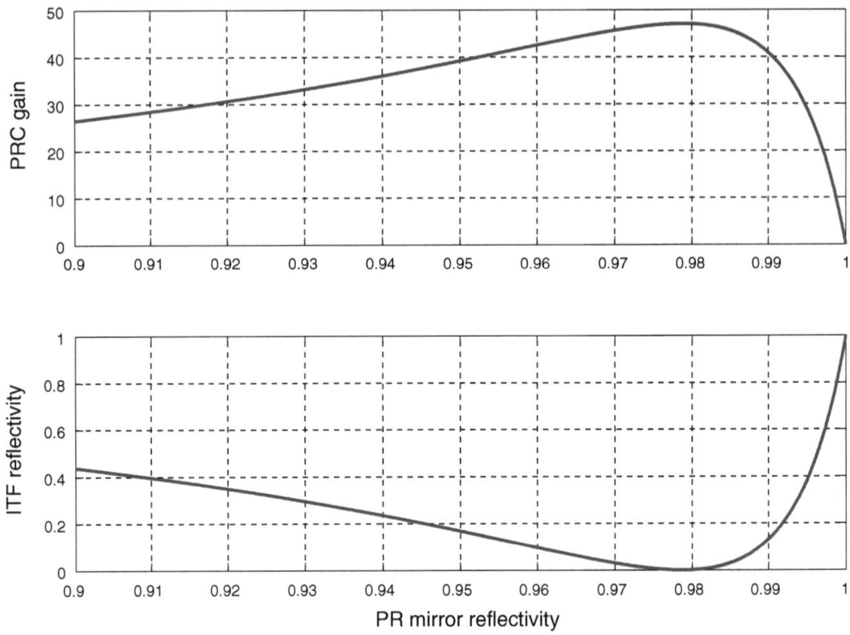

Fig. 3.9 Dependence of the power recycling gain on the PR mirror reflectivity. Here the arm Fabry-Perot cavities have a finesse of 450 and round-trip losses of 75 ppm, as requested in the Advanced Virgo design

and we see again that the smaller the losses the larger the gain in circulating power we can get by installing a power recycling mirror.

The last important point to understand is the effect of the power recycling cavity on the detector sensitivity, or in other words on the differential signal sidebands.

As explained in Sect. 3.2, a GW will create a differential displacement in the two arms, which will translate in a pair of signal sidebands in reflection of each cavity. The differential nature of GW will transfer to the sidebands: those coming from the north arm will have a π phase difference with respect to those coming from the west one. This implies that the signal sidebands interfere at the beam splitter with an additional dephasing with respect to the main beam, reversing the destructive and constructive interference. In other words, signal sidebands coming from differential displacements are completely transmitted to the antisymmetric port where they sum up. Therefore, the presence of the power recycling cavity is completely irrelevant for them, since they are not exiting through the symmetric port.

In conclusion, the addition of the power recycling cavity has the only effect of increasing the circulating power inside the interferometer and proportionally increasing the detector sensitivity.

The power-recycled Fabry-Perot Michelson configuration is the one that was chosen by most of the first generation interferometric detectors (Virgo, LIGO, and

TAMA) and will very likely be the first step also for the second generation of advanced detectors.

3.5 Shaping the Optical Response: Signal Recycling Technique

Here we are finally entering the realm of advanced gravitational wave detectors. As explained in the previous section, signal sidebands coming from differential displacements are completely transmitted to the antisymmetric port and unaffected by whatever happens in the symmetric part. On the contrary, the carrier field is completely reflected back to the symmetric port and not transmitted to the antisymmetric one. One can therefore add a mirror at the antisymmetric port [7], without affecting at all the carrier behavior inside the interferometer. Only differential signal sidebands will reach this mirror and will be partially reflected back into the interferometer, entering from the antisymmetric port. It is easy to see with symmetry considerations, that anything entering from the antisymmetric port will be transmitted only toward the two arms, with opposite signs, creating again differential sidebands inside the main interferometer. In conclusion, the newly installed *signal recycling* (SR) mirror will create an additional *signal recycling cavity* (SRC) which will be seen only by the signal sidebands. This cavity can be used to change the frequency response of the instrument.

The exact behavior of a dual recycled (signal and power) interferometer can be computed considering the power-recycled interferometer as a single box. Referring to Fig. 3.10, we consider the additional signal sidebands $\Psi_2'(\omega_s)$ that are exiting the power recycled interferometer at the antisymmetric port as a source. These sidebands are given by an equation that puts together all the elements we computed in the previous sections:

$$\Psi_2'(\omega_s) = -e^{i\pi/4 - ikl_+} \frac{2t_R}{1 - r_R R} \Psi_{in} \frac{t_i^2 r_e}{1 - r_i r_e} \frac{e^{-i\frac{\omega_s}{c}L}}{1 - r_i r_e e^{-2i\frac{\omega_s}{c}L}} \frac{2\pi}{\lambda} x_0 \tag{3.51}$$

In this equation, one can recognize the Fabry-Perot cavity frequency dependent response and the power increase due to the presence of the power recycling cavity. The phase term in front of everything comes from the constructive interference at the beam-splitter and takes into account the propagation of the signal sidebands from the arm cavity input mirror to the beam-splitter with the average length of the central Michelson part $l_+ = (l_N + l_W)/2$. This field is then recycled by the SR mirror:

$$\Psi_2(\omega_s) = \Psi_2'(\omega_s) + i r_S R_{\text{ASY}}(\omega_s) e^{-2i(k+\frac{\omega_s}{c})l_s} \Psi_2(\omega_s) \tag{3.52}$$

$$\Psi_{\text{ASY}}(\omega_s) = t_s e^{-i(k+\frac{\omega_s}{c})l_s} \Psi_2(\omega_s) \tag{3.53}$$

where $R_{\text{ASY}}(\omega_s)$ is the reflectivity of the entire power recycled interferometer for the signal sideband entering at the antisymmetric port. Considering again the interference

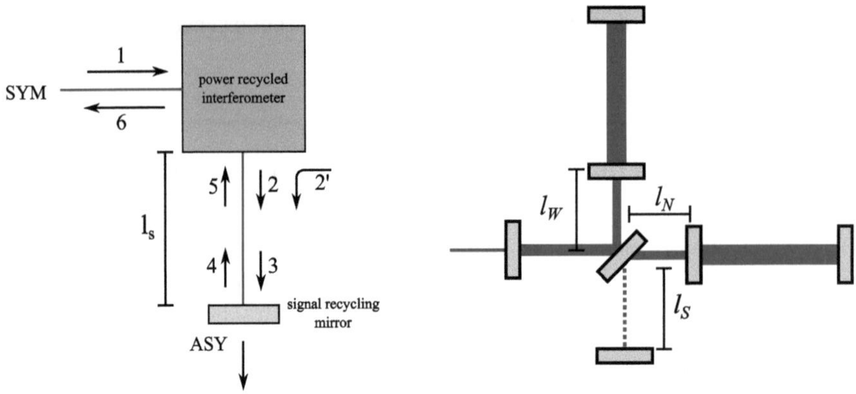

Fig. 3.10 Scheme of a dual recycled interferometer

conditions at the beam-splitter, this is simply given by the Fabry-Perot cavity reflectivity $r_{\mathrm{FP}}(\omega_s)$ at that frequency plus an additional dephasing due to the propagation from the beam-splitter to the input mirrors. The Fabry-Perot cavity reflectivity as a function of frequency can be easily obtained from Eq. 3.46. We are interested in the field transmitted by the SR mirror, which can be computed to be

$$\Psi_{\mathrm{ASY}}(\omega_s) = t_s \frac{e^{-i(k+\frac{\omega_S}{c})l_S}}{1 - i r_S r_{\mathrm{FP}}(\omega_s) e^{-i(k+\frac{\omega_S}{c})l_{\mathrm{SRC}}}} \Psi_2'(\omega_s) \tag{3.54}$$

where we introduced the total length of the signal recycling cavity $l_{\mathrm{SRC}} = l_S + l_+$. The microscopic position of the SR mirror is often expressed in terms of the phase the carrier field would acquire inside it, using the *signal recycling cavity tuning* which is conventionally defined as

$$\phi = k l_{\mathrm{SRC}} + \frac{\pi}{4} \tag{3.55}$$

We can now put everything together to compute the full response of a dual recycled interferometer. One additional approximation is considered here: the length of the signal recycling cavity is typically of the order of few tens of meters. Therefore, the additional propagation dephasing of the signal sidebands due to their small frequency (with respect to the carrier) is neglected.

$$\begin{aligned}\Psi_{\mathrm{ASY}}(\omega_s) = & -\frac{2 t_S t_R}{1 - r_R R} \Psi_{\mathrm{in}} \frac{t_i^2 r_e}{1 - r_i r_e} \frac{e^{i\phi}}{1 - r_S r_{\mathrm{FP}}(\omega_s) e^{2i\phi}} \\ & \times \frac{e^{-i\frac{\omega_S}{c}L}}{1 - r_i r_e e^{-2i\frac{\omega_S}{c}L}} \frac{2\pi}{\lambda} x_0\end{aligned} \tag{3.56}$$

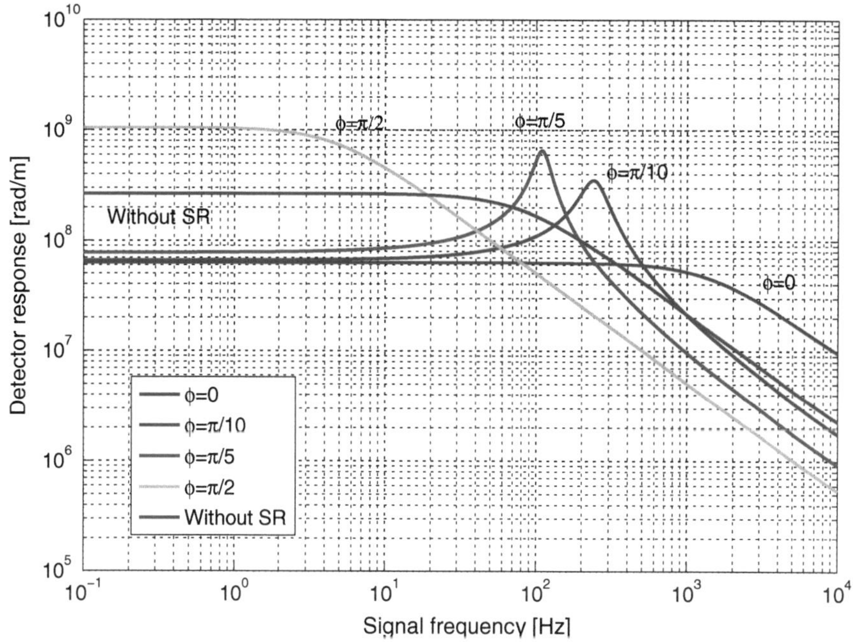

Fig. 3.11 Response of a dual recycled detector to differential displacement as a function of the signal frequency. Some different tunings of the signal recycling cavity are shown. For comparison, the purple curve shows the detector response without any signal recycling mirror

$$r_{FP}(\omega_s) = \frac{r_i - r_e(1-\mathscr{L})e^{-2i\frac{\omega_s}{c}L}}{1 - r_i r_e e^{-2i\frac{\omega_s}{c}L}} \tag{3.57}$$

$$R = \frac{r_i - r_e(1-\mathscr{L})}{1 - r_i r_e} \tag{3.58}$$

It is not obvious to understand from the previous equations what is the effect of adding a signal recycling mirror in terms of the interferometer response. The shape of the optical response of the detector clearly depends on the choice of the tuning ϕ. Figure 3.11 compares the response for some choices of the SRC tuning with the one of a detector without signal recycling. If the tuning is $\phi = 0$ the detector band-width is largely increased, at the price of a lower mean optical gain. This configuration is called *broad-band signal recycling*. In the opposite direction, if the tuning is set to $\phi = \pi/2$ the low frequency sensitivity is largely improved, but at the price of a narrower band-width. This is often called *tuned signal recycling*. All the intermediate values of ϕ introduce a peak frequency (function of the tuning) where the maximum gain is reached. The low frequency sensitivity is comparable to the broad-band case, while the high frequency is typically worsen than in the power-recycled configuration. These configurations are often called *detuned signal recycling*. Figure 3.12 shows how the shape of the detector response changes as a function of the signal recycling

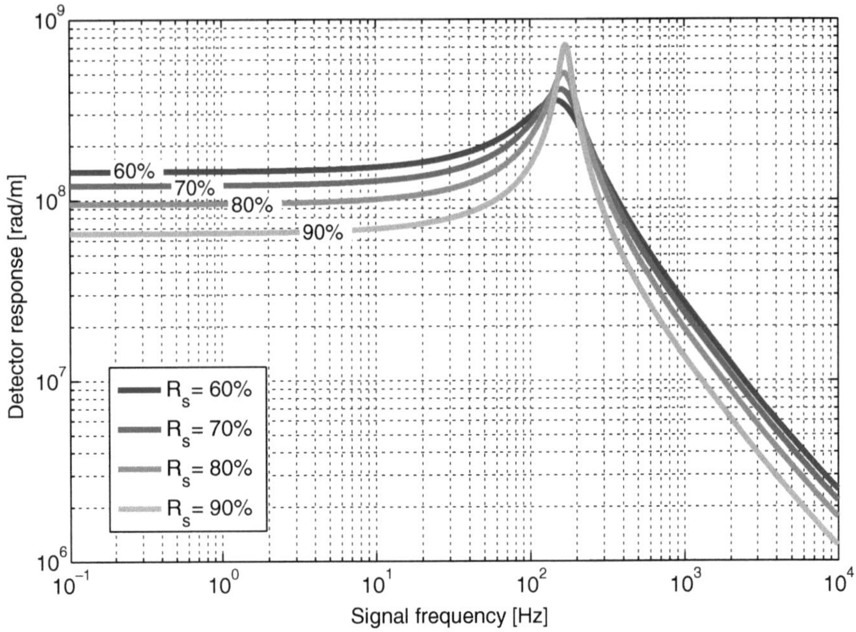

Fig. 3.12 Dependence of a dual recycled interferometer response on the reflectivity of the signal recycling mirror

mirror reflectivity: the larger the reflectivity, the narrower is the peak and the larger the peak gain.

Equations are very useful to proper design your detector. However, it is much more useful to get an intuitive understanding of what signal recycling is doing. Only differential signal sidebands are affected by the presence of the SRC. As explained in Sect. 3.3.2, signal sidebands are generated inside the arm cavities. Depending on their frequency they might be still inside the resonant peak and therefore optically amplified, or they might be outside and suppressed.

When a signal recycling mirror is present, the situation is slightly different. Indeed, differential sidebands will see a situation very close to the one depicted in Fig. 3.13. Signal sidebands at a given frequency are still generated inside the arm cavity with amplitude proportional to the field stored inside the arm. However, since they have a differential nature, when they exit the arm cavities they are completely transmitted to the antisymmetric port and inside the signal recycling cavity. For them, the beam splitter does not exist and they see a double resonant cavity made of the SR mirror, the arm cavity input mirror and the end mirror. The behavior of such a double cavity can be understood considering that the SR and the input mirrors act together creating a resonant cavity. We can model this cavity as an "equivalent" mirror with reflectivity and transmissivity which depend on the tuning of the signal recycling cavity:

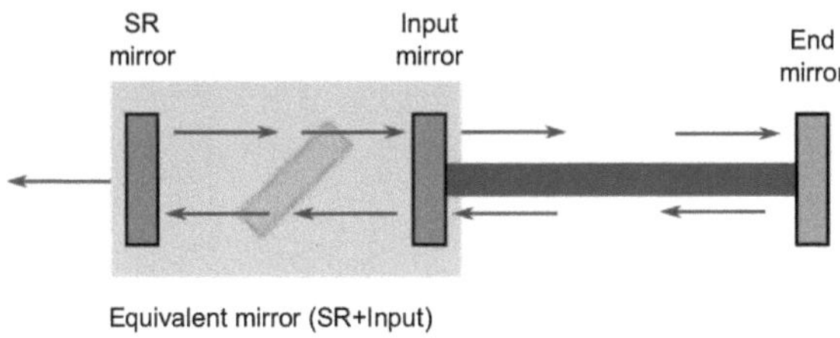

Fig. 3.13 Conceptual scheme of signal sideband fields propagation inside the signal recycling cavity. Input mirror and signal recycling mirrors acts together to provide an "equivalent" input mirror with tunable reflectivity

$$R = i\frac{r_i - r_s e^{2i\phi}}{1 - r_i r_s e^{2i\phi}} \tag{3.59}$$

$$T = \frac{t_i t_s e^{i\phi}}{1 - r_i r_s e^{2i\phi}} \tag{3.60}$$

For a general value of the tuning, this mirror will also introduce an additional dephasing in reflection with respect to the simple input mirror (see Fig. 3.14). This dephasing can be considered as independent of the signal frequency, since the length of the signal recycling cavity is much shorter than the wave-length corresponding to the signal frequency. This "equivalent" mirror will acts as the input one for the arm Fabry-Perot cavity to determine the behavior of the signal sidebands.

For $\phi = 0$ this additional cavity is resonant for the carrier field (which means for signal sidebands of zero frequency). In this case, the reflectivity of the equivalent mirror is the lowest possible (see Fig. 3.14) and the additional dephasing is null. This means that the signal sideband at (almost) zero frequency remains resonant inside the composite arm cavity. However, the input mirror of this cavity will have a lower reflectivity with respect to the real input mirror. In other words, signal sidebands will see an arm cavity with reduced finesse: the band-width is increased, but the maximum gain reduced. This corresponds to the already considered broadband configuration.

A similar behavior is obtained when $\phi = \pi/2$. In this case the additional dephasing is again null, but the reflectivity is almost one. Therefore, the zero frequency signal sidebands will be again resonant, but the equivalent arm cavity finesse will be very high. The bandwidth is largely reduced and the low frequency gain increased. This is the already discussed narrow-band or tuned configuration.

When we choose an arbitrary detuning of the SRC, the situation is clearly intermediate. The absolute value of the reflectivity is somewhere between the minimum and the maximum and increases with increasing detuning. Upon reflection on the equivalent input mirror the signal sidebands will get an addition dephasing which will move them out of resonance. However, there exists one signal frequency for which this additional dephasing is exactly compensated by one accumulated in propagation over the arm cavity length. This particular signal sideband will be again resonant inside the composite cavity and therefore enhanced. This is the origin of the gain peak in the response of a detuned dual recycled interferometer (see again Fig. 3.11).

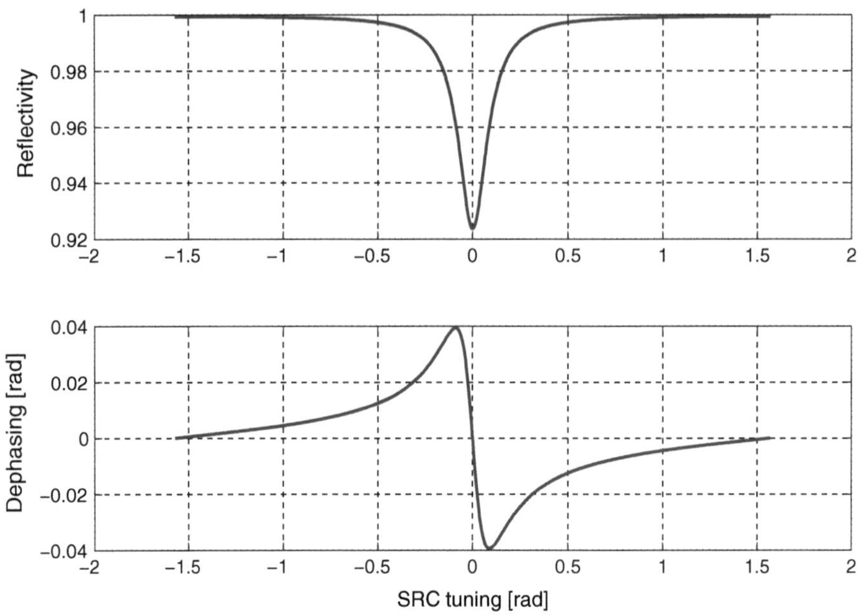

Fig. 3.14 Reflectivity and additional dephasing induced on reflection for the equivalent mirror composed of the SR and input ones. The behavior of this cavity is slightly different with respect to the arm Fabry-Perot considered for example in Fig. 3.5. In this case there is no π jump in the phase when passing from resonance to anti-resonance. This is due to the fact that the end mirror of this cavity has a lower reflectivity than the input one, making the cavity under-coupled [5]

It is maybe useful to add a comment here about positive and negative signal frequencies. When a differential displacement at a given physical frequency f_s happens, two signal sidebands are created at $+f_s$ and $-f_s$ around the carrier optical frequency. Since the entire interferometer is kept resonant for the carrier, only the frequency difference matters. For this reason, we often speak of positive and negative frequency sideband. When propagating inside the km-long arm cavity they will undergo a certain additional dephasing (with respect to the carrier) of opposite sign. This sign difference in the dephasing is relevant only when detuned signal recycling is used. Only the positive or the negative sidebands can get a dephasing in reflection of the equivalent mirror that completely compensates the propagation dephasing. This means that a given detuning of the SRC will make either the positive or negative sideband resonant, while the other one will remain not resonant. Another consequence is that inside detuned dual recycled interferometers signal sidebands are typically unbalanced, meaning that they have different amplitudes and phases.

Advanced gravitational wave detectors will use, at least in the final configuration, the signal recycling technique. This will not only improve the detector sensitivity at some frequencies, but will also allow building a tunable detector, capable of changing its sensitivity depending on the kind of astrophysical searches that are needed.

3.6 High Power Brings Radiation Pressure

Advanced gravitational wave detectors will implement a dual recycled configuration. The arm Fabry-Perot cavities will have a finesse of about 450, increased of a factor 3 with respect to first generation instruments. The round-trip losses inside the cavities are required to be not larger than 75 ppm, much lower than the 200–300 ppm of Virgo and Virgo+. According to the discussion of Sect. 3.4 this will limit the power recycling gain to about 46. The plan is to use high power lasers, in order to be able to inject at the input of the interferometer a maximum of 125 W, to be compared to the 20 W of LIGO and Virgo. This will give us about 6 kW of circulating power inside the PRC and finally about 800 kW inside the arm cavities. Such a large power will exert a force on the mirrors due to radiation pressure. Since the mirror reflectivities are very large, each photon bouncing on the mirror will transfer a momentum equal to the double of its own. The total force is therefore given by

$$F = \frac{2P}{c} \sim 5\ \text{mN} \tag{3.61}$$

This might seem a quite small force, but it is however of the same order of magnitude of the ones that are applied to control the mirror positions. Let us anticipate a bit of things that will be discussed in other chapters. Mirrors in advanced GW detectors will have typical masses of $m = 40$ kg and will be suspended to pendulums with length of about 70 cm. The static force computed above, applied to the mirror, will produce a displacement given by

$$\delta z = l \frac{F}{mg} \sim 10\ \mu\text{m} \tag{3.62}$$

which is ten times larger than the laser wavelength and therefore not at all negligible. However, this static force does not create problems provided you equip the system with a way to apply an opposite force to maintain the cavity length at resonance.

3.6.1 *Opto-Mechanical Couplings*

Let us now consider a Fabry-Perot resonant cavity as a simple but meaningful example of how radiation pressure can play a dynamical role. In GW interferometers, the mirror is suspended to pendulums with relatively low resonant frequency. If we apply a force F to the mirror, we obtain a displacement which is determined by the longitudinal equation of motion

$$\ddot{z} = -\omega_0^2 z - \gamma \dot{z} + \frac{F}{m} \tag{3.63}$$

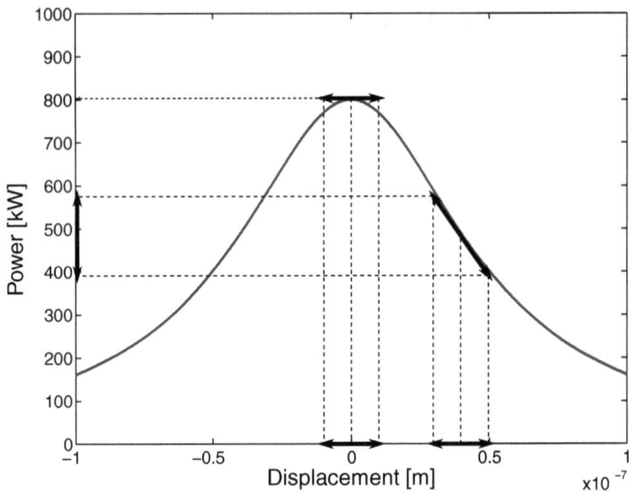

Fig. 3.15 When a cavity is moved around resonance, no first order power fluctuations are created. Instead, if the cavity is detuned from resonance, there is a direct linear conversion of displacement to power fluctuations

where ω_0 is the resonance frequency of the pendulum and γ a constant describing the amount of friction. The above equation is normally solved with a Fourier transform to compute the *mechanical transfer function*, which describes the dynamics of the mirror alone

$$z(\omega) = \frac{F(\omega)}{m} \frac{1}{\omega_0^2 - \omega^2 + i\gamma\omega} \tag{3.64}$$

Let's now imagine to have a Fabry-Perot cavity close to resonance. If we apply a very small sinusoidal force of frequency ω to one of the two mirrors, we obtain a small displacement determined by the above equation. As explained in the previous sections, this displacement creates two signal sidebands at the ω frequency which circulate inside the cavity. The main question here is: does the motion produce a fluctuation of the power inside the cavity? The answer to this question is clearly positive, however we are interested only in power fluctuations at the same frequency of our external disturbance. Referring to Fig. 3.15, it is clear that if the cavity is perfectly tuned at resonance, any motion around this point will not create a power fluctuation at the same frequency, but only at higher harmonics and with amplitude proportional to higher powers of the displacement. These nonlinear effects can be neglected, given the very small amplitude of the displacement we are considering.

If the cavity is, for some reason, detuned from the resonance, there is a linear conversion of the cavity length variation to power fluctuations at the same frequency, as clearly visible from Fig. 3.15.

Another way to understand this point is the following. Regardless of the cavity tuning, the length variation creates two signal sidebands at a given frequency ω with equal amplitudes, proportional to the displacement and to the stored field inside the cavity. These sidebands will be recycled inside the cavity. If the tuning is zero, the two sidebands will fall exactly at the same distance from the resonance center and they will maintain the same amplitude and phase relation. Instead, if the cavity is detuned, one of the two sidebands will be closer to the resonance than the other and will therefore be amplified more. In both cases, the field inside the cavity is described by three components:

$$\Psi = \Psi_0 + \Psi_+ e^{-i\omega t} + \Psi_- e^{i\omega t} \tag{3.65}$$

and the power is given by the squared modulus of the previous equation:

$$\begin{aligned} P = &|\Psi_0|^2 + |\Psi_+|^2 + |\Psi_-|^2 + \left(\Psi_0^* \Psi_+ + \Psi_0 \Psi_-^*\right) e^{-i\omega t} \\ &+ \left(\Psi_0 \Psi_+^* + \Psi_0^* \Psi_-\right) e^{i\omega t} \\ &+ \Psi_+ \Psi_-^* e^{-2i\omega t} + \Psi_+^* \Psi_- e^{2i\omega t} \end{aligned} \tag{3.66}$$

When the cavity is at resonance, Ψ_0 is real and the two sidebands are balanced: all terms oscillating at ω frequency cancel out and only the second harmonic terms remains. If instead the cavity is detuned, the two sidebands have different amplitudes and the cancellation is no more possible, meaning that power fluctuations at the ω frequency will be present.

If the power inside the cavity is oscillating, the radiation pressure force exerted to both the input and end mirrors will be also varying at the ω frequency. As visible from Eq. 3.66 the power fluctuation is proportional to the signal sideband amplitude which in turn is proportional to the cavity length motion. In conclusion, the additional force introduced by radiation pressure is linearly proportional to the cavity motion, or in other words the effect of radiation pressure is the introduction of an additional elastic term $F_{\mathrm{RP}} = -k_{\mathrm{RP}} z$.

Moreover, even if we are applying an external force only to the end mirror, the induced power fluctuation inside the cavity will exert a force to both end and input mirrors. The effect of radiation pressure is therefore two-fold:

- The end mirror displacement will not be the same as in free mirror case, since the force we apply causes a displacement which in turn causes an additional force to be applied to the end mirror. The mechanical response of the mirror (displacement as a function of applied force) is modified with respect to the simple pendulum by an *opto-mechanical effect* which adds another elastic term.
- The input mirror will also move due to the radiation pressure force caused by the power fluctuations induced by the end mirror motion. This introduces an *opto-mechanical coupling* between the two mirrors, which are no more dynamically disconnected as in the case of negligible power.

The important point to understand is the sign of the opto-mechanical stiffness k_{RP}. Let us first consider the case of a cavity detuned in the direction of increasing its length. When the oscillating length variation increases a bit more the length, the power inside the cavity decreases and therefore the radiation pressure force also decreases.

In this case, the stiffness is positive and radiation pressure introduces a restoring force which adds to the pendulum one. The total stiffness seen by the mirror remains positive and the resonance is simply shifted to higher frequencies. Let us now consider the opposite case, with a cavity detuned in the direction of decreasing its length. Now when the oscillating length variation increases the length, the cavity is moved toward the resonance and the power is increased. This has the effect of increasing the force. The stiffness introduced by radiation pressure is negative, introducing a kind of anti-spring. If the amplitude of the radiation pressure effect is small enough, the total stiffness remains positive. However, for increasing values of the stored power or of the offset from resonance, the k_{RP} might exceed in modulus the pendulum stiffness, resulting in a total negative stiffness: the opto-mechanical system become unstable.

A complete rigorous computation of opto-mechanical effects in a Fabry-Perot cavity is feasible, but the computations are far from being simple and straightforward. An example of one possible (very rigorous) approach can be found in [8].

Here we consider an approximate approach, that is however good enough for many applications. The dynamics of the input and end mirrors are described by their coordinate and the following equations:

$$\ddot{z_I} = -\omega_0^2 z_I - \gamma \dot{z}_I + \frac{F_{\mathrm{RP}}^I}{m} \tag{3.67}$$

$$\ddot{z_E} = -\omega_0^2 z_E - \gamma \dot{z}_E + \frac{F_{\mathrm{RP}}^E + F_{\mathrm{EXT}}}{m} \tag{3.68}$$

Here we assume that the two mirror have exactly the same mechanical behavior (same mass, same length of the pendulum and same friction). In general, the force induced by radiation pressure might be different for the two mirrors, since they are separated by few kilometers. However, if we consider only low frequency disturbances, we can neglect the propagation time inside the cavity and consider the two forces as equal in modulus but simply opposite in sign. This is valid as long as the frequencies we are considering are small with respect to the cavity pole (typically 60 Hz for a 3 km long cavity with a finesse of 450). We can therefore consider only the variation in the cavity length z, which is described by the equation one gets by subtracting the previous two:

$$\ddot{z} = -\omega_0^2 z - \gamma \dot{z} + \frac{2F_{\mathrm{RP}}}{m} + \frac{F_{\mathrm{EXT}}}{m} \tag{3.69}$$

We need now to compute the force exerted by radiation pressure. Assuming the reflectivity of both mirror to be very close to unity, Eq. 3.61 tells us that it is proportional to the power variation, which will be a function of the cavity length variation

$$F_{\mathrm{RP}} = \frac{2}{c}\,\delta P(z) \tag{3.70}$$

In general $\delta P(z)$ should contain a frequency dependence, due to the propagation of the signal sidebands inside the cavity. However, as already explained, we are neglecting this effect by considering only low frequencies. We know that the power stored inside the cavity is given by Eq. 3.37 which can be rewritten as:

$$P(z) = \frac{P_0}{1 + \left(\frac{2\mathscr{F}}{\pi}\right)^2 \sin^2 k\delta L} \tag{3.71}$$

where P_0 is the maximum power inside the cavity at resonance. We can express the cavity displacement from resonance as a static term $\phi = k\delta L_0$ plus the fluctuation kz we are interested in. Finally, the power fluctuation are given by the derivative of the previous equation

$$\delta P(z) = \frac{\mathrm{d}P}{\mathrm{d}z} z = -\frac{8\mathscr{F}^2}{\pi\lambda} P_0 \frac{\sin 2\phi}{\left(1 + \left(\frac{2\mathscr{F}}{\pi}\right)^2 \sin^2 \phi\right)^2} z \tag{3.72}$$

Substituting everything inside Eq. 3.69 we get:

$$\ddot{z} = -\frac{1}{m}\left[k + \frac{32\mathscr{F}^2}{\pi\lambda c} P_0 \frac{\sin 2\phi}{\left(1 + \frac{2\mathscr{F}}{\pi} \sin^2 \phi\right)^2}\right] z - \gamma\dot{z} + \frac{F_{\mathrm{EXT}}}{m} \tag{3.73}$$

The term inside brackets is the total stiffness of the system for the degree of freedom corresponding to the cavity length. The typical detuning ϕ is small, so we can approximate to first order:

$$k_{\mathrm{tot}} = k_{\mathrm{pend}} + \frac{64\mathscr{F}^2 P_0}{\pi\lambda c}\phi \tag{3.74}$$

The corresponding equation of motion can be solved again using the Fourier transform:

$$z(\omega) = \frac{\frac{F_{\mathrm{EXT}}(\omega)}{m}}{-\omega^2 + i\gamma\omega + \frac{k_{\mathrm{tot}}}{m}} \tag{3.75}$$

The behavior of this opto-mechanical oscillator is determined by the roots of the denominator which give the characteristic frequencies of the system

$$\omega_p = i\frac{\gamma}{2} \pm \sqrt{\frac{k_{\mathrm{tot}}}{m} - \frac{\gamma^2}{4}} \tag{3.76}$$

In typical GW application the friction is very small, making γ negligible inside the square bracket. If we call $\omega_0 = |k_{\mathrm{tot}}|/m$ we can distinguish the two cases

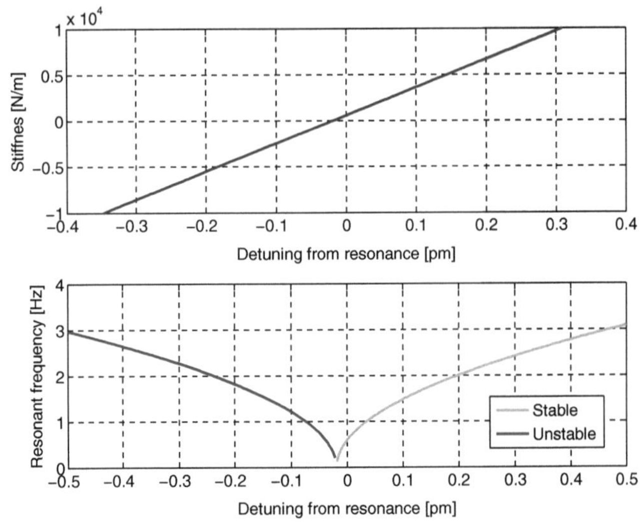

Fig. 3.16 Dependence of the opto-mechanical stiffness (*top*) and resonant frequency (*bottom*) on the detuning from resonance of Fabry-Perot cavity with a finesse of 450 and 800 kW of circulating power

$$k_{\text{tot}} > 0 \qquad \omega_p = \pm\omega_0 + i\frac{\gamma}{2} \tag{3.77}$$

$$k_{\text{tot}} < 0 \qquad \omega_p = \pm i\omega_0 \tag{3.78}$$

In the second case the system is unstable, being similar to an upside-down pendulum.

Let us take as an example an Advanced Virgo like Fabry-Perot cavity, with a finesse of 450 and a circulating power of about 800 kW. The single pendulum resonant frequency is about 600 mHz and the mirror masses 42 kg. The simple pendulum stiffness is therefore about 600 N/m. Figure 3.16 shows how the total opto-mechanical stiffness depends on the cavity detuning from resonance. A (relatively small) detuning of 1 pm is more than enough to make the cavity unstable, with a (negative) resonant frequency of about 6 Hz.

3.6.2 *Change in the ITF Optical Response*

We will discuss in Chap. 6 how Advanced Detectors will operate with a small differential detuning of the two arms from resonance, of the order of 10 pm. Moreover we already saw that when operating with a detuned signal recycling cavity the signal sidebands are typically unbalanced. We can therefore expect significant radiation pressure effects in the optical response of the interferometer to a differential displacement. Indeed, the low frequency response turns out to be quite different from the one obtained in Sect. 3.5.

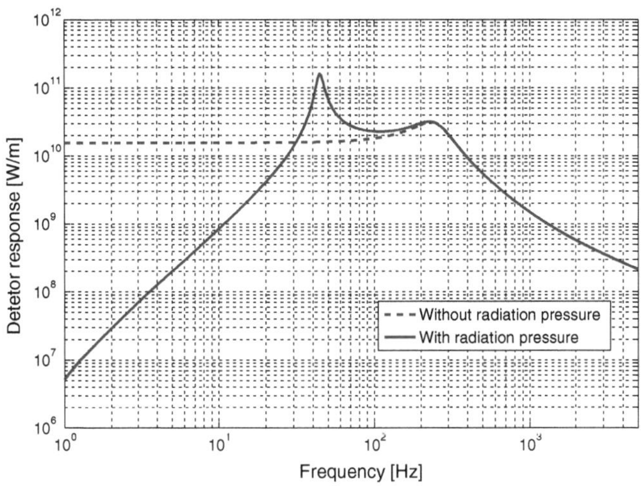

Fig. 3.17 Response of a dual recycled interferometer with Advanced Virgo parameters, compared to the (unphysical) situation when there are no radiation pressure effects

Computing the effect of radiation pressure on the response of a full dual recycled interferometer is a much more complex computation. For this reason, we enter the realm of numerical optical simulations. The equations linking the displacement of a given mirror with the surrounding fields are quite simple and can be easily implemented numerically.

Figure 3.17 shows the actual response of a dual recycled interferometer to a differential motion of the arm length. The dashed line is what we would measure in a world without radiation pressure, as computed in Sect. 3.5. The effect of radiation pressure is that of stiffening the arms in the differential degree of freedom. Any motion we try to impose on the arms causes a fluctuation of the power that exerts a force to the mirror such that the low frequency motion is largely suppressed. The additional peak around 40 Hz is the new (unstable) resonance of the opto-mechanical system, shifted up from the 600 mHz case.

3.7 Beam Optics and High Order Transverse Modes

So far we used the plane-wave approximation to describe the laser field inside the interferometer. This is however not completely correct, since our mirrors have finite sizes and our laser beams are limited by diffraction.

3.7.1 The Paraxial Diffraction Equation

In the general case, the propagation of light inside an optical system is described by the Maxwell equations. As already discussed in Sect. 3.1.1, the polarized nature of the laser beam allows reducing to a scalar wave equation

$$\left(\Delta + k^2\right) \mathcal{E}_k(x, y, z) = 0 \tag{3.79}$$

In the simpler case of a laser beam inside an interferometer, the propagation is well described in the *paraxial approximation* [5, 9]. The light field is supposed to propagate mainly along the cavity axis direction called z and all its transverse variations to be small. If the field is written as

$$\mathcal{E}(x, y, z) = e^{ikz} \Psi(x, y, z) \tag{3.80}$$

and the variation along z of the envelope is small $\left|\frac{\partial \Psi}{\partial z}\right| \ll k\Psi$, the wave equation simplifies to the *paraxial diffraction equation*:

$$\left(2ik\partial_z + \partial_x^2 + \partial_y^2\right) \Psi = 0 \tag{3.81}$$

which can be used to describe the propagation of an almost monochromatic beam concentrated along the z direction. Without entering too much into details (that can be found for example in [7, 9]), one can say that under the assumption of cylindrical symmetry the *fundamental Gaussian solution* of the paraxial diffraction equation is found:

$$\Psi(r, z) = \frac{1}{\sqrt{1 + \frac{z^2}{z_R^2}}} e^{-\frac{x^2+y^2}{w^2(z)}} e^{-ik\frac{x^2+y^2}{2R(z)}} e^{i \arctan \frac{z}{z_R}} e^{-ikz} \tag{3.82}$$

where

$$w(z) = w_0 \sqrt{1 + \frac{z^2}{z_R^2}} \tag{3.83}$$

$$R(z) = z\left(1 + \frac{z_R^2}{z^2}\right) \tag{3.84}$$

$$z_R = \frac{kw_0^2}{2} \tag{3.85}$$

The transverse intensity shape is a Gaussian function for every z. Its width is minimum when $z = 0$, at the *waist* of the beam, and increases for larger z according to the function $w(z)$ (see Fig. 3.18). The *Rayleigh distance* z_R gives the length scale

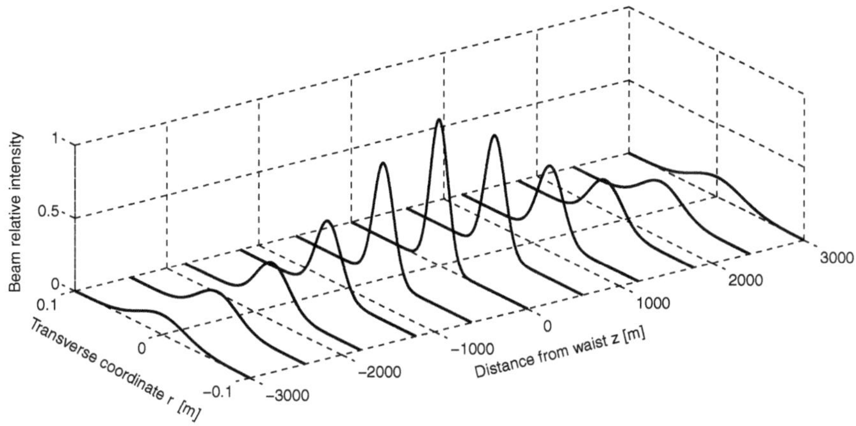

Fig. 3.18 Beam intensity profiles at different distances from the waist

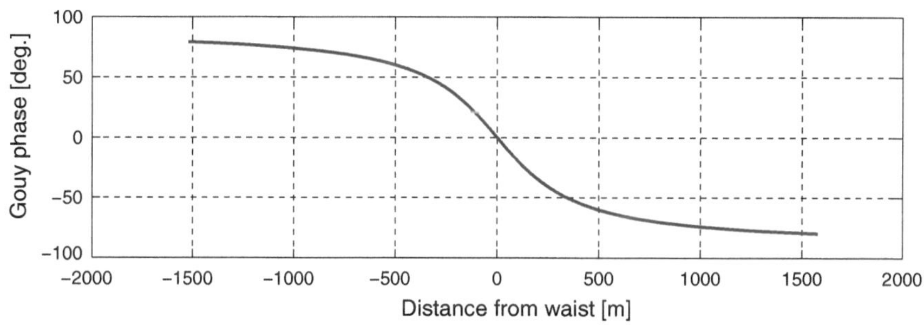

Fig. 3.19 Dependence of the Gouy phase on the distance the beam traveled from the waist. The Rayleigh length in this case is about 280 m

of the beam expansion: after propagating over a distance z_R along the axis, the beam is $\sqrt{2}$ times larger in the transverse direction.

The constant phase surfaces of this beam are parabolic, with a decreasing curvature given at $r = 0$ by $R(z)$. When propagating along z the field undergoes an additional dephasing with respect to the one predicted in the plane-wave approximation. This is the *Gouy phase*

$$\phi_G = -\arctan\frac{z}{z_R} \tag{3.86}$$

The behavior of the Gouy phase as a function of the distance the beam propagated from the waist is shown in Fig. 3.19.

Table 3.1 Hermite polynomials of lowest order

Order	Polynomial
0	1
1	$2x$
2	$4x^2 - 2$
3	$8x^3 - 12x$

3.7.2 Transverse Electro-Magnetic Modes

Starting from the fundamental solution of Eq. 3.82 it is possible to relax the requirement of cylindrical symmetry and search for solutions with different transverse shape. The most used set is given by the *Hermite-Gauss* functions, also called *electro-magnetic transverse modes* (TEM). They depends on two integer indexes n and m:

$$\mathrm{TEM}_{mn}(x, y, z) = N_{mn}(z)e^{ikz} H_m\left(\frac{\sqrt{2}x}{w(z)}\right) H_n\left(\frac{\sqrt{2}y}{w(z)}\right) e^{-i(n+m+1)\arctan(z/z_R)} e^{ik\frac{x^2+y^2}{2R(z)}} e^{-\frac{x^2+y^2}{w^2(z)}} \quad (3.87)$$

where the normalization coefficient $N_{mn}(z)$ is chosen in order to make this set orthonormal for any fixed z:

$$N_{mn}(z) = \sqrt{\frac{2}{\pi w(z)^2\, 2^{n+m} m!n!}} \quad (3.88)$$

and the $H_n(t)$ are the *Hermite polynomials* that can be defined for example as [10]:

$$H_n(t) = e^{t^2}\left(-\frac{d}{dt}\right)^n e^{-t^2} \quad (3.89)$$

Given a value of the beam size at the waist w_0, these solutions form a complete, orthonormal set, meaning that every beam can be expressed as a linear combination of these functions at fixed z. The first Hermite polynomials are listed in Table 3.1 and the transverse intensity pattern of the first TEM modes are shown in Fig. 3.20.

3.7.3 Resonant Cavity Stability and Mode Separation

If one excludes the waist position ($z = 0$) where the wave-front is flat, the surfaces of constant phase are curved. The mirrors we are using to build our detector therefore can not be all flat, otherwise they will not match the wavefront of the laser beam and will act as a lens. This is particularly true for Fabry-Perot cavities. Imagine that

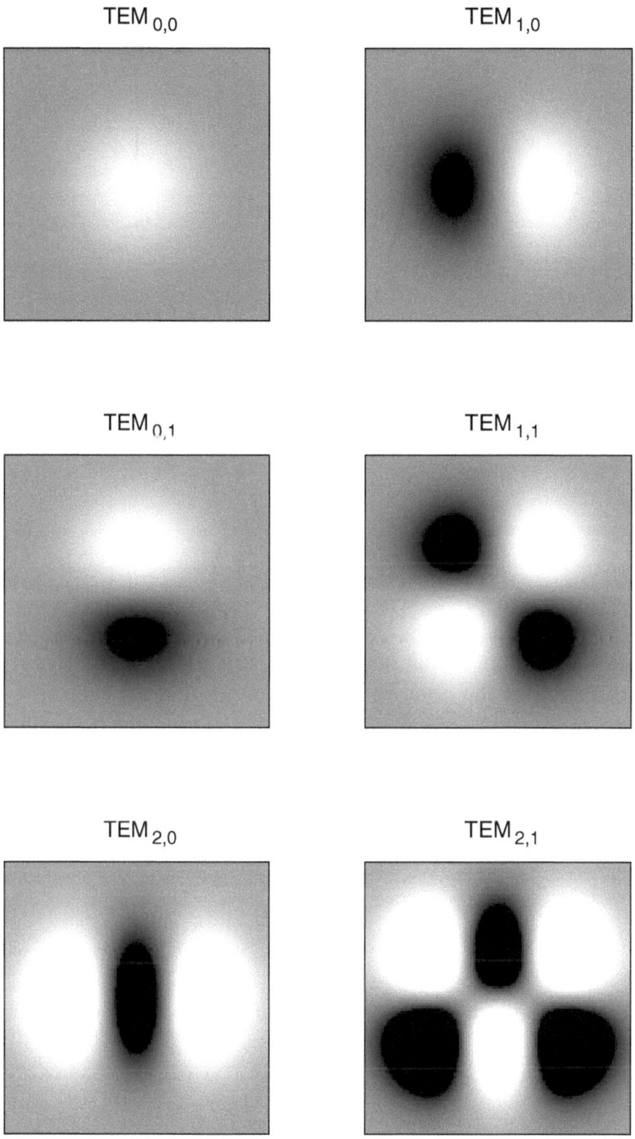

Fig. 3.20 Transverse intensity maps of the first TEM modes

one of the mirror surfaces does not match the laser wave-front: every time the beam bounces back from the mirror it will change its shape and therefore partially destroy the crucial constructive interference that must hold to take advantage of the cavity resonance.

Therefore in general the two mirrors that compose a Fabry-Perot cavity must be curved in such a way to match exactly the wave-front curvature at their position. Referring to Fig. 3.21, Eq. 3.84 tells us what is the curvature of the wavefront at the

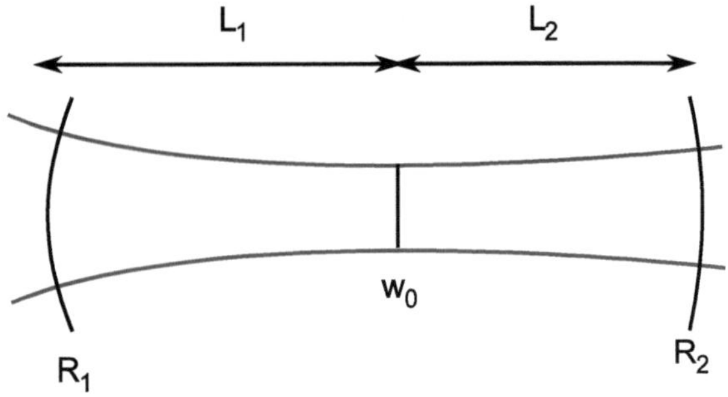

Fig. 3.21 Scheme of a Gaussian beam inside a Fabry-Perot cavity

two mirror positions, as a function of the beam waist size:

$$R_1 = L_1\left(1 + \frac{z_R^2}{L_1^2}\right) \tag{3.90}$$

$$R_2 = L_2\left(1 + \frac{z_R^2}{L_2^2}\right) \tag{3.91}$$

The convention used here is that both radii of curvature are positive if their center lays in the direction of the other mirror. Assuming that the radii of curvature are known, the previous two equations can be used to compute the properties of the laser beam which matches the two surfaces (the *resonant mode*). To solve these equations, we first define the cavity *g factors* [9]:

$$g_1 = 1 - \frac{L}{R_1} \tag{3.92}$$

$$g_2 = 1 - \frac{L}{R_2} \tag{3.93}$$

and then we can use these definitions to solve for the beam waist dimension and position:

$$L_1 = \frac{g_2(1-g_1)}{g_1+g_2-2g_1g_2}L \tag{3.94}$$

$$L_2 = \frac{g_1(1-g_2)}{g_1+g_2-2g_1g_2}L \tag{3.95}$$

$$w_0^2 = \frac{L\lambda}{\pi}\sqrt{\frac{g_1g_2(1-g_1g_2)}{(g_1+g_2-2g_1g_2)^2}} \tag{3.96}$$

From the last equation it appears that a resonant mode exists only if some conditions are satisfied. In particular, the cavity admits a resonant mode only if

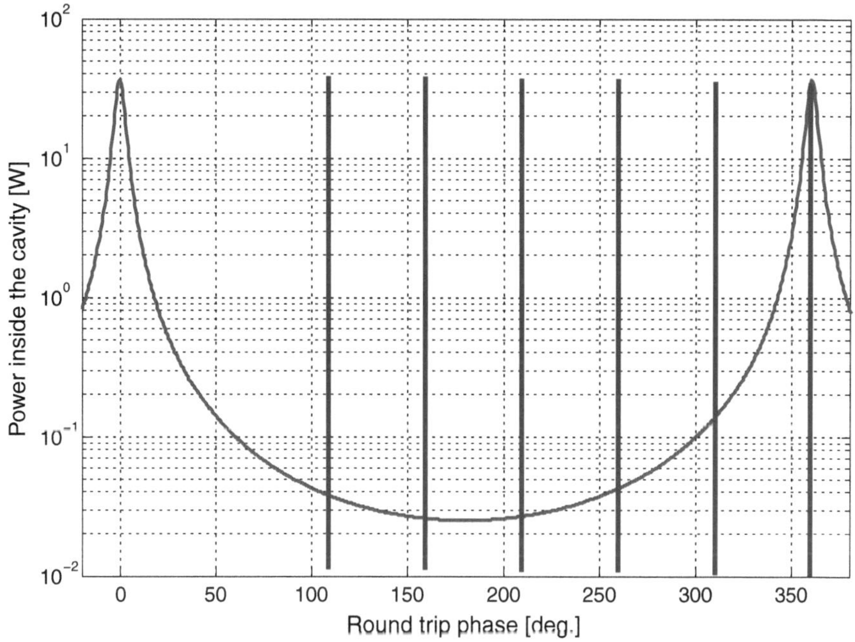

Fig. 3.22 Resonance conditions inside a stable Fabry-Perot cavity, with a Gouy phase of about 25°. High order modes are well separated with respect to the fundamental one

$$0 \leq g_1 g_2 \leq 1 \tag{3.97}$$

If this relation is satisfied the cavity is called *stable*, otherwise *unstable*.

Let us now consider a stable cavity. The fundamental resonant mode is well defined inside it. Let's assume that the cavity microscopic length is tuned in such a way to have the fundamental Gaussian mode (TEM00) resonant inside it. The actual microscopic position is slightly different from the one expected from the plane-wave approximation, because of the presence of the Gouy phase shift. This is however irrelevant for all practical purposes.

However, Eq. 3.87 shows that, when propagating inside the cavity, high order modes will acquire an additional dephasing given by the cavity Gouy phase multiplied by the mode order. If this phase shift is larger than the cavity line-width (which is typically true for the high-finesse arm cavities used in GW detectors) the high order modes will no more be resonant and therefore they will not be amplified by the cavity. This shows the *mode cleaning* effect of stable resonant cavity: only one mode can resonate inside them. All other modes will be rejected by the cavity. For this reason, stable cavities are often called *non-degenerate* (see Fig. 3.22).

Figure 3.23 shows what happens when we scan the microscopic position of a resonant cavity. The horizontal axis is rescaled in terms of the phase acquired by the propagating field when doing a complete round trip. Some defects have been added

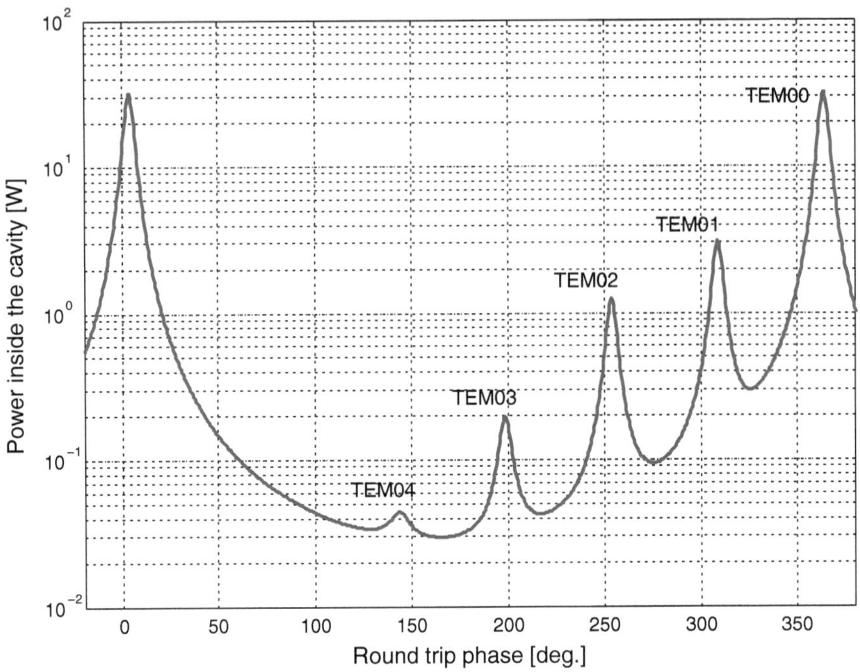

Fig. 3.23 Simulation of scanning a Fabry-Perot resonant cavity with some defects. High order modes are well visible when the length of the cavity makes them resonant

to the cavity in order to generate higher order modes. In this case, the cavity is stable and the Gouy phase is large enough to move all higher order modes out of resonance.

If instead the Gouy phase shift is smaller than the cavity line-width, there will be a finite (possibly large) number of high order modes that will fall inside the same resonance of the fundamental one. In this case the cavity will accept and amplify them in a way similar to the fundamental mode. These cavities are called *marginally stable*, meaning that, despite being optically stable and admitting a unique resonant mode, they are not able to completely separate the fundamental mode from a set of high order ones (Fig. 3.24).

3.8 Stability of Recycling Cavities

Before closing this chapter, we should consider the geometry of the entire interferometer. The kilometer-long arm cavities must be stable and with a good separation between high order modes. In this way, the beam shape resonating inside them is well defined. Advanced detectors will use a bi-concave geometry of the arm cavities: in this way the beam size on the mirrors will be as large as possible. This is useful, as will be briefly discussed in the next section, in order to reduce the impact of the

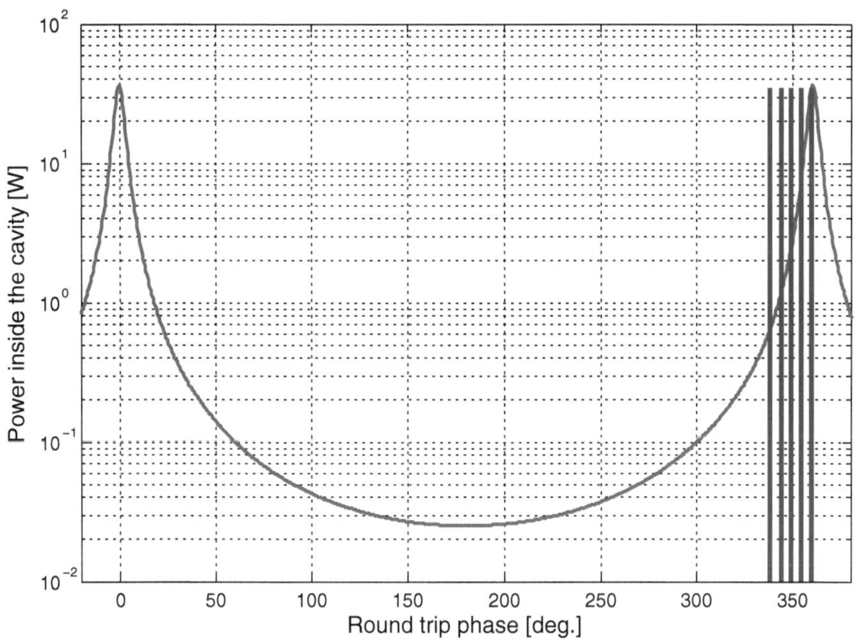

Fig. 3.24 Resonance conditions inside a stable Fabry-Perot cavity, with a small Gouy phase. Some high order modes are still inside the same resonant peak as the fundamental one

mirror coating Brownian noise (the bigger the beam spot, the larger the surface over which Brownian motion is averaged).

In the Advanced Virgo case the cavity is 3 km long. The two mirrors will have radii of curvature of 1420 m (input) and 1683 m (end). In this way the spot radii will be about 49 mm on the input test masses and 58 mm on the end test masses. The waist position will be close to the center of the cavity.

One must then choose the geometry of the recycling cavities. The most obvious choice would be to have them stable. However, this has some important implications. First of all the mode which is resonant inside the PRC and SRC must be equal to the one which is resonant in the arms, otherwise we will loose part of the laser power because of the *mismatch* between the two. Moreover, to make the recycling cavity stable one must ensure that the Gouy phase accumulated by propagation inside it is not negligible.

Looking at Eq. 3.86 we see that the Gouy phase increases with the length of the cavity. Therefore a simple way to make the cavity stable is to make it long: the total Gouy phase is given by the difference between the power recycling and the input mirror, which are placed a distances z_{PR} and z_I from the beam waist:

$$\Delta\phi_G = \arctan\frac{z_I}{z_R} - \arctan\frac{z_{PR}}{z_R} \tag{3.98}$$

Unfortunately, in the bi-concave configuration the input mirror is already placed quite far from the waist position, in the region when the arc tangent slope is quite small. This means that to accumulate enough phase the cavity should be very long. This solution is not feasible from an infrastructural point of view, since it would imply the construction of another km-long cavity.

A second way to ensure that the Gouy phase inside the recycling cavity is large is to insert focusing optical elements. In this way, one can change the shape of the mode and decrease significantly its Rayleigh distance. This approach is used in the Advanced LIGO and KAGRA detectors, by folding the two recycling cavities with mirrors having small radii of curvature (of the order of tens of meters). The main drawback is the increased complexity of the system, both from an optical and mechanical point of view.

The opposite choice would be to design marginally stable cavities. This was the configuration chosen by all first generation detector and by Advanced Virgo. The recycling cavities are simpler, since the power and signal recycling mirror could be on the same axis of the arm cavities and their curvature will be very close to the one of the input mirrors. The main drawback is the fact that a large numbers of high order modes can resonate inside them. However, this is not a real problem for the carrier fields. Indeed when operating at the working point, the fundamental mode of the carrier will be resonant simultaneously inside the arm and the power recycling cavities, as explained in Sect. 3.4. Let us assume we generate somehow a high order mode inside the recycling cavity. This will not be resonant inside the arm cavities, since they are stable. Therefore it will be reflected by the arm cavities being anti-resonant and therefore acquiring an additional π dephasing with respect to the TEM00 (as explained in Sect. 3.3). This large dephasing will for sure move the high order mode out of resonance also in the power recycling cavity. Therefore the stability of the arm cavity for the carrier field enforces also some kind of stability of the recycling cavities. This effect does not happen however for fields that are designed to be resonant inside the PRC but anti-resonant inside the arms. As we will see in the following chapter, all interferometers use radio-frequency sidebands for control purposes. For these fields, the power and signal recycling cavities remain marginally stable and therefore do not provide mode filtering. This is the origin of all the problems related to thermal deformations of the mirrors, which will be the topic of Chap. 9.

3.9 Exercises

3.9.1 Etalon Effect

Real mirrors are obviously made of a transparent substrate (typically of fused silica) with a high reflectivity (HR) coating on one surface. If the secondary face of the mirror is left untreated, it will reflect about 4%, creating lots of troubles in the form of losses and multiple beams. For this reason it is also coated with an anti-reflection (AR) treatment. Typical reflectivities of this AR coating can range from 50 to 500 ppm.

If the two faces of the mirror are parallel, the beam reflected back by the AR face will have the same shape of the main one. Therefore, a spurious resonant cavity is created: the two reflecting surfaces composing the cavity are the coatings on the two faces of the mirror itself. If the temperature of the mirror changes, the optical path length can vary due to a variation of the refractive index of the substrate. This is the so-called *etalon effect*.

1. Compute the dependence of the mirror reflectivity as a function of its temperature, assuming you know the dn/dt of fused silica at room temperature.
2. What are the maximum and minimum values of the reflectivity as a function of the HR and AR coatings? Assume a HR reflectivity of 98 % and an AR reflectivity of 200 ppm.
3. Imagine now that this etalon effect acts on the input mirror of a Fabry-Perot cavity. What is the effect on the cavity properties?
4. In Virgo and Virgo+ the input mirrors had parallel faces, thus implementing an etalon. Can you imagine why having an etalon effect was desirable in a power recycled interferometer? (Hint: the presence of the power recycling cavity is not important and you can neglect it).
5. Can you imagine other application to an Advanced Detector of the etalon effect, excluding the Fabry-Perot resonant cavities?

References

1. E.D. Black, R.N. Gutenkunst, An introduction to signal extraction in interferometric gravitational wave detectors. Am. J. Phys. **71**, 365 (2003)
2. M. Pitkin, S. Reid, S. Rowan, J. Hough, Gravitational wave detection by interferometry (ground and space). Living Rev Relat **14**, 5 (2011)
3. P.R. Saulson, *Fundamentals of Interferometric Gravitational Wave Detectors* (World Scientific, Singapore, 1994)
4. J.D. Jackson, *Classical Electrodynamics* (Wiley, New York, 1998)
5. J.-Y. Vinet, *The Virgo physics book, Optics and related topics* (2006), http://www.cascina.virgo.infn.it/vPB/
6. T.T. Lyons, M.W. Regehr, F.J. Raab, Shot Noise in Gravitational-Wave Detectors with Fabry-Perot Arms. Appl. Opt. **39**, 6761–6770 (2000) LIGO-P000014-00-D
7. G. Muller, T. Delker, D.B. Tanner, D. Reitze, Dual-recycled cavity-enhanced Michelson interferometer for gravitational-wave detection. Appl Opt **42**, 1257 (2003)
8. M. Rakhmanov, Dynamics of Laser Interferometric Gravitational Wave Detectors, Ph.D. thesis, California Institute of Technology, 2000
9. A.E. Siegman, *Lasers* (University Science Book, Mill Valley, 1986)
10. R. Courant, D. Hilbert, *Methods of Mathematical Physics* (Wiley-Interscience, New York, 1989)
11. F. Pampaloni, J. Enderlein, Gaussian, Hermite-Gaussian, and Laguerre-Gaussian beams: a primer, arXiv:physics/0410021 (2004)

Chapter 4
Pre-stabilized Lasers for Advanced Detectors

C.-Nary Man

Abstract Gravitational wave detectors need very stable continuous wave laser sources able to delivering high power beams. Realization of those lasers is a special R&D calling on very low noise controls on very reliable laser sources. After a brief introduction on the laser principles, we review the current laser sources for gravitational wave interferometric detectors, shortly describing the technologies of both solid-state and fiber lasers and amplifiers. A final section addresses the issue of laser pre-stabilization.

4.1 Brief Introduction on the Laser

Providing the laser source for interferometric detection of gravitational waves has always been an exciting challenge for the scientists. Differently from the usual lasers found in other experiments, this laser needs to be simultaneously incredibly stable while delivering a high power (HP) in a very stable spatial mode. In the evolution of the kilometric detectors Virgo and LIGO or the long-based subkilometric detectors GEO or TAMA, the request for the laser source has been an increasing output power from a few Watts to hundreds Watts and possibly higher for future projects like Einstein Telescope (ET). Besides that the wavelength of this source does not matter except for the fact that at some wavelengths, the mirrors can have better performances in terms of scattering as well as benefiting of more know-how in the coating deposition techniques.

Let's first take a glance at the laser basics. Theodore Maiman built the first laser in 1960 on the idea that a coherent beam can be created from an optical amplifier (the gain medium) surrounded by a regenerating feedback to build up an oscillation. The laser process occurs most often in a four-level system of the active ion. When

C.-N. Man (✉)
Dir UMR 7250, OCA-CNRS, Bd de l'Observatoire, BP 4229 06304 Nice cedex 4, France
e-mail: nary.man@oca.eu

M. Bassan (ed.), *Advanced Interferometers and the Search for Gravitational Waves*, Astrophysics and Space Science Library 404, DOI: 10.1007/978-3-319-03792-9_4,

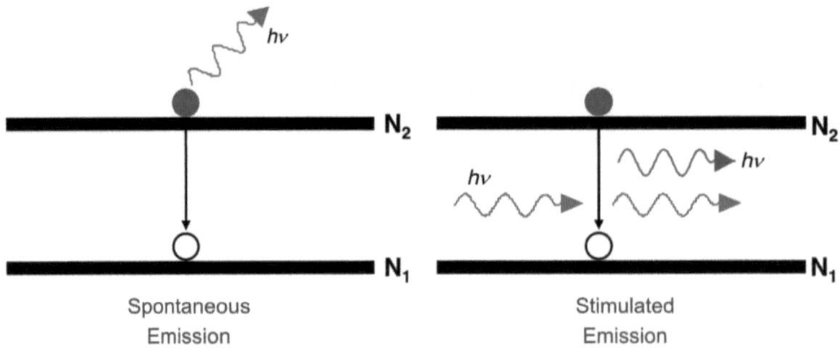

Fig. 4.1 Only the two levels of the transition have been represented here for sake of clarity

illuminated by appropriate light (the pumping process), the system absorbs the light energy and re-emits photons in a spontaneous emission and a stimulated emission forming the laser light itself. Those phenomena are represented schematically in Fig. 4.1.

The energy of the spontaneously emitted photon is related to the energetic difference of the two levels: $h\nu = E_2 - E_1$ and the change of the upper level population N_2 is given by the Einstein coefficient A which is the inverse of the lifetime τ_{sp}

$$\left(\frac{\mathrm{d}N_2}{\mathrm{d}t}\right)_{sp} = -AN_2 \qquad A = \frac{1}{\tau_{sp}}$$

(In 1917, Albert Einstein proposed that there are three processes occurring in the formation of an atomic spectral line. The three processes are referred to as spontaneous emission, stimulated emission, and absorption. With each is associated an Einstein coefficient that measures the probability of that particular occurring process).

A resonant photon $h\nu$ stimulates the active atom or ion to make a transition from the level N_2 to the level N_1 (population number N_1) with a creation of a second photon identical in direction (vector k), in frequency ν and in phase and polarization:

$$\left(\frac{\mathrm{d}N_2}{\mathrm{d}t}\right)_{st} = -W_{21}N_2 \qquad W_{21} = \sigma_{21}F$$

where W_{21} is the stimulated transition probability, σ_{21} the cross section of the transition and F is the photon flux. The photon stimulates absorption from E_1 to E_2 with a population variation of:

$$\left(\frac{\mathrm{d}N_1}{\mathrm{d}t}\right)_{abs} = -W_{12}N_1 \qquad W_{12} = \sigma_{12}F$$

where Einstein demonstrated that $\sigma_{21} = \sigma_1$. With sufficient pumping there is a population inversion when $N_2 > N_1$ and the differential between the incoming flux

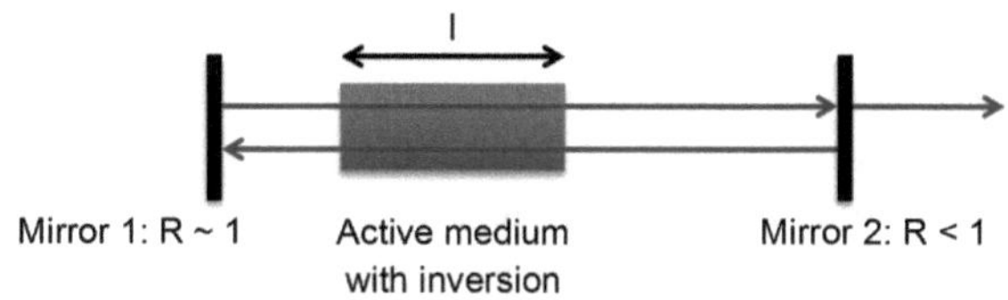

Fig. 4.2 Basic scheme of a laser

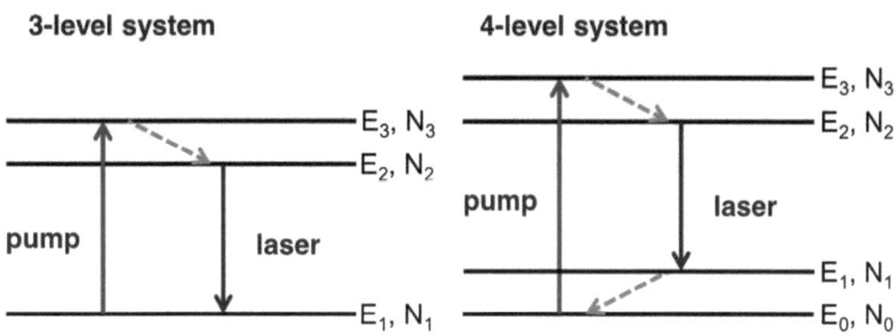

Fig. 4.3 Three-level and four-level systems are the most representatives of the usual lasers

and the out coming flux

$$dF = \sigma_{21} F(N_2 - N_1)dZ$$

is then positive. By adding now a resonator that feeds back the light with a leak rate (output coupler of the resonator) smaller than the gain, we end up with a laser (light amplification by stimulated emission radiation) on Fig. 4.2.

As mentioned previously, laser media are based on three- or four-level system, because a population inversion between two levels is impossible. A three-level system presents the advantage of higher efficiency in the pumping and a faster decay while in the four-level system the population inversion is easier because the thermal population of the lower level of the laser transition has a weaker thermal population (Fig. 4.3).

Let's have a quick look at the basic laser parameters based on the rate equations with the example of a four-level system of Fig. 4.3. Under adequate pumping process, the population density N is related to the photon number q by:

$$\frac{dN_2}{dt} = W_p N_0 - \frac{N_2}{\tau} - BqN_2$$
$$\frac{dq}{dt} = V_a BqN_2 - \frac{q}{\tau_c}$$
$$N_t = N_0 + N_2$$

N_i = population density (cm^{-3})
q = photon number inside the resonator
τ = lifetime of the upper laser level
V_a = mode volume in the active medium
τ_c = photon lifetime in the resonator
N_2/τ = spontaneous decay, $1/\tau$: spontaneous rate
W_pN_0 = pump process, W_p: pump rate
BqN_2 = stimulated decay process
V_aBqN_2 = growth rate of photon number in the resonator
q/τ_c = loss rate of photons = output coupling losses + internal losses

In a continuous wave operation, we set $d/dt = 0$ in the above equations. That yields the important parameters of the laser, where L is the length of the resonator and T is the output coupler transmission:

- the threshold pumping power for a four-level system:

$$P_{\text{thr}} = \frac{1}{\eta_p} \frac{h\nu_p}{\sigma_{21}(\lambda_1)\tau}(T+L)\frac{V}{2d}$$

- the threshold parameter for a three-level system

$$P_{\text{thr}} = \frac{1}{\eta_p} \frac{h\nu_p}{\sigma_{21}(\lambda_1)\tau}(T+L+2d\sigma_{\text{GSA}}(\lambda_1)(n-n_{\text{thr}}))\frac{V}{2d}$$

- the slope efficiency which is another relevant parameter that laser scientists uses when investigating a laser optimization, for small L and $T << 1$:

$$\eta = \eta_p \frac{\lambda_p}{\lambda_1} \frac{T}{T+L}$$

Figure 4.4 represents an example of the output power versus pump power, a line whose slope is the efficiency and P_{thr} is the minimum pump power for starting the laser oscillation.

4.2 Laser Sources for Gravitational Wave Interferometric Detection

The gravitational waves produce on the dark fringe of the Michelson such a tiny signal that it has to be almost free of any perturbation such as the power fluctuations of the laser, the frequency fluctuations and beam position fluctuations of the laser. When reading out a signal from a light source, one fundamental and unavoidable noise is the shot noise. In Michelson interferometers, the ratio of the potential gravitational wave signal to the readout shot noise scales as the square root of the

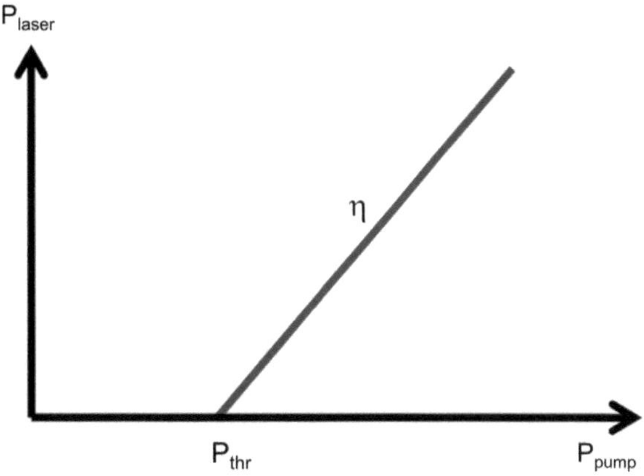

Fig. 4.4 Threshold and slope efficiency of a typical laser in a laser output power versus pump power

power in the interferometer. Then, to overcome the shot noise the laser power has to respect some requirements following the sensitivity planned for the detector. This sensitivity, when limited by readout shot noise is proportional to the square root of the laser power circulating inside the interferometer. Hopefully the combination of resonant optical cavities (Fabry–Perot) and recycling techniques (Chap. 3 of this book) will relax the constraints on the laser source from kilowatts or hundred of kilowatts in advanced generations, to hundred of watts. Recycling techniques have been invented for microwave oscillators and applied later to optical interferometry [1] it consists of enclosing the Michelson interferometer inside a resonant cavity, by adding at the input port, a mirror whose reflectivity compensates the losses of the Michelson's mirrors. Then the first generation of detectors used 10 W laser to end up with 1 kW inside the Michelson, while the advanced detectors Advanced LIGO (aLIGO) and Advanced Virgo (AdV) will need around 200 W to get 1 MW of circulating power.

The high power circulating inside the cavities produces another noise created by fluctuating radiation pressure forces on the mirrors, and the next question is when will that noise be equivalent to the readout shot noise. For a power P the mirror displacement caused by radiation pressure scales as $\sqrt{P}/f^2$, where f is the Fourier frequency above the pendulum resonances. Then induced radiation pressure noise is an issue at lower frequencies while readout shot noise is more relevant at high frequencies. The introduction of signal recycling which is a mirror placed at the output port that encloses the sidebands produced by the signal, will make this simple analysis more complex and a complete simulation of the detector is then necessary [2].

The Michelson interferometer needs to be illuminated by a single frequency polarized source in order to achieve a very good interference contrast in absence of a perfect symmetry in the Michelson and in the Fabry–Perot arms. If the contrast defect of the Michelson is $(1 - C)$, where C is very close to 1, the interference between the two arms reflects any frequency and power fluctuations of the laser. In order to be able to detect such a small gravitational wave signal as 10^{-23} in relative strain, the relative frequency and power fluctuations of the laser have to stick to the same order of magnitude. Then for a laser light emitting at 1 μm wavelength, we end up with the requirements of a few μHz/$\sqrt{}$Hz for the laser frequency fluctuations, while in comparison a free-running quiet laser presents a noise of a few kHz/$\sqrt{}$Hz. The current detectors use the TEM_{00} fundamental mode of the cavities and an efficient incoming laser beam has to contain as much as possible TEM_{00}; the higher order modes that are not rejected before will produce dangerous scattered light inside the interferometer. To minimize the influence of laser power fluctuations, the interferometer operates on the dark fringe but the DC readout method used in advanced detectors, requires a small deviation from the dark fringe. At this point again, a complete simulation of the interferometer behavior will help determining which level of light fluctuations is allowed to reach the desired sensitivity. The relative power noise, especially at low Fourier frequencies, where the radiation pressure coupling is most significant, has to be in the $10^{-9}/\sqrt{\text{Hz}}$ range and the beam pointing stability injected into the interferometer has to be below $10^{-11}\text{rad}/\sqrt{\text{Hz}}$ in angular jitter or of $10^{-10}\text{m}/\sqrt{\text{Hz}}$ in lateral shifts. The details of the specifications required can be found in different chapters of this book. Sophisticated stabilization techniques have been in for a few decades with the development of frequency standards metrology as far as long term (integration time higher than 1 s) performances are concerned; the gravitational wave field has extended their applied developments to the short-term performances with the venue of new design reference cavity and fast transducers. Let's define what kind of lasers could reach such performances.

4.3 Getting Stable HP Lasers

Getting a high power output with a highly stable laser is a serious challenge that is usually managed by separating the two functions of high power output and stabilization. Actually a low power stable laser called master laser is used to injection-lock a high power in a configuration called injection-locking technique [3], where the beam of the master laser controls the oscillation of the high power laser, called slave laser, with all the lasing gain redirected in a single frequency output. A second alternative is to just amplify the beam of the master laser in a single pass slave oscillator, this technique is called master oscillator power amplifier (MOPA). In this second case, the high power stage has no optical cavity around it. While in the first case the stable output of the low power laser is transferred automatically to the HP laser, the MOPA contains no filtering cavity for the amplifier noise and the transfer of stability to the HP beam can be achieved in saturation regimes.

4.4 HP Laser Candidates

Bulk Lasers

When the advanced detectors started to be developed, neodymium doped yttrium aluminum garnet (Nd:YAG) was the best choice for 100 W class lasers. The solid-state Nd:YAG or Nd:YVO4 (Yttrium Orthovanadate) laser emitting in the near infrared range (1.06 μm) has superseded the Argon ion laser emitting at 5,145 nm used in the first prototypes for its visible light. The solid-state laser (SSL) presents many advantages such as intrinsic stability of the solid medium, diode pumping with long lifetime, and last but not least, the longer wavelength implying smaller Rayleigh scattering in the optics which varies as $1/\lambda^2$.

To reach HP level, different configurations have been proposed in order to minimize or to remove the pumping heat; actually about 24–25 % of the pump power is converted into heat in the laser crystal. A few sophisticated cooling interfaces have been designed for HP Nd:YAG lasers as well as very clever laser zigzag paths [4] in the side-pumped crystal to average the thermal effects. Nevertheless, residual thermal induced effects such as stress-fracture limit of the crystal, nonisotropic thermal lensing spoiling the laser spatial mode, etc... limit the output power of those HP lasers to about 200 W today [5]. By managing correctly the thermal load on more than one amplifier stage, this SSL technology reveals to be a good basic solution for this first generation of advanced detectors where about 200 W is required.

The non-planar ring oscillator [6] has a very high intrinsic stability due to its compactness and is used as the master laser in all the projects today (Fig. 4.5). Designed in a small monolithic block the Nd:YAG crystal forms the laser cavity and the only fluctuations encountered by the laser are due to small temperature changes in the crystal, caused themselves by the power fluctuations of the pumping diodes. For frequency controls, a piezo actuator is mounted on top of the crystal and acts as a fast transducer of 1 MHz/V, while the slow drifts are fed back to the temperature control of the crystal with a tuning of 3 GHz/K. As far as the power noise is concerned, the NPRO is equipped with a noise eater that suppresses the 30 dB relaxation oscillation peak (at a few hundred kHz) intrinsic to any solid state laser, and leaves the relative intensity noise (RIN) below $10^{-6}/\sqrt{\text{Hz}}$.

The thermo-optical effects in this monolithic block limits the NPRO design to a few Watts [7] and make them good candidates for playing the role of master laser.

HP Lasers for Advanced LIGO

The laser concept chosen for the advanced LIGO project (aLIGO) is an end-pumped rod laser design [8] which presents the advantages of matching the pump light with the circular beam shape of the Gaussian laser mode in a very good spatial overlap between the gain medium and the laser mode. The thermal effects in rod lasers are

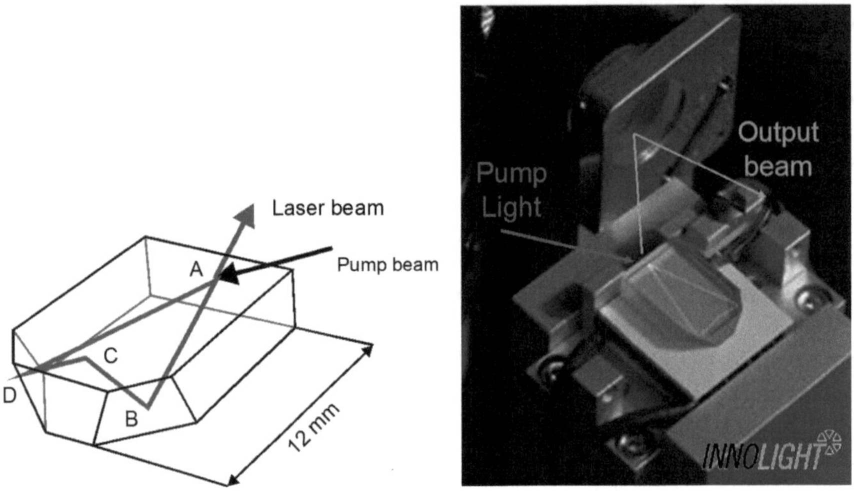

Fig. 4.5 The NPRO non-planar ring oscillator is robust, compact, and simple and has a high intrinsic stability. Dimensions of the crystal are 3 mm × 8 mm × 12mm

still strong compared to the other designs but it can either be compensated or used advantageously to achieve stable fundamental mode operation. One limiting effect of the rod lasers is the depolarization caused by stress-induced birefringence [9] and [10], then a linearly polarized laser mode senses different refractive indices on the horizontal and vertical directions and depolarization and a differential thermal focusing (bi-focusing) occurs in two transverse sections that limit the achievable output power of linearly polarized beams. To scale to HP the use of Nd:YVO_4 instead of Nd:YAG leads to a strong reduction of the depolarization and bi-focusing effects while emitting at the same wavelength 1.06 μm. Nevertheless to reach 200 W output without totaling intense thermal effects on one laser, it is unavoidable to use more than one amplification stage from the master to the final HP stage.

The single frequency laser for advanced LIGO (Fig. 4.6) is a compound system consisting of a Master low power laser, a first stage amplifier and a final HP stage. A Master oscillator which is a 2 W NPRO Nd:YAG is sent to a medium power amplifier bringing the laser power to 35 W: it is composed of a Nd:YV04 rods in series, each of them pumped axially by HP laser diodes. The HP stage contains 4 Nd:YV04 rods end-pumped laser diodes, but built inside a ring cavity able to lase up to more than 200 W. This final stage is then injection-locked by the 35 W beam of the first stage to yield a 200 W single frequency output.

Though thermal effects are strong inside laser crystals, the rods are suitably pumped in a very clever configuration such as to minimize those effects and the 220 W output of the amplifiers contain more than 88 % of fundamental mode. Then by spatial filtering, 165 W can be measured in the beam with only less than 2 % in high order modes [11].

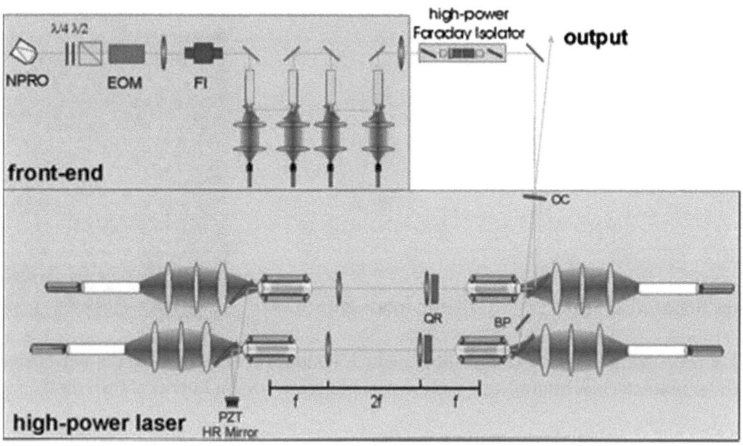

Fig. 4.6 Advanced LIGO laser concept: the master laser NPRO is amplified by a front-end first stage of four laser rods and delivers around 35 W which is sent into the HP stage combining two pairs of laser rods. Two lenses and a quartz rotator (QR) are inserted in the middle of each pair of rods to act as a depolarization unit [8]

Fiber Lasers and Amplifiers

A fiber laser is the laser in which the active medium is an optical fiber doped with rare-earth elements such as erbium, ytterbium, neodymium, etc... Fiber amplifiers are doped fibers providing amplification without lasing. Fiber medium has some intrinsic advantages over bulk medium: the large ratio of surface area to volume of fibers provides excellent heat dissipation, fiber laser is virtually alignment free due to the wave-guided laser cavity and pump laser path, and thus more robust against mechanical disturbance. The output beam quality of a fiber is determined mostly by the guiding technique because the thermal load is no more concentrated on a small volume but distributed along the fiber. At the same time as the gain, the thermal effect evacuation also scales up with the length of the fiber. Furthermore, the low number of optical components and the small size of the laser head also simplify the maintenance on site.

Rare-earth doped fiber amplifiers can offer single-frequency output at higher efficiencies (up to 50 %) and at lower cost than solid-state amplifiers for the similar power levels [12]. Over recent years, the powers that can be achieved from Ytterbium-doped silica fiber lasers have grown rapidly due to parallel advances in high power diodes, diode-to-fiber coupling schemes and doped fiber design and fabrication. To ensure single-mode guidance in the fiber, the core diameter of the standard fiber is at the same order of magnitude as the wavelength of the guided beam (5–10 microns). To transport higher optical intensities as well as to have higher damage thresholds, larger diameter cores of the order of 50 μm (large-mode-area or LMA fibers) have been invented while keeping single-mode guidance (Fig. 4.7). A straightforward approach

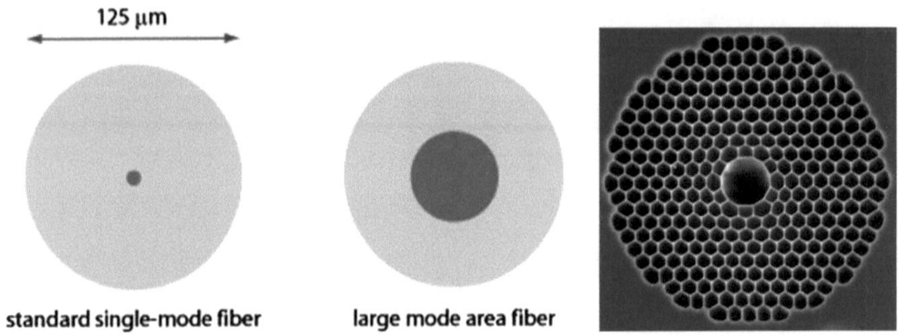

Fig. 4.7 Comparison of bare uncoated fibers with a standard core size (8 μm diameter) and a large core (50 μm diameter). A microscope picture of a photonic crystal fiber is shown on the *right*

is then to decrease the refractive index difference between the core and the cladding putting limitations to the bending capacity of the fiber. More sophisticated design based on photonic crystal fibers have been developed to address this drawback. Photonic crystal fibers (PCF) also called holey or microstructure fibers are optical fibers in which waveguide properties are done not from a spatially varying glass composition but from an arrangement of very tiny and closely spaced air holes which go through the whole length of the fiber. They typically guide light in a relatively narrow wavelength range but can be used for pulse compression with very high optical intensities as most of the power propagates in the hollow core. Those kind of fiber lasers have enabled output powers to exceed 1 kW while retaining excellent efficiencies [13].

Fiber lasers or amplifiers based on single-mode fibers can generate diffraction-limited output when it restricts the pump source to diffraction-limited beam and thus usually low power sources. The invention of double-clad fibers resolved that issue by extending the pumping to the cladding: the laser light propagates in a single-mode core, surrounded by an inner cladding in which the pump light propagates. Only the core is rare-earth doped. The pump light is confined to the inner cladding by an outer cladding with lower refractive index. Figure 4.8 shows a schematic of a double-clad fiber used as an amplifier.

Double clad fiber lasers with continuous-wave powers in excess of 100 W with diffraction-limited beam quality have been demonstrated in 2002 and a 246 W single-frequency master oscillator power amplifier have been already achieved with 91 % in TEM_{00} mode [14]. This has been achieved with a large mode field diameter fiber without any stimulated Brillouin (SBS) scattering effect. One of the major obstacles when propagating signals through fibers is the stimulated Brillouin scattering (SBS) effect, which is frequently encountered when narrow-bandoptical signals (e.g., from a

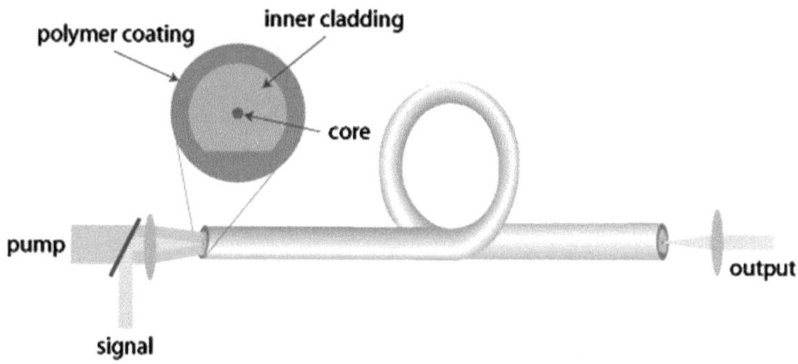

Fig. 4.8 Double-clad fiber with a pump light launched into the inner cladding while the signal to be amplified is launched into the doped core

single-frequency laser) are amplified in a fiber amplifier, or just propagated through a passive fiber. While the material nonlinearity of, e.g., silica is actually not very high, the typically small effective mode area and long propagation length strongly favor nonlinear effects. For silica fibers, the Brillouin frequency shift is of the order of 10–20 GHz, and the Brillouin gain has an intrinsic bandwidth of typically 50–100 MHz, which is determined by the strong acoustic absorption (short phonon lifetime). New fiber designs have been invented to lower successively the SBS threshold in doped fibers, but the issue remains open when using undoped fibers to transport HP narrow linewidth lasers in a few meters length, for instance the fiber length should be shorter than 2 m when 200 W frequency stabilized laser is launched through it without being spoiled by SBS. This is still an issue today when we want to transport the HP stabilized laser to the interferometer through fibers in order to avoid all the loss due to intermediate optics.

Another HP limitation of the rare-earth doped silica fibers come from the performance degradations due to photodarkening mechanism which involves the formation of color centers or other microscopic structural transformations in the medium [15] this process is a fundamental issue for fiber amplifiers involving long exposure to very high pump intensity levels. Though it is a permanent effect it can be reversed by heating the fiber or by co doping it with Cerium.

Rod fibers with large core (Fig. 4.9 and [16]) have been proposed in order to maximize the pumped intensity in the doped core while reducing the fiber length. Commercially available amplifiers exist on then 300 W range in pulsed operation [17], but again delivering cw HP output give rise to thermal gradients and then risk of mode instabilities due to mode competition.

Ytterbium-doped YAG (Yb:YAG), can lase at 1,064 nm in a four-level system and at 1,030 nm in a three-level system. This shorter wavelength could replace the 1,064 nm light as it has a smaller quantum defect and a better thermal management because it is pumped by 940 nm diodes instead of 890 nm for the 1,064 line. Furthermore, the 940 nm diodes have potentially longer lifetimes and the gain at 1,030 nm is two times larger than at 1,064 nm. The switch to this new wavelength is not a

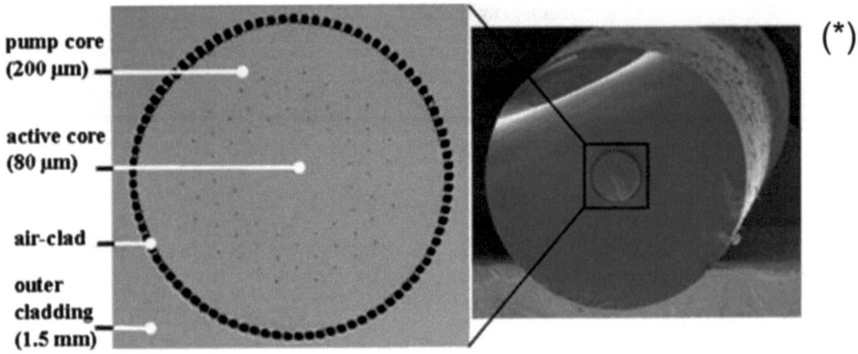

Fig. 4.9 Photonic crystal fiber, a section of the rod fiber

big issue for the mirror coatings performances as long as they are coated for normal incidence. For beamsplitters and other non-normal incidence mirrors, a check of reflectivity performances is necessary before going to 1,030 nm. Today, an impressive number of publications on HP fiber amplifiers exist but very few of them are useful for our applications as most often the given M^2 factor (Siegman's Book) give less information than the measurements on the modal content of the beam [18].

Fiber Amplifiers for Advanced Virgo

For Advanced Virgo (AdV), the laser concept has been defined more recently and to exploit a new solution which can be scaled up to higher power, Yb3+ doped fibers have been chosen for some relevant issues: beam quality, power tunability, efficiency, maintenance, and last but not least the cost. Adopting this new technology for the AdV has been an important challenge for the project and to minimize the potential risks in long-term operation, the reference solution adopted is the coherent addition of lasers to achieve the HP beam. The potential risks are the onset of mode instabilities and photodarkening effects in the fibers at high pump levels when they are used to amplify narrow linewidth lasers such as the ones specified for gravitational wave detection. Today a few options are still to be tested and compared for the R&D: combining the beams of many medium power amplifiers to get close to 200 W or have a multistage amplifier, each of them working with a gain low enough to avoid the mode instabilities and the photodarkening effect in the lifetime of the experiment.

In order to investigate the fiber laser behavior, a prototype has been designed in 2009 by the US company Nufern, a manufacturer of specialized fiber optics, delivering 180 W in single frequency operation. It is a two-stage amplifier fiber coupled to a seed source of 50 mW by a pre-amplifier delivering 10 W. Figure 4.10 shows the output power versus the diode current: the backward signal appears (SBS) appears above 150 W and a nominal power of 177 W could be achieved with 300 W pump power. We have checked (Fig. 4.11) that the frequency noise is due to the master frequency noise and with a standard servo-loop acting on the diodes current, we could get a noise floor close to photon shot noise at a photocurrent of 24 mA [18]. Unfortunately, an unexplained phenomenon that we suspect to be photodarkening effect combined to some SBS, decreases the power after a few weeks of running and damaged the fiber completely.

The GW laser source needs to be reliable in long-term operation and the need is beyond a publication's result. Prototypes and commercial amplifiers that claimed HP output have to be operated during a few months to be sure that no SBS or photodarkening or mode instabilities appear. It is obvious that commercial fibers have made large progress and will continue to be better and better in terms of HP behavior and in the time of the need for a full HP laser for Advanced Virgo, we are confident that commercial fiber amplifiers will appear on the shelves.

The combination or coherent addition of continuous wave lasers is not a big issue since it has been done almost 30 years ago with Argon ion lasers [20] and applied on fiber lasers in 2011 [21]. They obtained a complete summation of two fiber lasers without loss and a very good beam quality (97 % TEM_{00} content) without introduction of extra noise in the addition.

4.5 Laser Pre-stabilizations

The goal of the lasers pre-stabilization (PSL subsystem) is to achieve an intermediate level in the 100 $mHz/\sqrt{Hz}$ range on the laser beam before it is sent to explore the long baseline cavities. The laser heads are provided with high-bandwidth actuators for the last stage of frequency control loops that lock the laser frequency to the length of the kilometric cavities. The laser frequency is controlled in short term via the piezo actuator of the NPRO (low power stable laser) and by an electro-optic modulator (EOM) placed between the NPRO laser and the amplifiers.

The schematic layout of the aLIGO pre-stabilized laser [11] and AdV are represented respectively on Figs. 4.12 and 4.13. Both layouts have the same typical stabilizations: a pre-mode-cleaner (PMC) consisting of a rigid spacer in a ring three-mirror configuration cavity to avoid light feedback. The length of aLIGO's PMC is 2 m round-trip while in AdV it is about 50 cm length. The HP beam transmitted through the PMC has a pointing in angular and in transverse directions, filtered by a factor proportional to the finesse of the PMC. This step is a first "rough" filtering of the beam to ease the alignment of the incoming through the vacuum system containing the suspended input mode cleaners (IMC). This IMC acts as a low pass filtering for all the amplitude and beam jitters noise of the laser. The design of PMC takes into account the spatial filtering needs for the length and the power standing of its mirrors, determining the finesse of the cavity. If F is the finesse and L is the length of the cavity, the filtering transfer function of the cavity optical resonator can be described as a low-pass with a corner frequency given by:

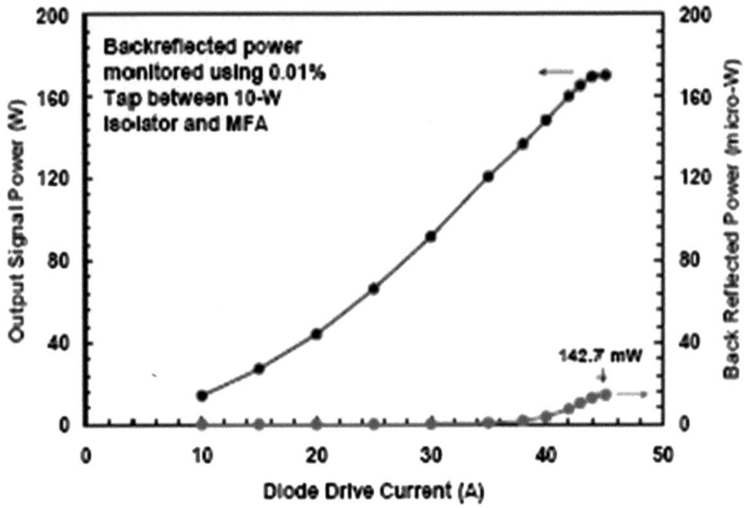

Fig. 4.10 Measured output power of the Nufern laser prototype and evidence of backward propagating SBS power through 0.01 % tap port (Nufern flyer) [19]

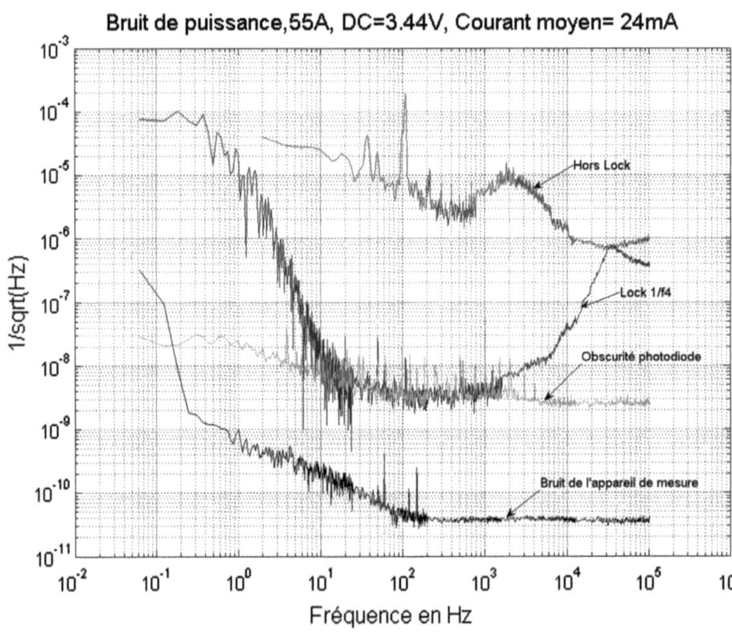

Fig. 4.11 Relative intensity noise (RIN) in units of $1/\sqrt{\text{Hz}}$ of the Nufern fiber amplifier prototype versus frequency in the GW detection range. In *red curve* of free-running RIN, in *black* electronic noise, in *green dark* noise of the photodiode and in *blue* the RIN in closed loop below $10^{-8}/\sqrt{\text{Hz}}$ limited only by the photodiode noise

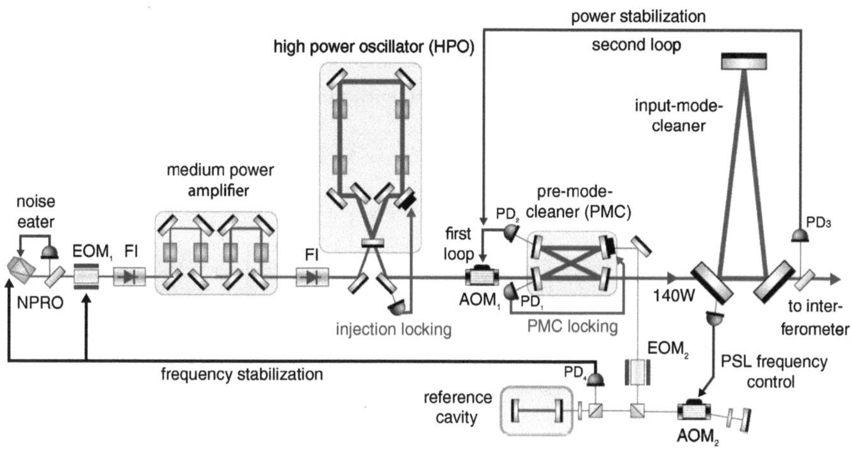

Fig. 4.12 Layout of the Advanced LIGO pre-stabilized laser system including a schematic of the nested frequency control scheme. *EOM* electro-optical modulator, *AOM* acousto-optical modulator, *FI* Faraday isolator, *PMC* pre-mode-cleaner resonator, *PZT* piezo-electric transducer, *FSS* frequency stabilization electronics

$$f_{\mathrm{FP}} = \frac{c}{4LF}$$

with $L = 0.5\,\mathrm{m}$ and $F = 10000$, we can get a cut-off frequency of 15 kHz which will give an attenuation of more than 100-fold for any frequency above a few MHz. Then, depending on the modulation frequencies used in the interferometer, the PMC can also be designed to filter the laser power noise at those frequencies. For the beam noises, it will filter out the high order modes not resonating in the mode-cleaner by a factor proportional to $1/\sqrt{F^2}$. The PMC length is kept resonant with the laser frequency by a piezo-actuator adequately mounted inside the resonator, i.e., without introducing any spurious drift by itself.

When investigating a laser's performances, one needs to measure precisely all the free-running noises such as the RIN, the beam pointing noise, the modes content, and the frequency fluctuations. Those measurements are mandatory to determine if active (mode-cleaners) and passive controls (servo-loops) will make it suitable for a GW interferometric detector.

4.6 Conclusion

The very prolific work done today on fiber lasers will possibly set the path to very high power needed for the third generation of interferometers, though their optical design could reduce significantly the requirements. Depending on the thermal noise limitations of mirrors, coatings, preferred wavelength could be shorter and beam

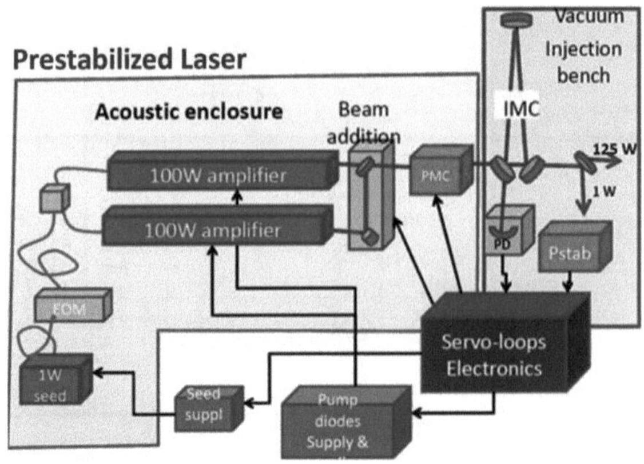

Fig. 4.13 Layout of the advanced Virgo PSL with the same kind of elements as for LIGO. The input mode cleaner IMC is a part of INJ (injection) subsystem but closely related to PSL

shape could be different from the fundamental TEM_{00} modes. High order Laguerre-Gaussian modes have been calculated to give less thermal noise and mainly less themal effects and their use might allow to matching the beam coming out of the laser and the beam exciting one resonance of the optical caivty. Other lasers medium could be exploited such as discs or multisegmented laser rods, etc …Though they already exist for industrial applications or pulsed applications, their operation on narrow linewidth remains to be experimented in terms of long-term reliability. In any case, the lasers will most likely be based on injection-locking or MOPA techniques or a combination of both. Furthermore, the lasers need to operate in single transverse mode with free-running noises compatible with the performances of the servo-loops to be added on for the future gravitational waves detectors.

References

1. R. Drever et al., A gravity-wave detector using optical cavity sensing, in E. Schmutzer (ed.) *Proceedings of the 9th International Conference on General Relativity and Gravitation*, Jena, 14–19 July 1980, Cambridge University Press (1983), pp. 265–267
2. K. Somiya et al., Phys. Rev. D. **73**, 122005 (2006)
3. O. Cregut, C.N. Man, D. Shoemaker, A. Brillet, A. Menhert, P. Peuser, N.P. Schmitt, P. Zeller, K. Wallmeroth, 18 W single-frequency operation of an injection-locked cw Nd:Yag laser. Phys. Lett. A **140**, 294 (1989)
4. A.K. Sridharan et al., Appl. Opt. **45**, 3340 (2006)
5. C. Basu et al., *n*° 14. Opt. Lett. **37**, 2862 (2012)
6. InnoLight. Mephisto Product Line (2009) http://www.innolight.de/
7. I. Freitag et al., Opt. Commun. **115**, 511 (1995)
8. M. Frede et al., Opt. Express **15**, 459 (2007)
9. W. Koechner, Appl. Opt. **9**, 2548–2 (1970)
10. S. Tidwell et al., IEEE J. Quantum Electron. **28**, 997–1009 (1992)
11. P. Kwee et al. Opt. Express **20**, 10617 (2012)
12. J. Limpert et al., Selected topics in quantum electronics. IEEE J **13**, 537–545 (2007)
13. Y. Jeong et al., Opt. Express **12**, 6088–6092 (2004)

14. M. Karow et al., *n*°5 Opt. Express **20**, 5319 (2012)
15. Manek-Honninger et al., Opt. Express **15**, 1606 (2007)
16. J. Boullet et al., Opt. Express. **16**(22), 17891–17902 (2008)
17. Eolite, Bordeaux, http://eolite.com/Octopus.html?wpid=39620
18. C. Greverie, A. Brillet, C. Man, W. Chaibi, J.P. Coulon, Felsik 2010; K, CLEO 2010, OSA technical digest paper JTuD36
19. Nufern Cie, http://www.nufern.com
20. C.N. Man, A. Brillet, Injection locking and coherent summation of argon ion lasers. Proceedings of the gravitation, geometry and relativistic physics, Lectures notes in Physics, Aussois **212**, p. 222 (1984)
21. H. Tunnermann et al., Opt. Express **19**, 19600 (2011)

Chapter 5
Input Optics System

Matteo Tacca

Abstract In the gravitational waves interferometric detectors, the properties of the light beam at the input of the interferometer are very different from the ones of a standard laser source. Therefore, an interface between the laser source and the interferometer is essential. In this chapter, the interface devoted to deliver a light beam characterized by the right properties is described. This interface is called Input Optics System. The Input Optics System of Advanced Virgo is considered and taken as example: all the subsystems are described, focusing on the involved physical phenomena. Some possible future upgrades are also introduced. At the end, the few differences between the Advanced Virgo and the Advanced Ligo Input Optics Systems are presented.

5.1 Introduction

In the gravitational waves interferometric detectors, the properties of the light beam at the input of the interferometer are very different from the ones of a standard laser source. Therefore, an interface between the laser source and the interferometer is essential. This interface is called Input Optics System and is devoted to deliver a light beam characterized by the right properties for what concerns power, dimensions, frequency stability, and position stability in order to ensure the right optical matching inside the interferometer. Since the Input Optics System has to satisfy many requirements, it is quite complex and is made of many subsystems. Most of

M. Tacca (✉)
European Gravitational Observatory (EGO), Via Amaldi, S. Stefano a Macerata—Cascina, 56021 Pisa, PI, Italy
e-mail: matteo.tacca@ego-gw.it

Laboratoire Astroparticule et Cosmologie (APC), 10, rue Alice Domon et Leonie Duquet, 75205 Paris Cedex 13, France
e-mail: matteo.tacca@apc.univ-paris7.fr

M. Bassan (ed.), *Advanced Interferometers and the Search for Gravitational Waves*, Astrophysics and Space Science Library 404, DOI: 10.1007/978-3-319-03792-9_5,

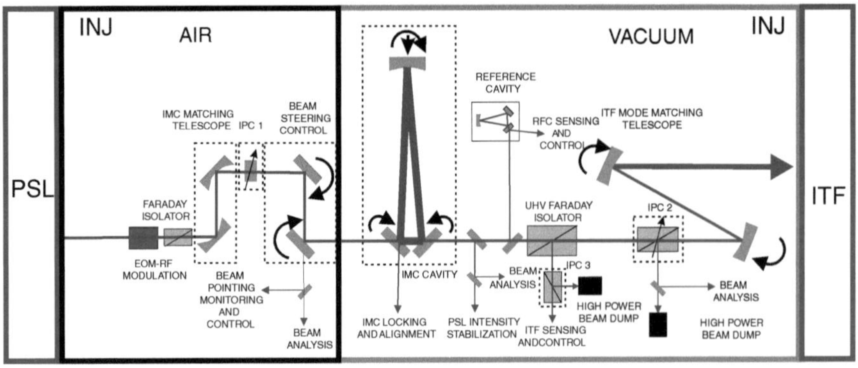

Fig. 5.1 Advanced Virgo Injection System

the aims of the Input Optics System are the same for all the large gravitational wave detectors, i.e., Advanced LIGO (aLIGO) [1], Advanced Virgo (AdVirgo) [2], and KAGRA [3]. Therefore, many features and the physical phenomena of this system are common in the various interferometers. In order to describe its various subsystems, the Advanced Virgo Input Optics System, called Injection System (INJ) [2], is considered and taken as example. At the end, a few differences between the AdVirgo and aLIGO Input Optics Systems are summarized.

The Advanced Virgo Injection System is divided into two large parts: the first part is assembled on an in-air optical bench (called External Injection Bench, EIB), the second is on an in-vacuum suspended optical bench (named Suspended Injection Bench, SIB). Figure 5.1 shows the general scheme of the Injection System.

In particular, the Radio Frequency (RF) modulation system, the Input Mode Cleaner (IMC) mode matching telescope, one Input Power Control (IPC) system, the Beam Pointing Control (BPC) system, and some steering optical elements are installed the in-air bench. On the other hand, a part of the Input Mode Cleaner Cavity, the High Power Faraday Isolator (FI), the Injection Mode Matching Telescope (MMT), the Reference Cavity (RFC), one Input Power Control system, and some steering optical elements are assembled on the in-vacuum bench.

The main troubles, common to almost all subsystems of the Input Optics System, are due to the thermal aberrations introduced by the very high input power laser beam that hits all the optical elements. In facts, some watts of power can be dissipated and absorbed in the various elements, resulting in temperature gradients. These temperature gradients introduce refractive index gradients and geometrical alterations of the elements, which, in the end, can affect their optical properties.

5.2 Radio Frequency Modulation System

A Radio Frequency modulation of the laser beam is needed in order to provide the different modulation frequencies exploited in both the longitudinal and angular control systems of the interferometer. In particular, five different modulation frequencies are needed for the automatic control system of Advanced Virgo. In order to modulate an optical beam two main modulation techniques are used: the electro-optic modulation [4, 5] and the acousto-optic modulation [4, 5].

5.2.1 Electro-Optic Effect

Electro-optic modulators are usually made with crystals that exhibit the electro-optic effect, which is a variation of the optical properties of a material as a consequence of the application of an external electric field. Many distinct phenomena are induced by the application of an electric field. They can be divided into two main effects: the variation of the absorption of the crystal and the change of its reactive index.

An electric field can transform an isotropic medium in an anisotropic one. In particular, the refractive index can be expressed as a function of the applied electric field:

$$n(E) = n + a_1 E + a_2 E^2 + \cdots \tag{5.1}$$

The coefficients a_1 and a_2 are, respectively, the linear and the second-order electro-optic coefficients. The refractive index variations induced by the linear term are named Pockels Effect, whereas the ones related to the second-order coefficient are called Kerr Effect. The Pockels Effect can occur only in certain crystals, the non-centrosymmetric materials. On the contrary, all the materials show the Kerr Effect with different magnitudes even if, in general, it is weaker than the Pockels Effect.

5.2.1.1 Pockels Effect

The electro-optic modulators are built with crystals that exhibit the Pockels Effect, such as Lithium Niobate ($LiNbO_3$), Gallium Arsenide (GaAs), Potassium Titanyl Phosphate (KTP), Potassium Dihydrogen Phosphate (KDP), or Rubidium Titanyl Phosphate (RTP). The Pockels Effect can be simply understood by considering the refractive indices ellipsoid and its deformation induced by the applied electric field. In particular, an isotropic medium changes into a birefringent one (like GaAs), whereas a uniaxial material turns into a biaxial one (like KDP).

When an electric field is applied, the non-perturbed refractive indices ellipsoid is:

$$\left(\frac{1}{n^2}\right)_1 x^2 + \left(\frac{1}{n^2}\right)_2 y^2 + \left(\frac{1}{n^2}\right)_3 z^2 + 2\left(\frac{1}{n^2}\right)_4 yz + 2\left(\frac{1}{n^2}\right)_5 xz + 2\left(\frac{1}{n^2}\right)_6 xy = 1 \tag{5.2}$$

if x, y, and z are parallel to the principal axes of the crystal characterized by the indices n_1, n_2, and n_3 and following notation is adopted (considering a 3×3 matrix): $11 \rightarrow 1$; $22 \rightarrow 2$; $33 \rightarrow 3$; $23, 32 \rightarrow 4$; $13, 31 \rightarrow 5$; $12, 21 \rightarrow 6$. The linear perturbation of the refractive index induced by the applied electric field E_j can be written as:

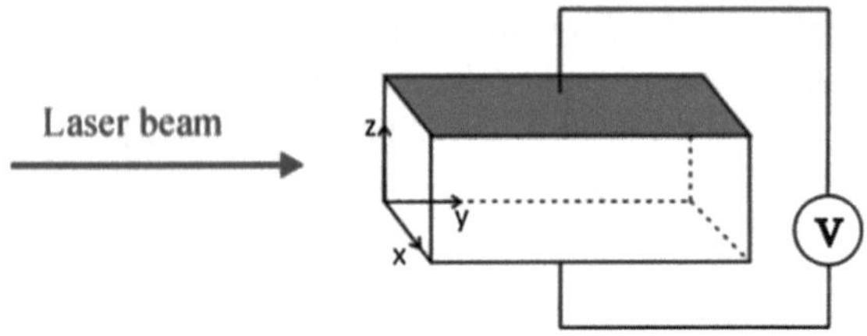

Fig. 5.2 Electric field applied on a uniaxial crystal

$$\Delta\left(\frac{1}{n^2}\right)_i = \Delta_i = \sum_{j=1}^{3} r_{ij} E_j \tag{5.3}$$

where r_{ij} are constants that depend on the material, named the Pockels coefficients. They are the components of the electro-optic tensor:

$$\begin{bmatrix} \Delta_1 \\ \Delta_2 \\ \Delta_3 \\ \Delta_4 \\ \Delta_5 \\ \Delta_6 \end{bmatrix} = \begin{bmatrix} r_{11} & r_{12} & r_{13} \\ r_{21} & r_{22} & r_{23} \\ r_{31} & r_{32} & r_{33} \\ r_{41} & r_{42} & r_{43} \\ r_{51} & r_{52} & r_{53} \\ r_{61} & r_{62} & r_{63} \end{bmatrix} \begin{bmatrix} E_1 \\ E_2 \\ E_3 \end{bmatrix} \tag{5.4}$$

Only a few of the Pockels coefficients are nonzero for most of the non-centrosymmetric materials. There are two different kinds of the Pockels Effect depending on the direction of the application of the external electric field: the transverse Pockels Effect when the electric field is orthogonal to the beam direction of propagation; the longitudinal one if the electric field is parallel to beam direction of propagation. Moreover, the electro-optic effect depends on the polarization of the light beam with respect to the crystal cross-section. For example, coefficient r_{13} nonzero constrains the application of the electric field along the z axis and the propagation of the light beam in the y direction with polarization along the x axis, as shown in Fig. 5.2.

In general, the principal axes of the perturbed refractive indices ellipsoid are changed by the application of the electric field; therefore the new axis and the new refractive index values have to be found starting from the ellipsoid equation. For example, the KDP is one of the most used materials in order to realize electro-optic crystals, thanks to its mechanical resistance and its transparency at visible wavelengths. Its electro-optic tensor is:

$$[r] = \begin{bmatrix} 0 & 0 & 0 \\ 0 & 0 & 0 \\ 0 & 0 & 0 \\ r_{41} & 0 & 0 \\ 0 & r_{41} & 0 \\ 0 & 0 & r_{63} \end{bmatrix} \tag{5.5}$$

The KDP crystals are uniaxial (refractive indices r_o and r_e) and they become biaxial after the application of an electric field (E_x, E_y, E_z). The new refractive indices ellipsoid is:

$$\frac{(x^2+y^2)}{n_o^2}+\frac{z^2}{n_e^2}+2r_{41}E_x yz+2r_{41}E_y xz+2r_{63}E_z xy=1 \tag{5.6}$$

When the applied electric field is parallel to the axis Oz, Eq. 5.6 becomes:

$$\frac{(x^2+y^2)}{n_o^2}+\frac{z^2}{n_e^2}+2r_{63}E_z xy=1 \tag{5.7}$$

Oz remains one of the principal axes of the indices ellipsoid characterized by its principal index $n_z = n_e$. On the other hand, the section at the plane $z = 0$ becomes an ellipse whose axes (Ox′, Oy′) are tilted of 45 deg. with respect to the crystal principal axes (Ox, Oy). Principal indices $n_{x'}$ and $n_{y'}$ relatives to the axes (Ox′, Oy′) can be calculated from the indices ellipsoid equation:

$$\begin{aligned} n_{x'} &= n_o - \Delta n \\ n_{y'} &= n_o + \Delta n \end{aligned} \tag{5.8}$$

where $\Delta n = \frac{1}{2}n_o^3 r_{63}E_z$.

If the Pockels Effect in a *KDP* crystal is used in the longitudinal configuration (applied electric field parallel to the direction of the propagation of the beam), the optical delay introduced by an electric field E in a crystal of length L is proportional to the difference between $n_{x'}$ and $n_{y'}$ and is equal to:

$$\Gamma = \frac{2\pi}{\lambda}n_o^3 r_{63}E_z L = \frac{2\pi}{\lambda}n_o^3 r_{63}V \tag{5.9}$$

where V is the tension applied to produce the electric field E. Therefore, a phase shift proportional to the applied tension is induced. If the crystal is put between two crossed polarizers, the phase modulation is transformed in an amplitude one, which is:

$$I_T = I_0 \sin^2\frac{\Gamma}{2} = I_0 \sin^2\frac{\pi}{2}\frac{V}{V_\pi} \tag{5.10}$$

where I_0 is the incident intensity, I_T the transmitted intensity, and $V_\pi = \lambda/(2n_o^3 r_{63})$ is the tension, independent from L, which has to be applied in order to obtain a half-waveplate ($\Gamma = \pi$).

On the other hand, if the *KPD* crystal is used in the transverse configuration (applied electric field perpendicular to the direction of the propagation of the beam), for a linearly polarized input beam, the introduced phase shift is:

$$\Gamma = \frac{2\pi}{\lambda} L \left[(n_e - n_o) - \frac{1}{2} n_o^3 r_{63} E_z \right] \tag{5.11}$$

Therefore, there is a natural birefringence and an electrical one. To compensate the natural birefringence the crystal length L can be chosen in order to have $2\pi/\lambda L(n_e - n_o)$ as a multiple of 2π. Between two crossed polarizers, the transmitted intensity becomes:

$$I_T = I_0 \sin^2 \frac{\pi}{2} \frac{V}{V_\pi} \tag{5.12}$$

where $V = E_z d$ and $V_\pi = (\lambda d)/(n_o^3 r_{63} L)$.

The RTP material is another most used crystal in the realization of the electro-optic modulators. It shows a stronger Pockels Effect. Its electro-optic tensor is:

$$[r] = \begin{bmatrix} 0 & 0 & r_{13} \\ 0 & r_{22} & r_{23} \\ 0 & 0 & r_{33} \\ 0 & r_{42} & 0 \\ r_{51} & 0 & 0 \\ 0 & 0 & 0 \end{bmatrix} \tag{5.13}$$

In particular, applying the electric field along the z-axis of this uniaxial crystal, the refractive indices ellipsoid becomes:

$$\frac{(x^2 + y^2)}{n_o^2} + \frac{z^2}{n_e^2} + r_{13} E_z x^2 + r_{23} E_z y^2 + r_{33} E_z z^2 = 1 \tag{5.14}$$

The crystal axis remains the same and the new refractive indices are

$$\begin{aligned} n_x &= n_o - 0.5 n_o^3 r_{13} E_z \\ n_y &= n_o - 0.5 n_o^3 r_{23} E_z \\ n_z &= n_e - 0.5 n_e^3 r_{33} E_z \end{aligned} \tag{5.15}$$

In a longitudinal configuration, since $n_x = n_y$, the light beam sees the same phase shift for any polarization. Therefore, a phase modulator, independent from the light polarization, can be implemented, whereas an amplitude modulation cannot be introduced.

On the other hand, in the transverse configuration, for an input beam linearly polarized along z, the induced phase shift is:

$$\Gamma = \frac{2\pi}{\lambda} n_z L = \frac{2\pi}{\lambda} L(n_e - 0.5 n_e^3 r_{33} E_z) \tag{5.16}$$

The tension necessary to obtain an amplitude modulation is $V_\pi = (\lambda d)/(n_e^3 r_{33} L)$.

5.2.1.2 Kerr Effect

The Kerr Effect is the change in the refractive index of a medium after the application of an external electric field. The Kerr Effect is different from the Pockels Effect because the induced refractive index variation is directly proportional to the square of the applied electric field. Moreover, it exists for all the symmetries of the crystals. The equation of the refractive indices ellipsoid is:

$$x^2\left(\frac{1}{n_1^2}+s_{11}E_x^2+s_{12}E_y^2+\cdots+2s_{16}E_xE_y\right)+y^2\left(\frac{1}{n_2^2}+s_{21}E_x^2+\cdots+2s_{26}E_xE_y\right)$$
$$+z^2\left(\frac{1}{n_3^2}+s_{31}E_x^2+\cdots+2s_{36}E_xE_y\right)+2yz(s_{41}E_x^2+\cdots+2s_{46}E_xE_y)$$
$$+2zx(s_{51}E_x^2+\cdots+2s_{56}E_xE_y)+2xy(s_{61}E_x^2+\cdots+2s_{66}E_xE_y)=1 \tag{5.17}$$

where the elements s_{ij} of the 6×6 matrix are the Kerr coefficients along the medium principal axis.

The induced birefringence is:

$$\Delta n=\frac{1}{2}n^3 s_{ij}E^2 \tag{5.18}$$

The Kerr effect is observed in almost every liquid. The induced birefringence is due to the molecules orientation after the application of the electric field; but it is present also in solid materials. The Kerr effect is exploited to control the polarization of an incident beam.

5.2.1.3 Electro-Optics Modulators

In the description of the Pockels Effect (Sect. 5.2.1.1), it has been shown that the application of an external electric field on a crystal induces a phase shift, which can be used as a direct phase modulator of the light beam. The polarization of the incident beam is parallel to one of the principal directions of the crystal. The application of the electric field linearly modulates the phase of the light beam that crosses the crystal. At the output of the crystal, a signal characterized by the spectral components $\omega_0 \pm n\Omega$ is obtained, where ω_0 is the laser frequency and Ω is the modulation frequency. The amplitude of the wave at the frequency $\omega_0 \pm n\Omega$ is proportional to the nth-order Bessel function: $J_n(2\pi/\lambda L\Delta n)$. Figure 5.3 shows the scheme of an electro-optic phase modulator used in the transverse configuration.

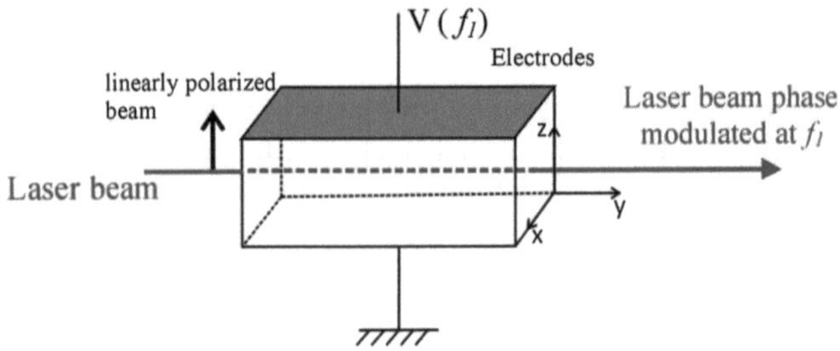

Fig. 5.3 Electro-optic phase modulator (transverse configuration)

5.2.2 Acousto-Optic Effect

An acoustic wave propagating in a medium produces mechanical stresses that cause a modification in the refractive index through the photoelectric effect. An acoustic wave that varies periodically can generate a refractive index grating of the same period. An optical wave that hits this grating will be diffracted in two different ways. If it is diffracted in only one order of diffraction, the grating is working in the Bragg regime; if it is diffracted in various orders of diffraction, the grating is working in Raman-Nath regime.

The interaction between acoustic and optical waves is exploited in the realization of modulation devices. It can be described with the photoelastic tensor of the materials. The effect of a stress on the optical properties of a medium is the modification of the refractive indices ellipsoid. If x, y, z are the principal axes of the ellipsoid and n_1, n_2, n_3 are the principal refractive indices, the non-modified ellipsoid is:

$$\frac{x^2}{n_1^2} + \frac{y^2}{n_2^2} + \frac{z^2}{n_3^2} = 1 \tag{5.19}$$

The refractive index and the stress S are related through the photoelastic tensor $[p_{ij}]$:

$$\Delta\left(\frac{1}{n^2}\right)_i = p_{ij}S_j \tag{5.20}$$

where $i, j = 1, 2, \ldots, 6$ and $S_1 = S_{xx}$, $S_2 = S_{yy}$, $S_3 = S_{zz}$, $S_4 = S_{yz}$, $S_5 = S_{zx}$, $S_6 = S_{xy}$ are the stress components. After the interaction between the acoustic and the optic waves, the modified ellipsoid is:

$$\begin{aligned} &x^2\left(\frac{1}{n_1^2} + \sum_{j=1}^{6} p_{1j}S_j\right) + y^2\left(\frac{1}{n_2^2} + \sum_{j=1}^{6} p_{2j}S_j\right) + z^2\left(\frac{1}{n_3^2} + \sum_{j=1}^{6} p_{3j}S_j\right) \\ &+2yz\left(\sum_{j=1}^{6} p_{4j}S_j\right) + 2zx\left(\sum_{j=1}^{6} p_{5j}S_j\right) + 2xy\left(\sum_{j=1}^{6} p_{6j}S_j\right) = 1 \end{aligned} \tag{5.21}$$

Therefore, the stress modifies the dimension and the form of the refractive indices ellipsoid. The relations among the coefficients p_{ij} of the photoelastic tensor are determined by the symmetries of the system.

For example, in one isotropic medium (water) $[p_{ij}]$ is characterized by two independent coefficients p_{11} and p_{12}. If a longitudinal acoustic wave propagates along z, the associated stress is:

$$S_3 = S\sin(\Omega t - Kz) \tag{5.22}$$

where K is the acoustic wave vector. The refractive indices ellipsoid becomes:

$$x^2\left(\frac{1}{n^2} + p_{12}S\sin(\Omega t - Kz)\right) + y^2\left(\frac{1}{n^2} + p_{12}S\sin(\Omega t - Kz)\right) + z^2\left(\frac{1}{n^2} + p_{11}S\sin(\Omega t - Kz)\right) = 1 \tag{5.23}$$

where n is the water refractive index. Since there are no crossed components, the principal axes do not change and the modified refractive indices are:

$$\begin{aligned} n_x &= n - 0.5n^3 p_{12}S\sin(\Omega t - Kz) \\ n_y &= n - 0.5n^3 p_{12}S\sin(\Omega t - Kz) \\ n_z &= n - 0.5n^3 p_{11}S\sin(\Omega t - Kz) \end{aligned} \tag{5.24}$$

Therefore, if an acoustic wave is applied, the water becomes a periodic medium equivalent to a grating with a period of $2\pi/K = \Lambda$.

An acoustic wave, which varies periodically, produces a refractive index grating characterized by a period equal to the acoustic wavelength. Refractive index variations around the mean value are of the order of 10^{-4}. An optical wave interacting with this refractive index grating is diffracted. The variation of the refractive index is:

$$\Delta n = \sqrt{M_2 10^7 P_a/2A} \tag{5.25}$$

where P_a is the total acoustic power, A is the section of the sound wave, and M_2 is the diffraction merit factor, defined as:

$$M_2 = \frac{n^6 p^2}{\rho v_a^3} = \eta \tag{5.26}$$

where n is the initial refractive index of the medium, p the corresponding component of the photoelastic tensor, ρ is the density, and v_a is the speed of sound in the medium. η identifies the ratio of the diffracted light.

In addition to M_2, there are two other merit factors, M_1 and M_3:

$$M_1 = \frac{n^7 p^2}{\rho v_a} \propto \Delta f$$

$$M_3 = \frac{n^7 p^2}{\rho v_a^2} \propto \Delta f \tau \tag{5.27}$$

where Δf is the modulation bandwidth and τ is the access time. Therefore, M_2 qualifies the efficiency of the diffraction, M_1 the modulation bandwidth and M_3 the access time.

The perturbation varies in time and space. It moves inside the medium with the speed of sound v_a, which is some orders of magnitude lower than the speed of light, thus it is nearly stationary. An optical wave, characterized by a wave vector $k = 2\pi n/\lambda$, that interacts with an acoustic wave, characterized by the wave vector $K = 2\pi/\Lambda$, is effectively reflected in the directions defined by:

$$2k \sin\theta = mK = m\frac{2\pi}{\Lambda} \tag{5.28}$$

where m is an integer and represents the diffraction order. The diffraction on several orders is possible only if the thickness of the medium is much smaller with respect to Λ: this is the condition for the thin gratings in the Raman-Nath regime. If the medium thickness is larger, the light beam will be diffracted only on the first-order, which that is the condition for the Bragg regime. In this situation Eq. 5.28 becomes:

$$2k \sin\theta = K \tag{5.29}$$

that is,

$$2\Lambda \sin\theta = \frac{\lambda}{n} \tag{5.30}$$

It can be shown that the frequency of the reflected beam is shifted to a factor $\pm\Omega$ with respect to the frequency of the incident one.

In order to distinguish between the two regimes, the parameter Q is used:

$$Q = \frac{2\pi\lambda L}{n\Lambda^2} \tag{5.31}$$

where L the length of the acousto-optic interaction. If $Q << 1$ the incident wave is diffracted in several orders that represents the Raman-Nath regime, which corresponds to a thin grating. If $Q >> 1$ the incident wave is diffracted only in the first-order that is the Bragg regime, which corresponds to a thick grating. In both regimes, the diffracted waves are characterized by a frequency that is different with respect to that of the incident wave.

5.2.2.1 Raman-Nath Regime

In the Raman-Nath regime, the phase of a light beam that crosses along z a medium excited by an acoustic wave is varied by the factor:

$$\Delta\phi = \frac{\Delta n 2\pi nL}{\lambda} \sin \frac{2\pi y}{\Lambda} \tag{5.32}$$

where Δn is the refractive index variation induced by the acoustic wave, L the length of the acousto-optic interaction, and Λ the acoustic wavelength. By replacing Δn with the expression 5.25, the phase shift becomes:

$$\Delta\phi = \frac{2\pi nL}{\lambda}\sqrt{M_2 10^7 P_a/2a} \sin \frac{2\pi y}{\Lambda} \tag{5.33}$$

where a is the thickness of the acoustic beam.

The incident beam is diffracted in several orders defined by the angle θ

$$\sin\theta_{RN} = \frac{m\lambda}{n\Lambda} \tag{5.34}$$

If I is the intensity of the diffracted beams and I_0 is the transmitted optical intensity without acoustic modulation, the diffraction efficiency η is given by:

$$\begin{aligned} \eta &= \frac{I}{I_0} = \frac{[J_m(\Delta\phi_M)]^2}{2} \qquad |m| > 0 \\ \eta &= \frac{I}{I_0} = [J_0(\Delta\phi_M)]^2 \qquad m = 0 \end{aligned} \tag{5.35}$$

where J are the Bessel functions of order m and $\Delta\phi_M$ is the maximum-induced phase shift.

5.2.2.2 Bragg Regime

In order to fulfill the Bragg condition, the light beam incident on a medium excited by an acoustic wave must have an angle of incidence defined by:

$$\sin\theta_B = \frac{\lambda}{2n\Lambda} \tag{5.36}$$

The diffracted beam (order 1) is characterized by an angle of $2\theta_B$ with respect to the direct beam (order 0). The diffraction efficiency is:

$$\eta = \frac{I}{I_0} = \sin^2\left(\frac{\Delta\phi}{2}\right) \tag{5.37}$$

5.2.2.3 Acousto-Optic Modulators

The acousto-optic effect can be exploited in the realization of various devices as light modulators, beam deflectors, tunable filters, and spectrum analyzers. In particular a variety of light modulators, for both amplitude and frequency modulation, can be produced. Either the Raman-Nath regime or the Bragg regime can be used for the modulation, but the Raman-Nath regime is characterized by a small interaction length, too small for practical application, that limits the frequency operation and the bandwidth. Therefore, the Bragg modulators are the most used, since they are characterized by a high frequency operation and a wide bandwidth.

Equation 5.37 shows that the Bragg diffraction efficiency is proportional to the acoustic power. Therefore, if the acoustic power is modulated, so is the diffracted light beam. Thus, by exploiting the Bragg diffraction an information can be given to a light beam. The efficiency of the diffraction is acceptable only when the incident and the refracted beams are almost symmetric with respect to the acoustic wave. Since the angle of incident is small, θ_B can be written as:

$$\sin\theta_B \simeq \theta_B = \frac{\lambda}{2n\Lambda} = \frac{\lambda f_a}{2n v_a} \tag{5.38}$$

where f_a is the sound frequency. Therefore, for small Bragg angles θ_B, the deflection angle $2\theta_B$ is linearly proportional to the frequency of sound.

The bandwidth Δf of an acousto-optic modulator depends on the angular spread of the light beam. Since both the acoustic and the light waves are finite, also the angular spread is finite. Therefore, the Bragg diffraction can occur over a range of acoustic frequencies. The relation between the finite bandwidth and the variation of the angle of incidence and diffraction, due to the angular spread of the acoustic and light waves, is:

$$\Delta f = \frac{2n v_a \cos\theta}{\lambda} \Delta\theta \tag{5.39}$$

$\Delta\theta$ is the variation of the Bragg angle needed to satisfy the Bragg condition when the acoustic wave is characterized by a frequency bandwidth Δf. Moreover, $\Delta\theta$ is the sum of the angular spreads of both the light wave $\delta\theta$ and the acoustic wave $\delta\phi$:

$$\Delta\theta = \delta\theta + \delta\phi \tag{5.40}$$

where $\delta\theta \simeq \frac{2\lambda}{\pi n \omega_0}$ and $\delta\phi \simeq \frac{\Lambda}{L}$. ω_0 is the waist of the light beam. For all the angular components of the light beam, the Bragg condition can be satisfied over a range of acoustic frequencies. The diffracted beam corresponding to a particular incidence angle is characterized by an angular spread of $2\delta\phi$. Every direction corresponds to a different frequency shift. The condition that permits to recover the intensity modulation of the diffracted beam is $\delta\theta \simeq \delta\phi$.

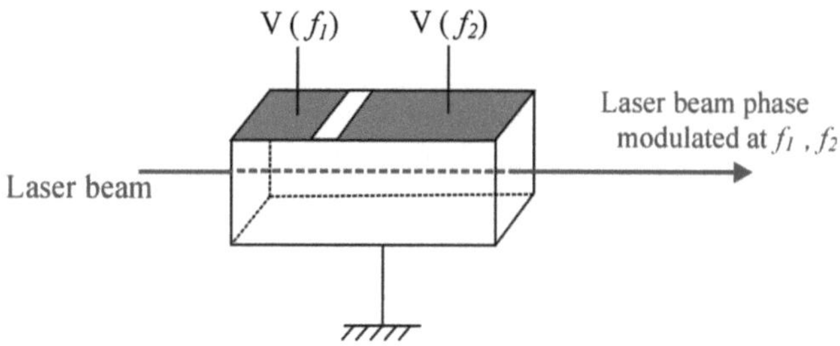

Fig. 5.4 Scheme of the 2-section modulator

5.2.3 Advanced Virgo Modulator

Among the various modulation methods, in the gravitational waves interferometers the phase modulation is needed, since it is the basis of the Pound Drever Hall technique [6]. Therefore, an electro-optics modulator, exploiting the Pockels Effect, is the chosen technology to introduce a modulation. Due to the high incident power, thermal effects can become significant in the electro-optic crystal [7]. Therefore, the choice of the appropriate material is crucial to limit as much as possible these effects. Indeed, it is important to select the right material not only to limit wavefront aberrations but also to limit heating that can induce slow variations of the modulation index and a disturbance of the interferometer control. In addition to the five modulation frequencies mentioned before, other requirements are put on the modulator that will be exploited in Advance Virgo: it has to withstand 200 W of input power, the maximum modulation depth approved is 0.2, and thermal lensing effects have to be quite limited. Among all the possible electro-optic crystals, the RTP has been chosen because it satisfies the requirements and it is characterized by a low absorption value [8].

An electro-optic modulator prototype, made in RTP, has been realized. In order to minimize the losses and to avoid as much as possible spurious reflections from input and output faces of the modulators, a 2-section modulator of the length of 20 mm is implemented. The proposed scheme is shown in Fig. 5.4.

With this scheme it is possible to obtain two modulation frequencies for each crystal. Therefore, three different EOMs will be installed on the high power beam in Advanced Virgo: two of them will have two sets of electrodes, the last one will be a simple electrode modulator.

5.3 Faraday Isolation

A source of problem in the correct working of the interferometer can be the light that propagates back from the interferometer itself and from the power recycling mirror into the Input Optics System. Therefore, an isolation system, which isolates the Input

Mode Cleaner Cavity from the ITF back-reflected beam, is needed. This isolation system has to satisfy some requirements: the isolation factor must be >40 dB for a power of 150 W, its total transmission must be >95 %, its residual thermal focal lens must be >100 m, and it must be ultra-high vacuum (UHV) compatible (pressure $< 10^{-6}$ mbar).

The adopted isolation system exploits the Faraday Effect.

5.3.1 Faraday Effect

The Faraday Effect is the most important magneto-optic effect [4]. This effect is the rotation of the polarization of a light beam that crosses a medium affected by the presence of a magnetic field. Considering an infinite medium characterized by a magnetization M, the interaction between a light beam and M can be described by two principal effects: the Faraday Effect and the Voigt Effect (or Cotton-Mouton Effect). The former appears when the magnetization and the direction of propagation of the light beam are parallel and it origins the rotation of the polarization, also called magnetic circular birefringence (MCB). The latter appears if the magnetization and the direction of propagation of the light beam are perpendicular and it origins the magnetic linear birefringence (MLB).

The linear polarization of a light beam can be decomposed into the superposition of two opposite circularly polarized components with the same amplitude and different phases. When a linearly polarized beam crosses a medium of length d affected by a magnetic field, its two circularly polarized components propagate with different speeds. Therefore, a relative phase shift between them is induced at the output of the medium originating a rotation of the orientation of the linearly polarized wave. The relation between the rotation angle of the polarization and the magnetic field is:

$$\beta = \nu B d \tag{5.41}$$

where B is the magnetic flux density in the direction of propagation and ν is the material Verdet constant, which varies with wavelength and temperature. The rotation of polarization is not reciprocal, i.e., it depends on the direction of transmission. If ν is positive and the magnetic field and the light beam propagation are parallel, a counterclockwise rotation is induced. On the other hand, if the magnetic field and the light beam are antiparallel, the induced rotation is clockwise. Therefore, if a light beam crosses a magnetized material in both directions (back and forth), the induced angle of rotation is doubled. Some materials, as the Terbium Gallium Garnet (TGG), are characterized by a very high Verdet constants (−40 rad / (Tm)). If a high magnetic field is applied on a small piece of these materials, large angles of rotation can be obtained. Therefore, these materials are exploited in the construction of the Faraday rotators, which are the main components of the Faraday isolators.

5.3.1.1 Faraday Isolator

An optical isolator is used to protect a light source from the undesired back reflections, which can make the source unstable. In the particular situation of a gravitational waves detector, an isolator protects the source from the light back reflected by the interferometer. A simple Faraday Isolator depends on the polarization of the input beam. In general, it is made of three parts: an input polarizer (aligned for a vertical polarization), a Faraday Rotator, and an output polarizer (usually named analyzer, aligned at 45° with respect to the input polarizer). In the forward direction, the input light beam crosses the polarizer and becomes vertically polarized. The Faraday Rotator rotates the polarization of the beam by 45°. The output analyzer enables the transmission of the light at the output of the isolator. In the backward direction the input beam becomes polarized at 45° after the analyzer. The Faraday Rotator adds a rotation of 45° and the light becomes horizontally polarized. Since the polarizer is aligned in the vertical direction, the transmission of the light is blocked.

5.3.2 Advanced Virgo Faraday Isolator

The isolation requirements have been defined considering that no fringes, coming by the spurious cavity made by IMC and Power Recycling mirror, with amplitude larger than 0.5 % are present in reflection of the IMC cavity. The full power considered in the definition of the requirements is 150 W (125 W of input power and 25 W of interferometer reflected power). With the high input power foreseen in Advanced Virgo, a standard Faraday Isolator will not work properly, since it will be characterized by loss of optical isolation due to the thermally induced birefringence and by modification of the mean angle of rotation. Moreover, it will also exhibit very high thermal lensing. All these troubles have to be corrected at the Faraday Isolator level itself. Therefore, a custom designed Faraday Isolator has to be used in order to reach all the requirements [9–11].

Since high power hits the crystal, a large amount of energy is absorbed inside the TGG medium inducing a variation of the internal temperature, which introduces also a modification of the refractive index of the medium. As a consequence, a change of the optical path inside the medium arises causing a thermal lensing effect. Figure 5.5 shows the experimental measurements of the thermal lensing for a laser power going from 10 to 180 W for two different TGG crystals, TGG1, and TGG2, characterized by two different absorption values (2300 ppm/cm for TGG1 and 2600 ppm/cm for TGG2) [11, 12]. It can be noticed that a thermal focal length of about 14 m is induced in both crystals at full power.

A passive compensation is exploited to correct the thermal lensing. In particular, a plate of the right length made of Deuterated Potassium Dihydrogen Phosphate (DKDP), a material characterized by a large negative thermo-optic coefficient, is inserted in the Faraday Isolator in order to compensate the thermal lensing created in the TGG crystal. Figure 5.6 shows the residual thermal lens power, evaluated

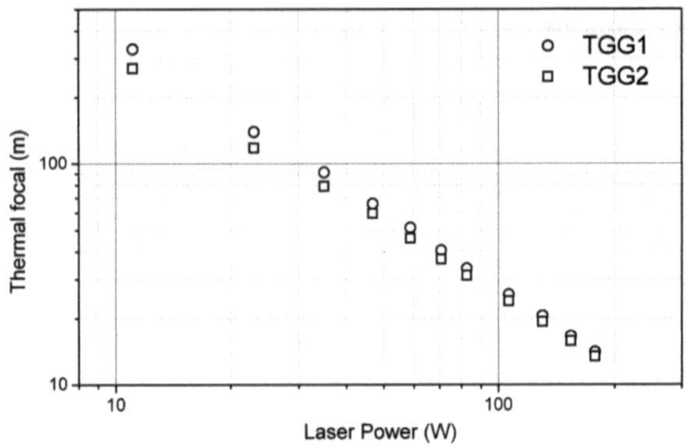

Fig. 5.5 TGG crystals thermal focal length

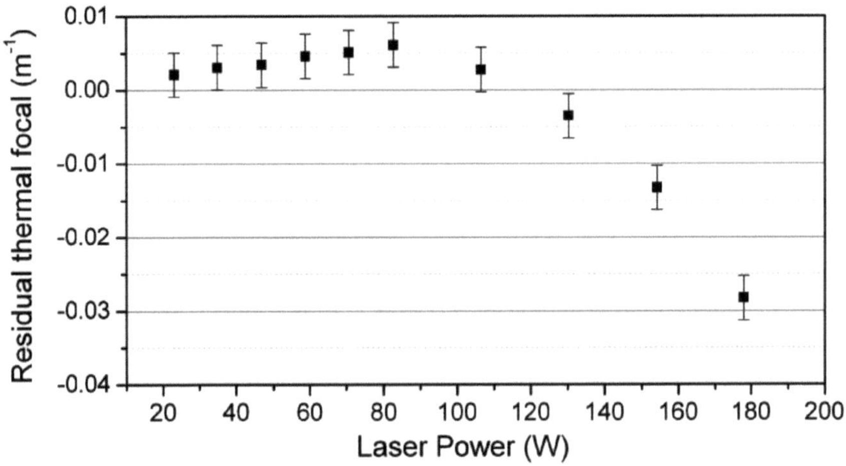

Fig. 5.6 Faraday isolator residual thermal lens power

in diopters (m^{-1}) after the installation of the DKDP plate, as a function of the incident power for a range from 20 to 180 W [11, 12]. It can be noticed that the compensation is very efficient up to 150 W. Then the compensation starts to get worse. The requirement (residual focal length >100 m (0.01 diopters) at 125 W) is satisfied.

The thermal isolation value changes because the angle of rotation is modified when the TGG crystal is heated. This modification is due to the temperature dependence of the Verdet constant:

$$\frac{\Delta\beta}{\beta} = \frac{1}{\nu}\frac{\mathrm{d}\nu}{\mathrm{d}T}\Delta T \tag{5.42}$$

Therefore, the thermal isolation varies when the power sent inside the Faraday Isolation is modified or when the thermal condition is changed (e.g., a Faraday

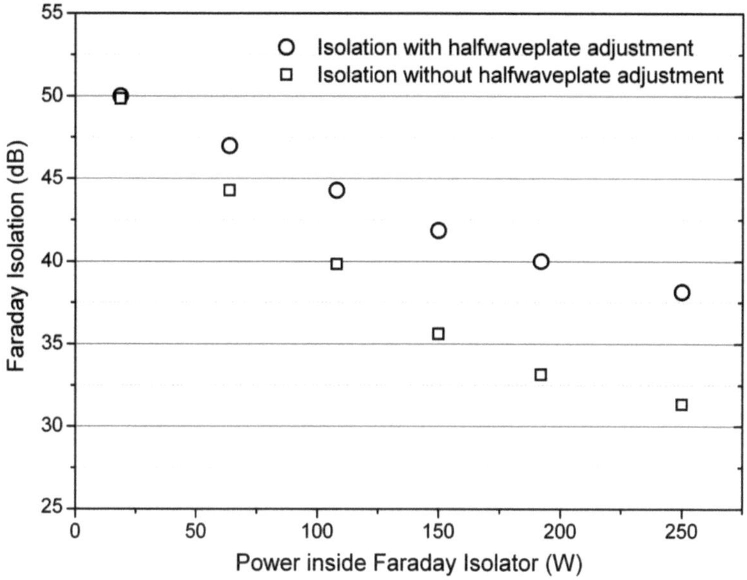

Fig. 5.7 Faraday isolator optical isolation

Isolator used in air or in vacuum). In order to compensate this effect, a motorized half-wave plate is introduced in the Faraday Isolator [13]. By adjusting the half-wave plate, the mean angle of rotation can be tuned and the optimum level of the optical isolation is recovered. Figure 5.7 shows the optical isolation reached with and without the introduction of the half-wave plate [11]. An improvement of about 7 dB is reached by introducing the wave plate.

Another important problem induced by the high power hitting the TGG crystal of the Faraday Isolator is the thermal depolarization. In fact, when the TGG crystal is heated, it acts as a birefringent medium and changes the polarization of the incident beam. In particular, the thermal depolarization depends also on the alignment of the incident polarization with respect to the crystal axis. The depolarization introduced by the two TGG (TTG1 and TGG2) crystal has been experimentally evaluated. The crystal is rotated in order to find the axis of minimum depolarization, which will suggest the right orientation to mount it in the Faraday Isolator. As example, Fig. 5.8 shows the obtained result for the thermal depolarization of one of the TGG crystals as a function of its orientation with an impinging power of 200 W [11].

It can be noticed that the crystals are characterized by orientations that minimize or maximize the thermal depolarization: there is a difference in the polarization state of 7 dB between these two axis orientations. After the identification of the two main orientations, it is possible to measure the depolarization as a function of the input power P for the minimum and maximum conditions. As example, the obtained experimental results for one of the TGG crystals are shown in Fig. 5.9 for impinging power going from 10 to 200 W [11]: thermal depolarization increases with the input power and an important gain on thermal depolarization is obtained when the best

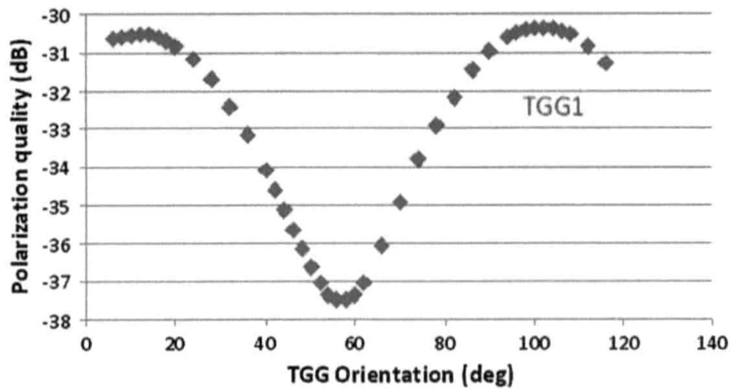

Fig. 5.8 Measured thermal depolarization in one TGG crystal as a function of the axis orientation

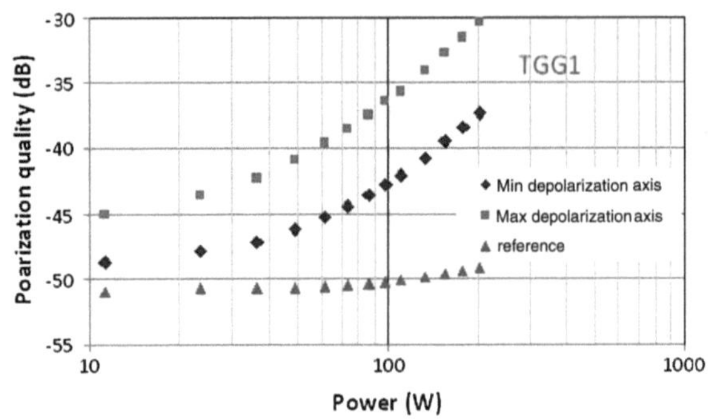

Fig. 5.9 Measured thermal depolarization in one TGG crystal as a function of the input power

orientation is chosen. In particular, at high power the trend of the depolarization is proportional to P^2. The behavior of the second TGG crystal is almost the same.

The physical mechanism at the origin of the limiting depolarization at high power is a self-induced conversion from a spin angular momentum (SAM) to an orbital angular momentum (OAM) of the photon [14]. In particular, the depolarization of the beam is due to the occurrence of maximal entanglement between the SAM and the OAM. The high power of the input Gaussian beam induces in the material temperature gradients that create a birefringence with a radial symmetry, which can transform the impinging light beam in a vortex one, which carries OAM. Therefore, the heated material can be considered as a q-plate, an optical device developed for optical angular momentum manipulation.

In order to limit the effect of the thermal depolarization and to compensate it in the backward direction, a solution which foresees two TGG crystals rotating the polarization by 22.5° each separated by a −67.5° quartz rotator, instead of the classical solution of a single TGG crystal rotating the polarization by 45°, is adopted [15]. With this configuration, it is mandatory to find the orientation giving the minimum depolarization for each TGG crystal and align them correctly inside the Faraday Isolator. The orientation of the polarization of light inside the Faraday Isolator is shown in Fig. 5.10. The red and the blue arrows describe the orientation of the polarization of light at different stages in the Faraday Isolator, respectively, in the backward and forward directions. The total polarization rotation is −22.5° in the forward direction and +112.5° in the backward direction.

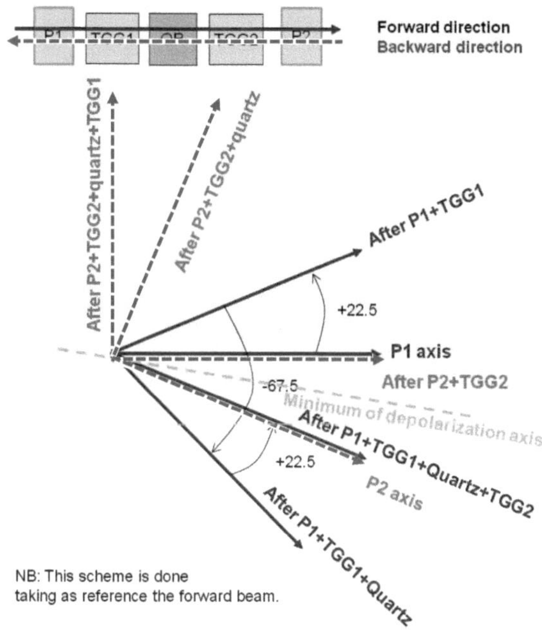

Fig. 5.10 Orientation of the polarization of light inside the Faraday Isolator

Fig. 5.11 Advanced Virgo Faraday Isolator scheme

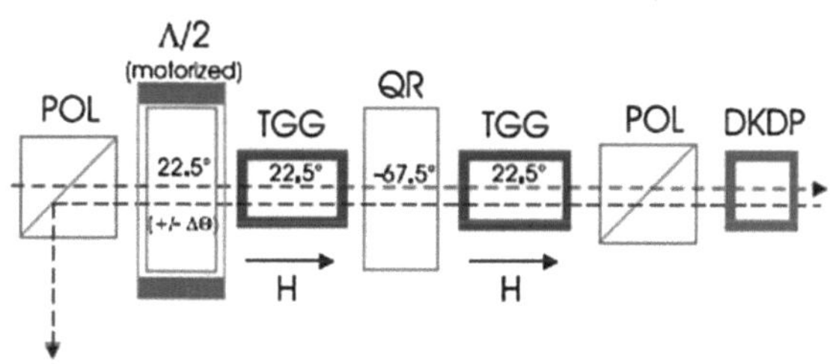

Considering all the compensators (the half-wave plate, which compensates the isolation change, the quartz rotator, which compensates the depolarization and the DKDP crystal, which compensates the thermal lensing) that are introduced, the final design of the system is shown in Fig. 5.11 and the assembled prototype is shown in Fig. 5.12.

5.4 Input Mode Cleaner Cavity

At the input of a gravitational waves interferometer, a pure fundamental Gaussian mode (TEM_{00}) (see Chap. 3) is needed in order to match the resonant mode of the arm cavities, but the light beam at the output of the laser source can be characterized by the presence of some transverse higher order modes. Therefore, the Input Optics

Fig. 5.12 Advanced Virgo Faraday Isolator assembled prototype

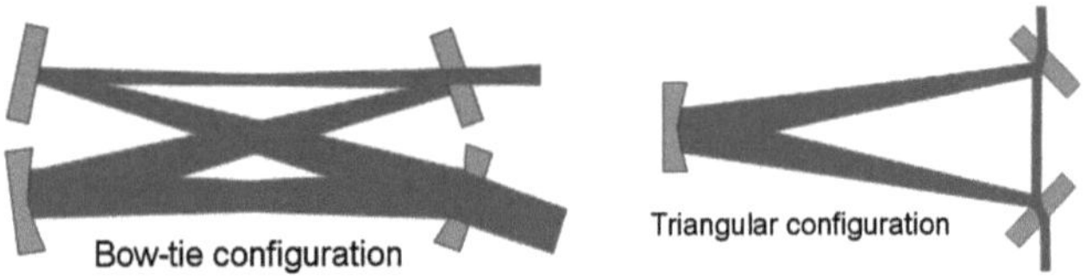

Fig. 5.13 Two possible different configurations for the Input Mode Cleaner Cavity

System has to guarantee that the laser beam is the proper one, not characterized by the presence of higher order modes, before entering in the interferometer. In order to reach this purpose a resonant cavity, whose role is to suppress all the higher order modes other than the fundamental one, is added on the Injection System. This cavity is called Input Mode Cleaner Cavity and the laser beam has to pass through it before entering the interferometer. Moreover, the laser frequency has to be very stable inside the interferometer and the Input Mode Cleaner Cavity conducts an important role also in the stabilization of it. In the end, thanks to the suppression of the higher order modes, this cavity is used also in the filtering of the residual beam jitter. Summarizing the Input Mode Cleaner Cavity acts as passive spatial filter (reducing the beam jitter by suppression of the higher order modes), as a filter of the laser frequency and amplitude fluctuations (reduction of the fluctuations above the cut-off frequency around 500 Hz) and as part of the laser frequency stabilization loop at high frequency (above some tens of kHz).

A ring geometry has been chosen for the Input Mode Cleaner Cavity in order to avoid the problem of back reflection on different optical elements of setup placed before it [16]. Two different configurations of the cavity have been considered: a so-called bow-tie cavity, characterized by four mirrors and a triangular cavity, characterized by three mirrors, as shown in Fig. 5.13.

Fig. 5.14 Input Mode Cleaner Dihedron

The bow-tie configuration is interesting because it should be compliant with possible future upgrades of Advanced Virgo, which could foresee the use of a higher order Laguerre-Gauss mode instead of the fundamental Gaussian one [17], but this technology is not yet mature to allow the choice of this kind of cavity. Moreover, a four-mirror cavity would be much more complicated to control and its performances in terms of small angles scattered light, which is one of the main problems of the Input Mode Cleaner Cavity, are worse than the ones of the triangular cavity [16]. Therefore, the triangular configuration has been chosen for the Input Mode Cleaner Cavity, which is the same solution that was adopted in Virgo. In the final design two of the three mirrors, the input and the output flat mirrors, are rigidly fixed in a dihedral shape (Fig. 5.14) and are installed on the Suspended Injection Bench. The other mirror, which is parabolic and is called IMC end mirror, is on the other side of the cavity and is mounted on a payload (Fig. 5.15) suspended to a Virgo Super-attenuator.

The design of the Input Mode Cleaner triangular cavity foresees a cavity length of 144 m, a cavity finesse of 1,200, and a cavity throughput greater than 90 %. Moreover, it has not to suffer radiation pressure effects introduced by the high power incident on the end mirror.

As almost all the other parts of the Input Optics System, also the Input Mode Cleaner Cavity can suffer from thermal effects due to the high input power. In particular, the thermal lensing induced on the input substrate affects the matching coupling of the input beam with the cavity. This effect, which is independent on the cavity geometry, depends on the properties both of the materials of the substrate and of the coating of the mirror: in particular, it relies on the absorption of the coating, on the variation of the refractive index with temperature and on the conductivity of the substrate. Figure 5.16 shows the optical coupling in a cavity as a function of the power absorbed in the coating of the first mirror. Since the requirement on the Input

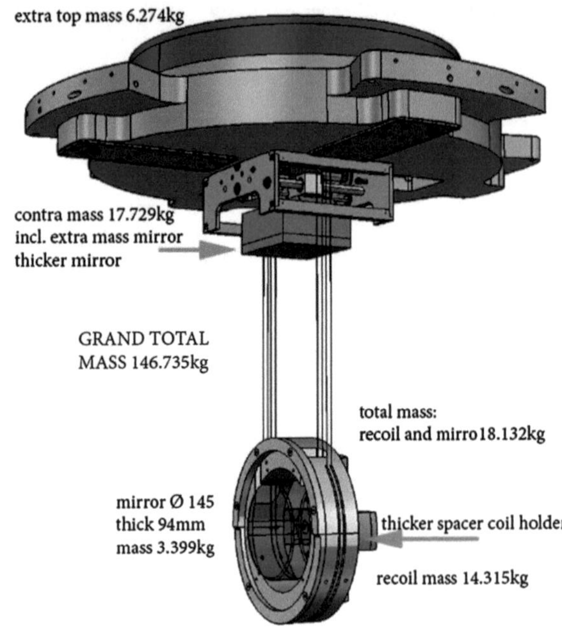

Fig. 5.15 Input Mode Cleaner Payload

Mode Cleaner Cavity is a coupling greater than 90 %, the total power absorbed in the coating should be lower than 100 mW.

This result can be achieved, in the case of high finesse (around 1,000) and high input power (around 180 W), if the coating absorption is lower than 2 ppm, which is a very stringent requirement on its quality.

Another stringent requirement that affects the quality of the mirrors of the cavity is set on the total throughput, which can be defined as the total amount of power transmitted by the cavity and has to be greater than 90 %. In fact, for the triangular Input Mode Cleaner Cavity, it is linked to the transmission coefficients of the mirrors (T_1 and T_2) and to the total round trip losses (L_{RT}) [18]:

$$T = \frac{4T_1T_2}{(T_1 + T_2 + L_{RT})^2} \tag{5.43}$$

In particular, the total round trip losses, which determine the part of the power that becomes unusable at each round trip, are related to the absorbed and the scattered light. These two phenomena depend on the quality of the coating of the mirrors in terms of absorption and roughness.

There is another important property of the cavity that depends on the same parameters (T_1, T_2 and L_{RT}), the finesse [18]:

$$T_1 + T_2 + L_{RT} = \frac{2\pi}{F} \tag{5.44}$$

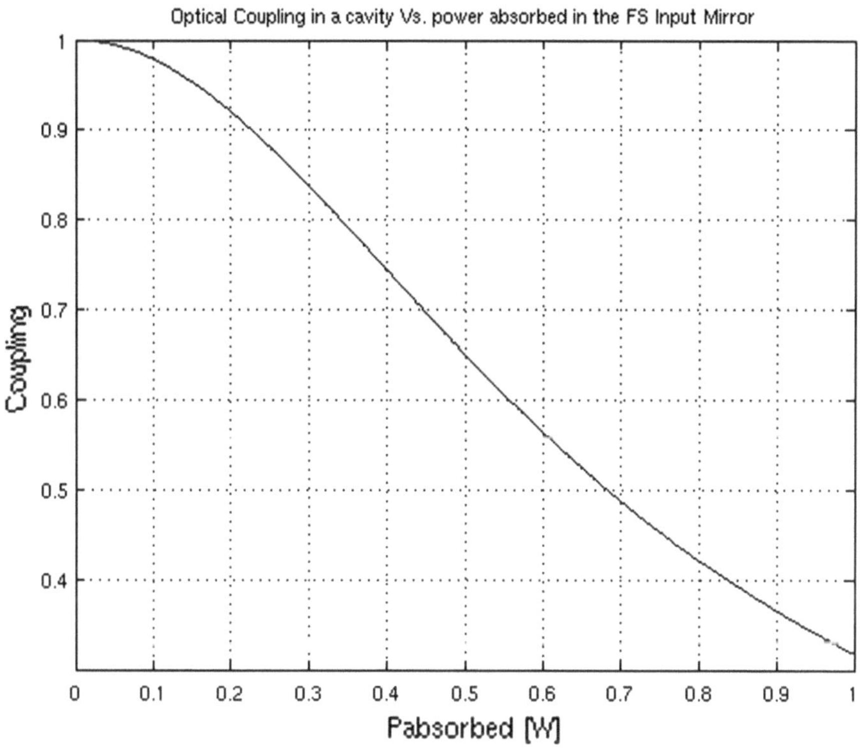

Fig. 5.16 Cavity optical coupling versus absorbed power in the coating of the first mirror

Therefore, round trip losses and finesse are linked in order to have high throughput. Figure 5.17 shows the dependance of the throughput on the finesse and the round trip losses for the Input Mode Cleaner triangular cavity [16].

Therefore, a compromise should be found for the choice of the finesse of the cavity. In fact it has to be as high as possible to increase the filtering effect of the cavity, but too high finesse will put very stringent constraints (very difficult to achieve) on the losses of the mirror and, thus, on their quality. Moreover, there is a large constraint also on the absorbed power in order to limit the thermal effects. The compromise has been found on a value of 1,200 for the finesse of the cavity, which means a throughput of 95 % for 100 ppm of round trip losses.

The choice of the length of the cavity has been determined considering the constraint of the interferometer modulation frequency. In fact, for a length L of about 150 m, the Free Spectral Range (FSR) of the cavity is $FSR = c/2\pi L = 1$ MHz. It means that all the multiples of 1 MHz can be used as modulation frequencies for the control of the interferometer. Having a shorter cavity could affect in a not negligible way the constraints on the modulation frequency. Moreover, the same filtering properties can be achieved by increasing the finesse, affecting the requirements on all the other components of the cavity.

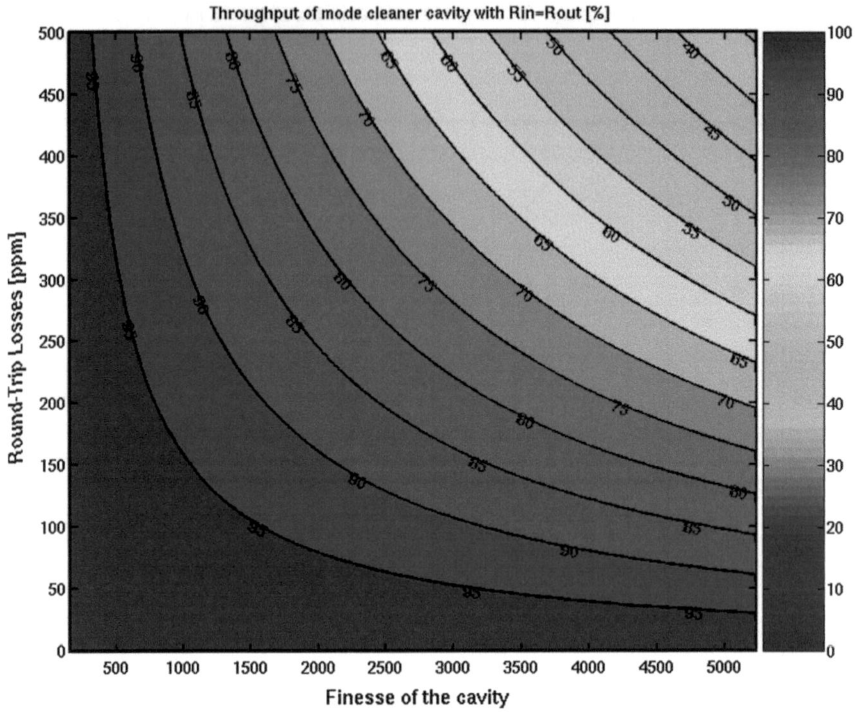

Fig. 5.17 Cavity throughput as a function of cavity finesse and round trip losses

5.5 Reference Cavity System

As mentioned in the previous section, the frequency of the laser has to be very stable. The Input Mode Cleaner Cavity is used as reference at high frequency. On the other hand, a Reference Cavity System (RFC) is exploited in order to provide a low frequency stabilization reference [16]. The monolithic Reference Cavity is the reference for the laser frequency pre-stabilization: the laser frequency will be locked on the Input Mode Cleaner Cavity and the IMC length will be locked on the Reference Cavity. At the beginning, it is the low frequency (<10 Hz) reference for the pre-stabilized beam, at the end it is used as main interferometer low frequency reference in order to control the Interferometer common mode. The residual frequency fluctuation requirement provided by the Reference Cavity is $\delta\nu < 1$ Hz rms up to 0.1 Hz.

In order to fulfill all the tasks, the Reference Cavity has to be quite independent from the other parts of the interferometer. It means that the alignment and the matching of the Reference Cavity must be performed independently from the interferometer beam.

Since the Reference Cavity has to be isolated from the ground seismic noise, it is installed on the lower part of the Suspended Injection Bench. The Reference Cavity is mounted inside its separate vacuum tank. A maximum of 30 mW of power is picked up from the main interferometer beam to feed the Reference Cavity. The beam sent to the Reference Cavity is picked off just after the IMC, before entering in the Faraday Isolator. In this way two main advantages are obtained: the beam entering in the RFC is geometrically stable in shape and lateral jitter, improving the matching with the cavity and the RFC alignment is decoupled from the one of the main beam and from matching adjustments on the interferometer.

5.6 Beam Pointing Control

The beam at the input of the interferometer should be always stable at the right position, but it could move because of the residual motion of the mechanical benches that hold the optical elements. Therefore, a subsystem, which reads and adjusts the position of the beam at the input of the Input Mode Cleaner and, in consequence, at the input of the interferometer, is implemented. This subsystem is named Beam Pointing Control System (BPC) and it also monitors and stabilizes the jitter, both lateral and angular, of the input beam.

5.6.1 Beam Jitter

In the field of the gravitational wave interferometry, the term "beam jitter" is used to indicate the time-dependent fluctuations in the location of the impinging laser field. It mainly originates at the level of the unsuspended optical components that steer the laser beam into the interferometer. It is shown in the spectrum of the interferometer output as peaks or bumps at particular frequencies (in some range under 500 Hz), which are the typical resonance frequencies of the benches and of the mounts that hold the optical components.

Starting from the hypothesis that the input beam of the interferometer is a TEM_{00} mode (zero-order Hermite-Gaussian mode) matched to the spatial mode of the interferometer, any jitter, i.e., tilt or displacement of the phase front of the beam, can be described as a pair of sidebands separated from the carrier by the jitter frequency [19]. The spatial order of the sidebands is the first-order Hermite-Gaussian mode ($TEM_{10} - TEM_{01}$), which propagates along with the zero-order mode. The complex amplitude of the sidebands $a_1 = (y + i\beta)$ can be calculated from the displacement y and the tilt β. In particular, the modulus of a_1 is:

$$|a_1| = \sqrt{\left(\frac{y}{w}\right)^2 + \left(\beta\frac{\pi w}{\lambda}\right)^2} \tag{5.45}$$

where w is the beam waist of the TEM_{00} mode and λ is the laser wavelength.

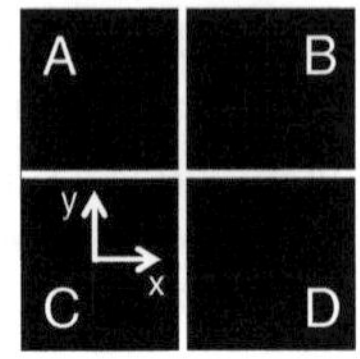

Fig. 5.18 Scheme of a quadrant photodiode

5.6.2 Beam Jitter Control System

In order to measure the jitter, quadrant photodiodes are used as sensors [20]. These photodiodes are discrete components that usually feature four optically active areas separated by a small gap, as shown in Fig. 5.18.

In order to understand the functioning of the position sensing quadrant consider a uniform round light spot that impinges on the active surface of the detector. The position of the spot, relative to the center of the photodiode, can be determined by comparing the signal acquired by the four separate quadrants. In particular, x and y displacements can be calculated as:

$$x = \frac{(B + D) - (A + C)}{A + B + C + D}$$
$$y = \frac{(A + B) - (C + D)}{A + B + C + D} \qquad (5.46)$$

The incident light spot must be smaller than the detector total active area and larger than the gap between the single quadrants. Moreover, the detection range increases with the spot size, while the positional resolution decreases.

The Beam Pointing Control System is composed of position sensors and steering mirrors. In particular, the signal acquired by the quadrants is used as an error signal for the actuators, mounted on the steering mirrors, which control the position of the beam and reduce the jitter at low frequency. The signal on the quadrant photodiodes has to be maintained null in order to guarantee the right alignment of the beam. Its scheme is shown in Fig. 5.19.

This system should allow to lower the beam jitter at low frequency and keep the beam at the right position. In order to measure the tilt and the displacement of the beam, two detectors are needed. These two sensors have to be installed at right positions and to detect beams of the proper dimension. In order to decouple the measurement of tilt and displacement, a Gouy phase shift of 90 deg. between the field detected by the two quadrants is mandatory. Therefore, one sensor is installed to detect the so-called Near Field (named QN in Fig. 5.19) and the other one has to acquired the Far Field (named QF in Fig. 5.19). In particular the Near Field Quadrant measures a pure displacement, while the Far Field Quadrant acquires a pure tilt. In order to have the right beam size on the two quadrants, different systems of lenses are installed on the path of the beam before both quadrants. Starting from the beam

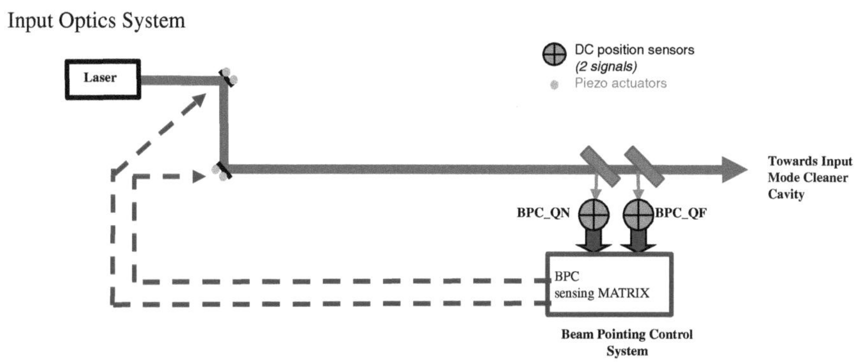

Fig. 5.19 Scheme of the Advanced Virgo Beam Pointing Control System

jitter measured by the quadrants of the Beam Pointing Control System it is possible to estimate the jitter on the beam at the input of the interferometer. In parallel to the sensors used to control the beam position, the whole system is composed of two other quadrants, which are used to monitor the jitter over the entire detection bandwidth.

5.7 Input Power Control System

The Input Optics System should provide a way to change the interferometer input power without changing the incident power on elements that can experience possible thermal effects, such as the Input Mode Cleaner or the Faraday Isolator. Moreover, it should also guarantee that the power impinging on the different detectors devoted to control the whole system does not change. For these purposes a subsystem composed of a few polarizers and a wave plate, named Input Power Control System (IPC), is installed right before the Input Mode Cleaner cavity. Two other IPC systems are installed in the Input Optics System: the first one is placed just after the Faraday Isolator in order to modify the interferometer input power without inducing other changes on the laser and injection systems; the second one is installed on the interferometer reflected beam to simplify the adjustment of the laser power on the quadrant photodiodes used for the interferometer sensing and control.

The Input Power Control System should be able to modify the laser power by two orders of magnitude (from 1 to 99 % of the full power) without spoiling the polarization state, which should be better than 30 dB for the whole power range. Moreover, it should induce no significant thermal distortions at full power. The overall thermal lens power of the system should be below 0.0045 diopters at full power.

The most convenient system consists of a half-wave plate placed between polarizers. It should handle both light going to and reflected back from the interferometer. Both situations are shown in Fig. 5.20.

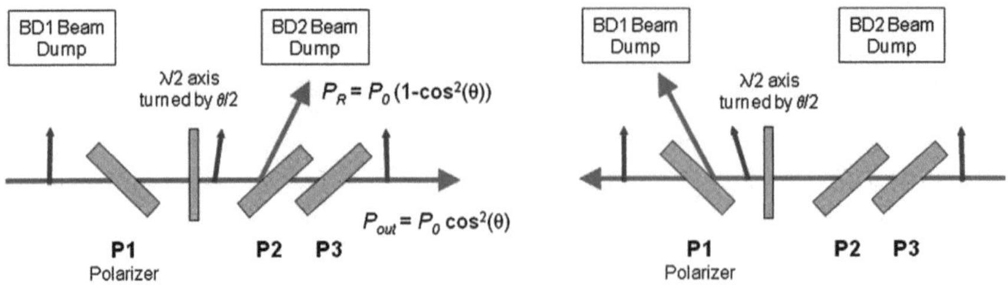

Fig. 5.20 Input Power Control System scheme

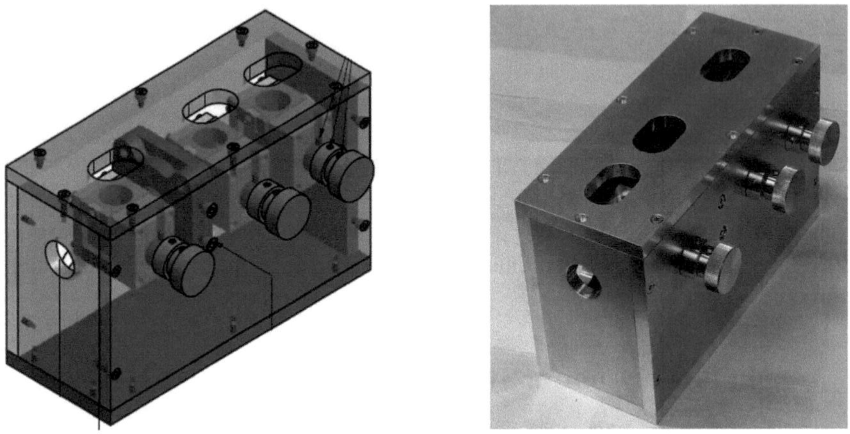

Fig. 5.21 Advanced Virgo Input Power Control System

Light going to interferometer (S polarization) passes through P1 and its polarization is rotated by an angle θ by the half-wave plate. After crossing P2 it recovers an S polarization and it goes toward the interferometer with an attenuated power of $P_{\text{out}} = P_0 \cos^2(\theta)$. Light emerging from P2 goes through a third polarizer, P3, aligned along the same direction to better clean the polarization. P3 is mandatory to assure a good polarization in low IPC transmission configuration. The light reflected by P2 should go to a high power beam dump.

Light reflected back from the interferometer should pass through P3 and P2 and its polarization is rotated by an angle θ by the half wave plate. Therefore, it goes back toward the laser system with a power $P_{\text{out}} = P_{\text{back}}(1 - \cos^2(\theta))$ after crossing P1. This last polarizer fixes the polarization of the light coming back whatever the position of the wave plate. The light reflected by P1 should go to a high power beam dump.

The scheme and the assembled prototype of the Advanced Virgo Input Power Control System are shown in Fig. 5.21.

It is composed of three optical contacted cubes that can be easily tilted and an adjustable half-wave plate. Figure 5.22 shows the polarization quality of the beam

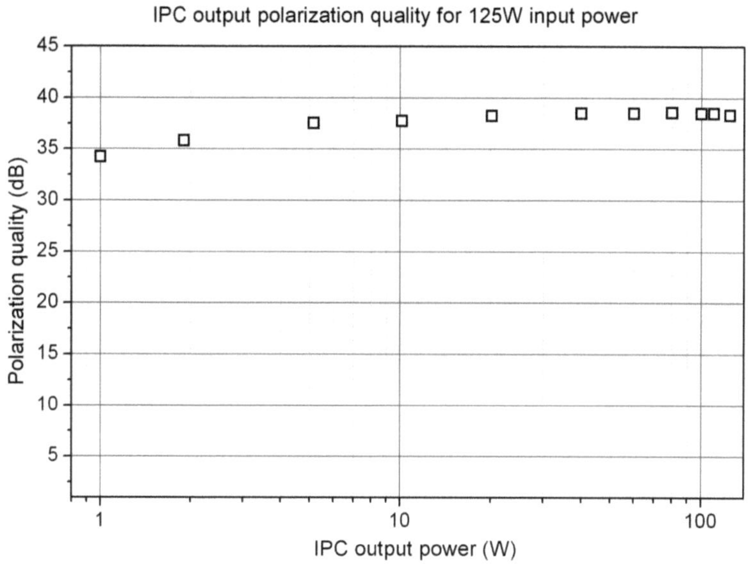

Fig. 5.22 Polarization quality of the light emerging from the Input Power Control System prototype

emerging from the prototype for an input power range from 1 to 125 W. The polarization quality remains better than 35 dB for all the range of power.

5.8 Mode Matching Telescope

In the gravitational waves interferometers some resonant cavities are installed. The main are the Fabry–Perot cavities that characterized the two arms of the Michelson interferometer. Another important cavity is the Input Mode Cleaner cavity, which is part of the Input Optics System. In all the cavities a light beam characterized by a proper shape, dimension, and phase can resonate. The beam that is produced at the output of the laser source is very different for what concerns the main properties of the mode, from the one that can resonate inside the Input Mode Cleaner Cavity and, most important, inside the interferometer. In order to obtain the right light beam two Mode Matching Telescopes are part of the Injection System: the first is installed on the External Injection Bench and allows to manipulate the laser beam to make it resonant in the Input Mode Cleaner Cavity, the second, the most important, is installed on the Suspended Injection Bench and permits to transform the light beam in order to make it resonant in the Fabry Perot Cavities. A similar system of telescopes is present also on the Detection System, in order to collect the light beam at the output of the interferometer and to steer it in the Output Mode Cleaner Cavity.

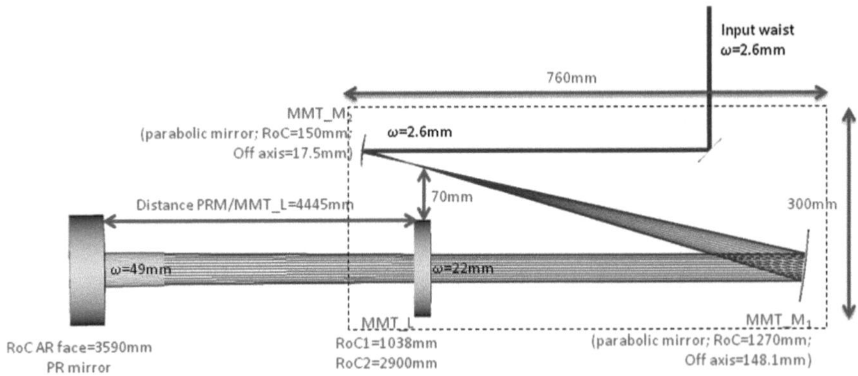

Fig. 5.23 Advanced Virgo Input Mode Matching Telescope scheme: Input waist = 2.6 mm, Output waist = 49 mm

5.8.1 Optical Telescope

A telescope is an optical system that collects a light beam coming from a source, transforms it, and produces an image of it. The optical telescopes can be classified on the basis of the optical components that they use to manipulate the input light beam [21]. In particular, there are two main classes of telescopes: the refractive ones, which exploit the refraction of a system of lenses and the reflective ones, which use the reflection of a system of mirrors. There are also some mixed schemes made by both mirrors and lenses. For example the catadioptric telescopes exploit both the reflection and the refraction of the light.

In the Advanced Virgo Input Optics System, the Mode Matching Telescope that has to steer the beam in the Input Mode Cleaner Cavity is reflective. It is composed of four parabolic mirrors characterized by different radii of curvature. It manipulates the laser beam in order to obtain the proper one at the input of the Input Mode Cleaner Cavity (waist of the cavity, whose size is 5.17 mm, located on the input/output flat mirrors).

The most critical telescope is the one installed on the Suspended Injection bench, which recovers the light from the Faraday Isolator, transforms it, and provides the right beam for the Fabry Perot Cavities. It is a catadioptric telescope composed of two parabolic mirrors and one meniscus diverging lens in combination with the Power Recycling Mirror, whose input face is characterized by the proper radius of curvature, as shown in Fig. 5.23. The great advantage of this configuration is its compactness, mandatory due to the lack of room on the bench.

This telescope is the most critical one, because the requirements are very stringent. In fact, it has to provide a mode matching greater than 99 %, a very low noise coming from the diffused and back-reflected light, very small aberrations, and a magnification sufficient to separate the secondary beams. Moreover, it is characterized by quite small dimensions because of the space constraints on the Injection Suspended Bench.

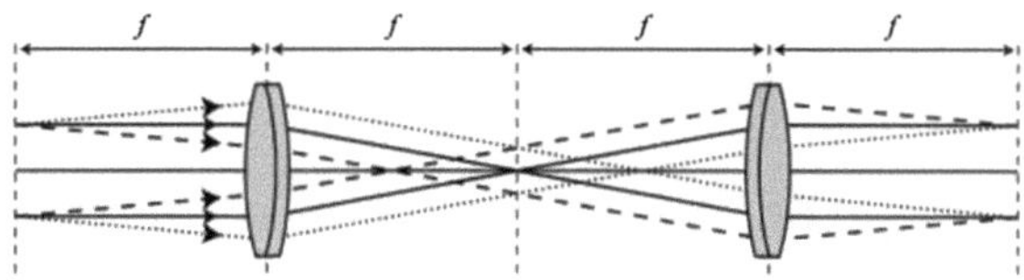

Fig. 5.24 Simple afocal system

The two parabolic mirrors form an afocal off axis parabolic telescope, which increases the beam size by a factor of 8.5 (i.e., from 2.6 to 22 mm). An example of a general simple afocal system is shown in Fig. 5.24.

It is an optical system that produces no net convergence or divergence of the input beam. In practice, it is a system without a focus, or better, characterized by an infinite effective focal lens. This kind of system can be implemented with a pair of optical elements where the distance between the element is equal to the sum of the focal lengths of each element. An important property of these systems is that it does not alter the divergence of a collimated beam, while altering the width of it, i.e., it introduces a magnification on the beam.

The diverging lens is used in combination with the Power Recycling mirror to match the beam to the interferometer, where the input beam waist has to be 49 mm. The meniscus shape of the diverging lens has been optimized in order to compensate the spherical aberrations introduced by the recycling mirror.

The Mode Matching Telescope can introduce a noise due to the diffused light created by imperfection on the surfaces of all the components, which can spoil the interferometer sensitivity. Moreover, also the noise coming from the back reflection of the faces of the meniscus lens can limit the sensitivity. Therefore, all the specifications on the roughness and radii of curvature of the optics elements of the telescope have been made in order to have a maximum noise, even in the worst case, due to scattered and back-reflected light that will not affect the interferometer sensitivity.

5.9 Future Improvements

In the future the Input Optics System could be upgraded by adding some other subsystems or adopting new technological solutions. In particular, two main upgrades that could improve the performance of the system, are under evaluation. The first one is a subsystem, called Input Beam Monitoring System (IBMS) that permits to monitor the laser beam properties in several points of the injection system [22]. The second one is a new technique, called Thermally Deformable Mirror (TDM), that could correct the thermal lensing effects for example in the Input Mode Cleaner Cavity or in the Faraday Isolator [23].

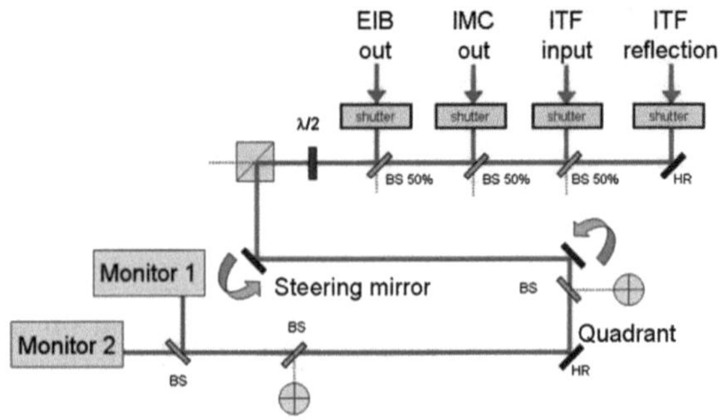

Fig. 5.25 Input Beam Monitoring System Scheme

5.9.1 Input Beam Monitoring System

The input beam monitoring system is used to monitor the laser beam properties in several points of the INJ system. The beam monitoring system will measure the intensity distribution and wavefront shape for the different fields of the beam at different points of the Injection System. It is important because it could permit to have a tool that can be quickly and easily used to assess the beam quality at all stages of the Input Optics System. Moreover, it should monitor the residual thermal effects and provide the error signals for tuning and acting compensation systems.

The Input Beam Monitoring System should assess the beam in particular points of the Injection System, specifically at the output of the External Injection Bench, at the output of the Input Mode Cleaner Cavity, at the input of the interferometer, and in reflection of the interferometer, as shown in Fig. 5.25.

The check of the beam at the output of the External Injection Bench is important because it permits to measure the thermal effects on it and to ensure the right quality of the beam that is entering in the Input Mode Cleaner Cavity. At the output of the IMC, it is important to verify the thermal effects of the cavity and to provide an error signal for a possible active compensation. At the input of the interferometer the thermal effects in the Faraday Isolator and the quality of the beam are checked. In reflection of the interferometer the thermal effects in it are measured. To perform all the relevant measurements, a single beam is shared among all the installed devices. Due to its conception, the Input Beam Monitoring System should be considered like a "black box" that is installed in the Input Optics System, with a single input beam that is injected in the system and it will be automatically adjusted in terms of power, alignment and size for all the receiving devices. A full measurement of the intensity and the relative phase of the beam is needed. In particular, the aim is to detect the intensity and the phase distribution for the global beam and for each field separately.

In order to make all the foreseen measurements, five diagnostic tools will be installed in this system: two CCD cameras (one put in the Near Field and the other in the Far Field), a wavefront sensor (WFS) [24], a Scanning Fabry–Perot Interferometer (SFPI), and a Phase Camera (PC) [25]. Two CCD cameras are exploited to acquire high-resolution intensity images both in the Near Field and in the Far Field. These

Fig. 5.26 Input Beam Monitoring System prototype

images are useful to know the distribution, the position, the size, and the divergence of the global input beam. Therefore, an imaging device with a sufficient resolution, a sufficient frame ratio of acquisition, and a sufficient active area to collect the entire beam has to be used and the CCD cameras have been chosen as sensors for these purposes. A multilateral shearing interferometer is used as wavefront sensor to acquire a global intensity and wavefront map of the beam: it splits the beam into several replicas that will interfere by overlapping after translation, magnification, or rotation. The interferogram is acquired by a CCD camera in order to be analyzed. To complete the analysis the intensity and phase distribution for each field of the input beam, composed of a carrier and two sidebands, have to be acquired. The evaluation of the relative power of each field is performed by a Scanning Fabry Perot Interferometer. It is based on the Fabry–Perot etalon principle: an incident beam is either reflected or transmitted by two successive reflective surfaces that are separated by free space and form a resonant cavity. One part of the incident light is transmitted and the other part is reflected, creating multiple reflections between surfaces, therefore the light interferes (both in a constructive and in a destructive way). In the case of constructive interference, the light is transmitted at the output of the device. In order to separately measure the different fields of the incident beam, a Phase Camera is adopted. It would be the ideal device to accurately determine the amplitude and phase distribution for each field of the input beam. In fact, it is a high-resolution wavefront sensor that can measure the complete spatial profile and phase of any frequency component of a light beam containing multiple frequencies. It is based on the heterodyne technique to measure the beat between the beam under test and a reference one that spatially overlaps it. The interference is acquired by two photodiodes, which permit to measure the amplitude and the phase of the carrier and the sidebands separately.

A prototype of the Input Beam Monitoring System is shown in Fig. 5.26.

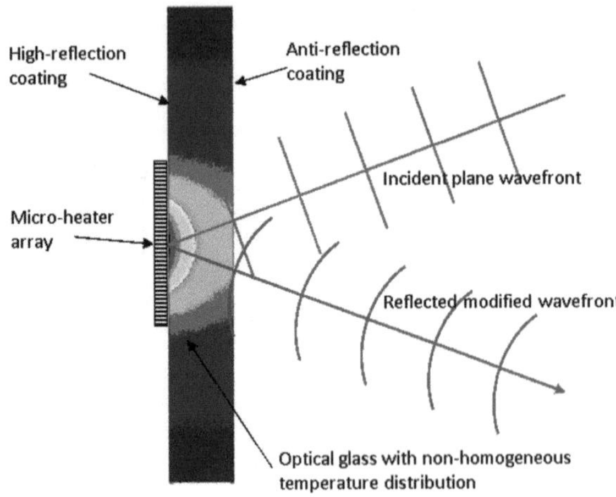

Fig. 5.27 Thermally Deformable Mirror working principle

5.9.2 Thermally Deformable Mirrors

As described in the previous sections, the Input Optics System is affected by thermal effects and nonoptimal matching in the cavities. Thermally induced beam-wavefront distortions can be compensated using deformable mirrors driven by thermal actuators. A new device, called Thermally Deformable Mirror [23], in which the heating actuators are placed in direct contact with the reflecting surface of the mirrors, is under test in order to be installed in the Injection System to correct, for example, the thermal lensing effects in the Input Mode Cleaner Cavity or in the Faraday Isolator.

The working principle of the TDM is shown in Fig. 5.27: an incident beam passes through the antireflective (AR) surface of the mirror and is reflected back by the high-reflective (HR)-coated surface. An array of micro-resistors is placed directly in contact with the HR surface and allows to locally control its temperature. By heating the surface, it is possible to change its shape and glass refractive index inducing a variation in the optical path length (OPL) of the beam, which results in a modification of the reflected wavefront. The variation of the optical path length is:

$$OPL = \int_C \left(\Delta T(s) \left[\frac{\mathrm{d}n}{\mathrm{d}T} + \alpha_T (1 + \nu) n \right] + n \right) \mathrm{d}s \tag{5.47}$$

where $\Delta T(s)$ is the local increase in temperature as a function of distance s along the path C, ν is the Poisson ratio, n is the glass refractive index, $\mathrm{d}n/\mathrm{d}T$ is the thermo-optic coefficient, and α_T is the coefficient of thermal expansion.

This system, exploited in the correction of the thermal effects, has some important advantages: it is simple and cheap, it does not introduce any extra noise, and it is ultra-high vacuum compatible. A prototype has been realized by Stevenage Circuits

Fig. 5.28 Thermally Deformable Mirror ptototype

Ltd. (Stevenage, UK) and is under test [26]: it is composed of a micro array of 61 independently controlled actuators ($0.9 \times 0.9\ mm^2$ each). Figure 5.28 shows the prototype installed on a test mirror.

This prototype has been design to correct the aberrations of a beam with a waist of 2.6 mm. The phase modifications induced on the beam by the TDM are measured with a wavefront sensor. Each actuator subjected to a constant current produces a phase deformation that is the individual spatial response of this actuator. Figure 5.29 shows examples of this response for three different actuators. The difference in amplitude responses among the actuators is mainly duc to thc diffcrcncc in thc resistors' values.

It can be seen that the wavefront phase modification is in good correlation with the switched on actuator. Therefore, the first tests are satisfying and this kind of device, after the proper development, could be installed in the future on the Input Optics System of the interferometer. Moreover, it could also open a way to introduce more adaptive optics devices in the correction of various aberrations in the next generations of gravitational waves interferometers.

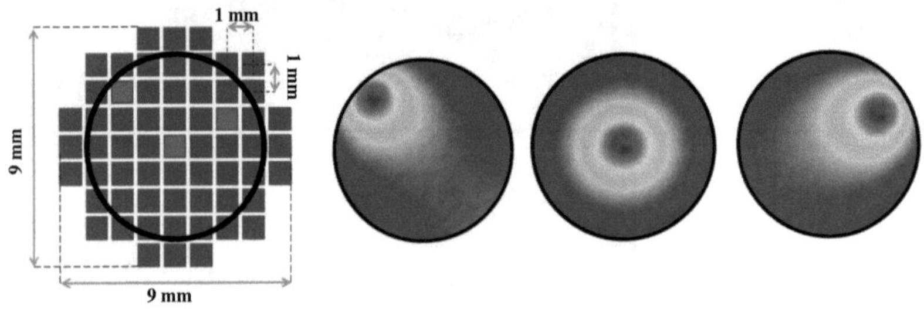

Fig. 5.29 Thermally Deformable Mirror ptototype results: on the *left* scheme of three switched on actuators (one at a time), on the *right* the obtained responses

5.10 Advanced Ligo Input Optics

In this section a short description of the Advanced Ligo Input Optics System is presented [27]. Its role is the same as in Advanced Virgo: it has to transform the laser beam in order to guarantee that it satisfies all the requirements at the input of the interferometer. But there are two main differences in the design of Advanced Ligo with respect to that of Advanced Virgo: the source system exploits a solid-state laser instead of a rod fiber one and the recycling cavities are stable instead of marginally stable. The Input Optics System is the connection between the laser source and the power recycling cavity, therefore there are some differences between its implementation in Advanced Virgo and in Advanced Ligo. As in Advanced Virgo, also the Advanced Ligo Input Optics System is divided into two main parts: one in air and the other under vacuum. In particular, the Electro-Optic Modulation System, the Input Mode Cleaner mode matching telescope, the Input Power Control System, the Beam Pointing Control System, and a Beam Analysis System are installed on the in-air bench. Therefore, the main tasks and, in consequence, the main components of the in-air optics system are the same as in Advanced Virgo. On the other hand, the differences between the Input Optics Systems of the two interferometer are mainly on the in-vacuum bench. In fact, some systems as the Input Mode Cleaner and the Isolation System are almost equal, but some other systems, as the Mode Matching System or the Alignment System to the Power Recycling Cavity are different. In fact, the Power Recycling Cavity in Advanced Ligo is formed by three mirrors instead of one and systems to align and to match the beam to this cavity are mandatory. These systems are installed on the in-vacuum bench of the Input Optics System. Another difference is that the Reference Cavity System in aLigo is not installed in the Input Optics System but is part of the Laser System.

Some parameters and some requirements among the devices of the two interferometers can be different, such as the values of the modulation frequencies required to the electro-optic crystals, but the main tasks are almost the same.

References

1. G.M. Harry, [LIGO Scientific Collaboration], Advanced LIGO: The next generation of gravitational wave detectors. Class. Quant. Grav. **27**, 084006 (2010)
2. Virgo Collaboration, Advanced Virgo Technical Design Report, VIR0128A12, https://tds.ego-gw.it/ql/?c=8940
3. K. Kuroda, [LCGT Collaboration], Large-scale gravitational wave telescope (LCGT). Int. J. Mod. Phys. D **20**, 1755 (2011)
4. A. Carenco, Composants actifs, Collection SFO **27**, (1992)
5. A. Yariv, *Quantum Electronics* (Wiley, New York ,1989)
6. E. D. Black, An introduction to Pound Drever Hall laser frequency stabilization. Am. J. Phys. **69**, 79–87 (2001)
7. J.D. Mansell et al., Evaluating the effect of transmissive optic thermal lensing on laser beam quality using a Shack-Hartmann wavefront sensor. Appl. Opt. **40**, 366–374 (2001)
8. B. Canuel et al., High power input optics R&D: Final report, Virgo internal document VIR-0296A-10 (2010)
9. E.A. Khazanov et al., AdV Faraday isolator design study, Virgo internal document VIR-0245A-10 (2010)
10. E.A. Khazanov et al., AdV INJ: TheFaraday isolator prototyper for AdV Description and assembling procedure, Virgo internal document VIR-0283A-10 (2010)
11. B. Canuel, E. Genin, AdV INJ: Faraday isolator prototype characterization, Virgo internal document VIR-0455A-11 (2011)
12. Virgo Collaboration, In-vacuum optical isolation changes by heating in a Faraday isolator Appl. Opt. **47**, 5853–5861 (2008)
13. Virgo Collaboration, In-vacuum optical isolation remote tuning. Appl. Opt. **49**, 4780 (2010)
14. S. Mosca et al., Photon self-induced spin-to-orbital conversion in a terbium-gallium-garnet crystal at high laser power. Phys. Rev. A **82**, 043806 (2010)
15. E.A. Khazanov et al., Suppression of self-induced depolarization of high-power laser radiation in glass-based Faraday isolators, JOSA B **17**, 99–102 (2000)
16. B. Canuel et al., Advanced Virgo: Inj subsystem Preliminary Design study, Virgo internal document VIR-023A-09 (2009)
17. J.Y. Vinet, On special optical modes and thermal issues in advanced gravitational wave interferometric detectors. Living Rev. Relativity **12**, 5 (2009)
18. E. Genin et al., Virgo Input Mode Cleaner: Optical characterization, Virgo internal document VIR-0232A-10 (2010)
19. G. Muller, Beam jitter coupling in advanced LIGO. Opt. Exp. **13**, 7118–7132 (2005)
20. S. Donati, *Electro-Optical Instrumentation: Sensing and Measuring with Lasers* (Prentice Hall, USA, 2004)
21. D.J. Schroeder, *Astronomical Optics* (Academic Press, New York, 2000)
22. M. Kasprzack et al., *Laser Beam Characterization System Development and Wavefront Correction System Prototype Validation for Advanced Virgo*. M.Sc Thesis, 2011
23. B. Canuel et al., Wavefront aberration compensation with a thermally deformable mirror. Class. Quant. Grav. **27**, 084006 (2010)
24. S. Velghe, Wave-front reconstruction from multidirectional phase derivatives generated by multilateral shearing interferometers. Opt. Lett. **30**, 245–247 (2005)
25. K. Goda, Frequency-resolving spatiotemporal wave-front sensor. Opt. Lett. **29**, 1452–1454 (2004)
26. M. Kasprzack et al., Performance of a thermally deformable mirror for correction of low order aberrations in laser beams. Appl. Opt. **52**, 2909–2916 (2013)
27. D. Feldbaum et al., Advanced LIGO Input Optics —status update, LIGO internal document G1200911–v2 (2012)

Chapter 6
Readout, Sensing, and Control

Gabriele Vajente

Abstract Suspending the mirrors is one of the most crucial tasks in gravitational wave interferometer technology. The performance of the suspensions must provide the required attenuation of seismic noise and reduction of thermal noise, two fundamental limits to the sensitivity of any gravitational wave detector. Moreover, the suspension system must be equipped with sensors and actuators which are used to actively control some relevant degrees of freedom, so to be able to keep the interferometer at its working point (i.e., "locked"). In the first part of this chapter we deal with the basic principles behind the super attenuator chains developed in Virgo to reduce the seismic noise. In the second part, we illustrate the techniques to suspend the mirror reducing the thermal noise in the detection bandwidth, according to the theory illustrated in Chap. 8.

6.1 Foundations of the Problem

In the previous chapters we learned how to build an interferometric gravitational wave detector with high sensitivity. Advanced GW detectors will be dual recycled Michelson interferometers with Fabry–Perot cavities in the arms. To take full advantage of the presence of all the resonant cavities we must maintain their length close to the correct working position. This is a two-step problem: we must first find a way to measure the microscopic deviation of all lengths from the resonance point and then find a way to correct them. Typically, the accuracy needed ranges from 10^{-15} m for the most critical arm differential length motion to some 10^{-12} for the other degrees of freedom.

G. Vajente (✉)
INFN Sezione di Pisa, Largo B. Pontecorvo 3, 56124 Pisa, Italy

California Institute of Technology, MC 100-36, 1200 E. California Blvd, Pasadena, CA 91125, USA
e-mail: vajente@caltech.edu

M. Bassan (ed.), *Advanced Interferometers and the Search for Gravitational Waves*, Astrophysics and Space Science Library 404, DOI: 10.1007/978-3-319-03792-9_6,

Moreover, we showed that an interferometer is a good transducer, in the sense that it is able to convert gravitational waves into phase changes in the carrier field at the anti-symmetric port. We however completely neglected the problem of how to measure this phase variation, or in other words how to convert them into some kind of electrical signal that can be measured, digitized, and stored for the analysis.

6.2 How to Measure the Laser Field Phase Change

Let us consider the following situation: our precious GW signal is encoded in the phase $\phi(t)$ of the laser carrier field at the dark port:

$$\Psi(t) = Ae^{2\pi i\nu t + i\phi(t)} \tag{6.1}$$

where ν is the laser optical frequency ($\nu \sim 3 \times 10^{14}$ Hz). When this field is detected at a photo-diode, only the power is accessible and our phase signal is lost:

$$P(t) = |\Psi(t)| = A^2 \tag{6.2}$$

We must therefore do something to our field before sending it to the photo-diode in order to be able to extract the phase information. One possible strategy was already discussed in Chap. 3, when describing the Michelson configuration.

6.2.1 Heterodyne Detection

One of the most used techniques is the *optical heterodyne*. Referring to Fig. 6.1, a laser source is used to probe an optical system. The laser field acquires a dephasing when traveling through the system

$$\Psi_{\text{SYS}} = Ae^{2\pi i\nu t + i\phi(t)} \tag{6.3}$$

To detect this phase we take a pick-off of the input beam, before it enters the system. This beam is then sent to an *acousto-optical modulator* (AOM)[1] which is driven by a sinusoidal signal at a heterodyne frequency f_H. The AOM generates a transmitted beam which has a frequency shifted by f_H:

$$\Psi_{\text{AOM}} = Be^{2\pi i(\nu + f_H)t} \tag{6.4}$$

This beam is then recombined with the one transmitted by the optical system and finally probed by the photo-diode

[1] See Chap. 5 for more details on AOMs.

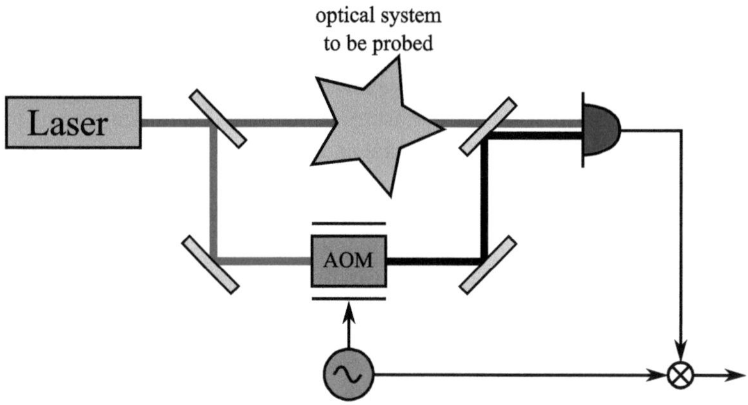

Fig. 6.1 Optical heterodyne scheme for measuring the phase the laser field has acquired inside an optical system

$$\Psi_{PD} = \Psi_{SYS} + \Psi_{AOM} \tag{6.5}$$

$$P(t) = |\Psi_{SYS} + \Psi_{AOM}|^2 \tag{6.6}$$

$$= A^2 + B^2 + AB\left[e^{-i\phi(t)}e^{2i\pi f_H t} + e^{i\phi(t)}e^{-2i\pi f_H t}\right] \tag{6.7}$$

$$= A^2 + B^2 + 2AB\cos(2\pi f_H t + \phi(t)) \tag{6.8}$$

The mixing of the frequency shifted beam with the probe one creates a beat note at the heterodyne frequency f_H which is visible in the power detected by the photo-diode. The phase of this beat note is exactly the phase acquired by the laser beam through the optical system. Since this signal is at a frequency f_H, which is typically of the order of few MHz, it is possible to precisely measure its phase. The simplest approach would be to acquire digitally the signal with fast converters, running at a frequency larger than f_H and then fit the sinusoidal motion. This approach is however rather new, since in the past such fast converters and digital system were not available. The standard method, still used in many cases, is to multiply the output of the photo-diode signal with synthesized sinusoid and cosinusoid again at f_H:

$$c(t) = P(t)\cos(2\pi f_H t) = \left(A^2 + B^2\right)\cos(2\pi f_H t) + AB\left[\cos(4\pi f_H t) + \cos\phi\right] \tag{6.9}$$

$$s(t) = P(t)\sin(2\pi f_H t) = \left(A^2 + B^2\right)\sin(2\pi f_H t) + AB\left[\sin(4\pi f_H t) - \sin\phi\right] \tag{6.10}$$

These signals contain components at high frequencies (f_H and $2f_H$) and at low frequency. The latter can be separated with a proper low pass filter. These low frequency components are the sine and cosine of the dephasing we are interested in and therefore can be used to reconstruct ϕ.

The heterodyne method is quite simple, but however it has some disadvantages. First of all, we need to properly recombine two beams at the photo-diode. Moreover, so far we considered that only the probe beam accumulates a dephasing, but this is not true in the real world. Any variable dephasing accumulated by the beam going through the AOM will show up in the output signal and it will not be possible to distinguish it from ϕ. Therefore, this system is only capable of measuring the phase difference of the probe beam with respect to the reference one. This kind of measurement is not what we need, for example, to control the length of a Fabry–Perot cavity: in this case we need to measure the relative distance variation and not the absolute positions.

6.2.2 Pound-Drever-Hall Technique

The base of all the methods used to measure the length variations in a GW interferometer is the *Pound-Drever-Hall technique* (PDH) [1, 2]. This technique was first developed with the goal of stabilizing the laser frequency on reference cavities.

Referring to Fig. 6.2, the input laser beam is passing through an *electro-optical modulator* (EOM).[2] This typically consists of a Pockels cell, which is a crystal with an optical length that can be changed by driving an applied voltage [3]. The EOM in our application is driven with a radio-frequency sinusoidal signal which is converted in a variation of the phase of the transmitted beam:

$$\Psi = \Psi_0 e^{i(\omega t + m \cos \Omega t)} \tag{6.11}$$

where ω corresponds to the laser optical frequency and Ω to the radio-frequency modulation. The amplitude of the phase modulation m is called the *modulation depth* or *index*. The exponential can be expanded in terms of Bessel functions [4]

$$\Psi = \Psi_0 e^{i\omega t} \left[J_0(m) + i J_1(m) e^{-i\Omega t} + i J_1(m) e^{i\Omega t} \right] + \cdots \tag{6.12}$$

We have therefore found a description of the modulated beam in terms of radio-frequency sidebands around the main carrier field, space by the modulation frequency. The dots in the above equation are there just to remind us that higher order sidebands are also generated, spaced by multiples of the modulation frequencies. These high-order sidebands have however a much smaller amplitude and are often neglected, since the Bessel functions behave, for small m as $J_n(m) \propto m^n$ [5]. Just for reference, the low m approximation of the first Bessel functions are

$$J_0(m) \sim 1 - \frac{m^2}{4} \tag{6.13}$$

$$J_1(m) \sim \frac{m}{2} \tag{6.14}$$

[2] See Chap. 5 for more details on EOMs.

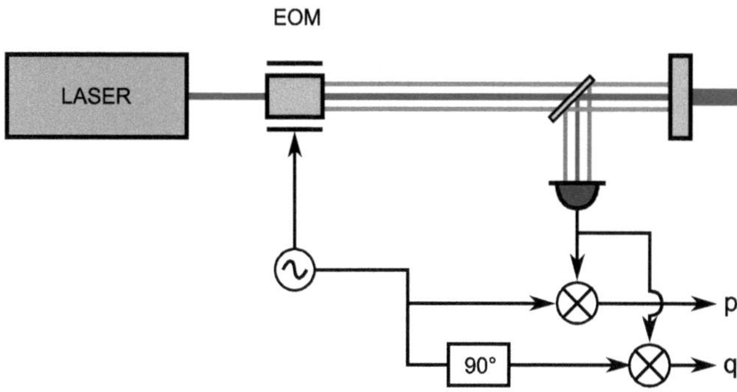

Fig. 6.2 Scheme of the Pound-Drever-Hall technique to measure the length of a Fabry–Perot cavity

Finally, if assuming m to be small and substituting these two approximations into Eq. 6.12, we found the small m expansion of Eq. 6.11.

Equation 6.12 describes the beam impinging on the Fabry–Perot resonant cavity as composed of three independent fields. The basic idea of the PDH technique is that if we choose properly the modulation frequency we can ensure that the sidebands are not resonant inside the cavity when the carrier is. Recalling what we learned in the previous chapter, when a field is resonant inside the cavity, the phase of its reflection changes very fast as a function of the cavity length. Instead, when a field is not resonant, its phase is roughly constant even when the cavity length varies slightly around the working point. We can therefore use the sidebands as static phase references for the measurement of the carrier phase. In this sense they play more or less the same role of the reference beam we introduced in the heterodyne configuration.

Let us assume that the cavity is tuned at resonance and allow a displacement z from this position. For the moment we are not requiring z to be small. As we learned in the previous chapter, the reflectivity of the Fabry–Perot cavity for the carrier field is given as

$$R_0 = i\frac{r_i - r_e e^{-2ikz}}{1 - r_i r_e e^{-2ikz}} \tag{6.15}$$

where we neglected the losses inside the arm. The reflectivity for the sidebands is slightly different because of their frequency shift which introduces an additional dephasing (with respect to the carrier) in propagation along the arm length given as

$$\phi_{\text{sb}} = \pm 2\frac{\Omega}{c}L \tag{6.16}$$

There should also be a term proportional to Ωz but this is for sure negligible, since z is of the order of 1 wavelength and the modulation frequency is typically of a few MHz. We can then write the Fabry–Perot arm reflectivity for the sidebands as

$$R_{\pm} = i\frac{r_i - r_e e^{-2ikz \mp 2i\frac{\Omega}{c}L}}{1 - r_i r_e e^{-2ikz \mp 2i\frac{\Omega}{c}L}} \tag{6.17}$$

Since we want to use the sidebands as phase reference, we want them anti-resonant inside the cavity when the carrier is resonant. This means that the additional dephasing should be π. This gives us a relation between the cavity length and the modulation frequency

$$f_{\text{mod}} = \left(N + \frac{1}{2}\right)\frac{c}{2L} \tag{6.18}$$

which means that the modulation frequency must be a semi-integer multiple of the free spectral range of the cavity. With this choice the reflectivity simplifies to

$$R_{\pm} = i\frac{r_i + r_e e^{-2ikz}}{1 + r_i r_e e^{-2ikz}} \tag{6.19}$$

As shown in Fig. 6.2 we use an optical isolator to send the reflected beam to the photo-diode which as usual detects the total power:

$$\Psi_R = \Psi_0 e^{i\Omega t}\left[J_0 R_0 + iJ_1 R_- e^{-i\Omega t} + iJ_1 R_+ e^{i\Omega t}\right] \tag{6.20}$$

$$\begin{aligned} P = |\Psi_0|^2 \Big[& J_0^2 |R_0|^2 + J_1^2 |R_+|^2 + J_1^2 |R_-|^2 \\ & + ie^{-i\Omega t} J_0 J_1 \left(-R_0 R_+^* + R_0^* R_-\right) - ie^{i\Omega t} J_0 J_1 \left(-R_0^* R_+ + R_0 R_-^*\right) \\ & + e^{-2i\Omega t} J_1^2 R_- R_+^* + e^{2i\Omega t} J_1^2 R_-^* R_+ \Big] \end{aligned} \tag{6.21}$$

The terms in the first row of the power equation give the low frequency contribution to the photo-diode signal, which corresponds to the static power in reflection of the cavity. The second row instead contains a contribution which oscillates at the modulation frequency, while the last one at twice the modulation frequency. The situation is therefore quite similar to the heterodyne case. We want to extract the beat note at the modulation frequency. For this reason we again multiply the output of the photo-diode with a sinusoidal signal at the modulation frequency

$$s(t) = \frac{1}{2}\left(e^{i\Omega t + i\phi} + e^{-i\Omega t - i\phi}\right) \tag{6.22}$$

Here we added the possibility of selecting the *demodulation phase* of the signal. This is equivalent to multiply with both a sine and a cosine and then combine the results.

The photo-diode output signal is therefore multiplied with the *local oscillator signal* defined above. The result will contain several components, each one oscillating at different frequencies: quasi static, Ω, 2Ω and 3Ω. However, we are interested only in the quasi static component and therefore we apply a low pass filter to the result.

The surviving terms give us the *demodulated signal*:

$$d = P_{\text{in}}\frac{J_0 J_1}{2}\left[i\left(R_0^* R_- - R_0 R_+^*\right)e^{i\phi} + c.c.\right] \tag{6.23}$$

$$= P_{\text{in}} J_0 J_1\, \mathscr{R}\left[i\left(R_0^* R_- - R_0 R_+^*\right)e^{i\phi}\right] \tag{6.24}$$

where in the last equation we are taking the real part of the result. The argument in square brackets of the last equation is a complex number and the demodulation phase ϕ just rotates it in the complex plane. However, we do not have an infinite number of independent components but only two, corresponding to the real and imaginary part of the complex number. Therefore, we can consider the two cases $\phi = 0$ and $\phi = \pi/2$ and construct the *in-phase* and *quadrature* signals:

$$p = P_{\text{in}} J_0 J_1\, \mathscr{I}\left(R_0^* R_- - R_0 R_+^*\right) \tag{6.25}$$

$$q = P_{\text{in}} J_0 J_1\, \mathscr{R}\left(R_0^* R_- - R_0 R_+^*\right) \tag{6.26}$$

We can use Eqs. 6.15 and 6.19 to substitute into the previous equations to find

$$p = -4P_{\text{in}} J_0 J_1\, r_e(1 + r_e^2) r_i t_i^2 \frac{\sin 2kz}{1 + r_i^4 r_e^4 - 2r_i^2 r_e^2 \cos 4kz} \tag{6.27}$$

$$q = 0 \tag{6.28}$$

The behavior of the in-phase signal as a function of the cavity displacement from resonance is shown in Fig. 6.3. The interesting feature is that the signal shows a linear dependence on the cavity displacement when close enough to the resonance. One can indeed show that the width of the linear region is proportional to the cavity line-width. The slope can be computed with the derivative of Eq. 6.27 in the origin. If we assume the cavity finesse to be large with respect to unity we finally obtain

$$\frac{\mathrm{d}p}{\mathrm{d}z} = J_0(m) J_1(m) P_{\text{in}} \frac{2\mathscr{F}}{\pi}\frac{4\pi}{\lambda} \tag{6.29}$$

This tells us that the larger the finesse is, the more sensitive our signal is to the cavity length variation. This however comes at the price of a reduced linear range of the error signal. The signal is also proportional to the input power, while the shot noise increases only with the square root of the power. We can therefore gain in signal-to-noise ratio with the square root of the power we are injecting into the cavity.

Finally, a clear advantage of the PDH technique with respect to the simple heterodyne one is that the signal is insensitive to any dephasing happening outside of the cavity, i.e., along the path from the laser to the input mirror and from the input mirror to the photo-diode. Indeed, any variation of these optical lengths induces dephasings which are common to the carrier and the sideband fields and therefore cancel out in the detection and demodulation process.

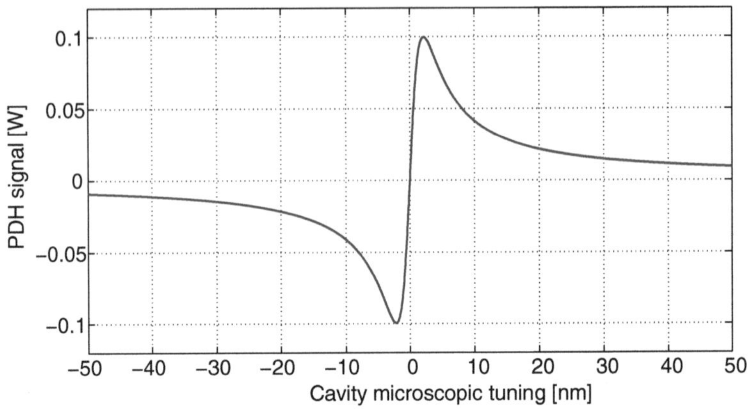

Fig. 6.3 Pound-Drever-Hall signal in reflection of a Fabry–Perot cavity with Advanced Virgo parameters. The input power is 1 W

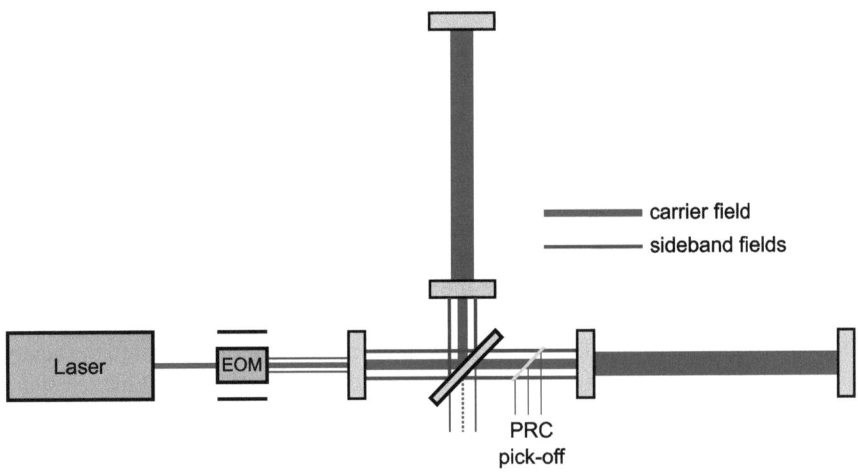

Fig. 6.4 Scheme of the frontal modulation technique applied to the control of a power-recycled interferometer

6.2.3 Frontal Modulation

The Pound-Drever-Hall technique is quite simple when applied to a single resonant cavity. However, it is possible to extend it to the control of a full interferometer for GW detection with the *frontal modulation* approach, see Fig. 6.4. Here the word frontal refers to the fact that the laser beam is phase modulated with an EOM before entering the main optical system to control.

The laser beam is modulated at a radio-frequency f_{mod} which is chosen as before in order to be anti-resonant inside the arm Fabry–Perot cavities. However, now we have a power-recycling cavity which must be maintained resonant for the carrier field. The point here is that we must make also the sidebands resonant inside the PRC, otherwise they will be almost completely reflected by the interferometer.

The two sidebands will undergo an additional dephasing with respect to the carrier when propagating over one entire round-trip inside the PRC. This dephasing is made of two contributions. The first is due to their anti-resonance inside the arms, which adds a π in reflection of the cavities. The second contribution comes clearly from the propagation over twice the PRC length. In total

$$\phi_{SB} = \pi + 2\frac{2\pi f_{mod}}{c} l_{PRC} \tag{6.30}$$

To have the sidebands resonant, we must ensure that this phase is a multiple of 2π:

$$\pi + 2\frac{2\pi f_{mod}}{c} l_{PRC} = 2N\pi \tag{6.31}$$

which turns out to be

$$f_{mod} = (2N-1)\frac{c}{4l_{PRC}} \tag{6.32}$$

If the power recycling cavity has a length of about 12 m as in Advanced Virgo, the modulation frequency must be an odd multiple of about 6.25 MHz. This condition must hold together with Eq. 6.18 that determines the anti-resonance inside the arm cavities. These two requirements together imply a relationship between the arm and the power recycling cavity lengths.

6.2.4 The Schnupp Asymmetry

We already explained in the previous chapter that in this configuration the GW signals can be read out with the carrier field at the anti-symmetric port. However, since the Michelson interferometer is tuned in order to be in the dark fringe condition, it is not possible to simply use the power read by a photo-diode to read the GW signal. The frontal modulation technique can be used for this purpose, in a way quite similar to the Pound-Drever-Hall technique for a single resonant cavity. We indeed need to tune the interferometer in such a way as to have some reference field at the dark port, in order to be able to read the beat note with the carrier and extract the GW signal with a demodulation technique.

The simplest way to have some sideband fields leaking through the antisymmetric port is to introduce a macroscopic difference between the two short Michelson arm: in other words we make the two distances between beam-splitter and input mirrors slightly different. This will have no impact on the carrier field, since only the microscopic length difference is relevant to ensure the destructive interference. The story is different for the two sidebands. When recombining at the beam-splitter, the sidebands that traveled nort will have accumulated a different phase with respect to those traveling west, due to the different distances:

$$\Delta\phi = 2\frac{\Omega}{c}(l_N - l_W) = 2\frac{\Omega}{c}\Delta l \tag{6.33}$$

where Δl is called the *Schnupp asymmetry*. This phase must be inserted in the equation that gives the field transmitted to the anti-symmetric port. The net result is that the Michelson interferometer is no more perfectly reflective for the sidebands entering from the symmetric port, but transmits a fraction to the anti-symmetric port

$$R_{\rm sb} = i r_{\rm FP} \cos\left(\frac{\Omega\Delta l}{c}\right) \tag{6.34}$$

$$T_{\rm sb} = \pm i r_{\rm FP} \sin\left(\frac{\Omega\Delta l}{c}\right) \tag{6.35}$$

If we consider a modulation frequency of 6.25 MHz as was done in Virgo and Virgo+ and a Schnupp asymmetry of 0.8 m we obtain that 1 % of the sidebands that circulate inside the power recycling cavity are transmitted to the anti-symmetric port, with equal and opposite amplitudes. Without entering too much into detailed computations, the antisymmetric port field will be composed of the two sidebands and a carrier field which is directly proportional to the differential motion of the two arm lengths. The photo-diode will detect a power which contains a beat note coming from the interference of the carrier and the two sideband fields. The amplitude of this beat note is proportional to the differential displacement and can be extracted using a demodulation technique as was done in the PDH scheme. It is clear that the optical gain of the system is proportional to the amplitude of the sideband fields, or in other words to the square root of the sideband power leaking to the antisymmetric port.

Moreover, the average power reaching the dark port will be dominated by the sidebands, assuming that the Schnupp asymmetry is large enough and that the contrast defect of the interferometer is small enough. This means that the shot noise which will limit the photo-diode output signal will have an amplitude proportional to the square root of the sideband power reaching the antisymmetric port. Putting all together, we see that in this configuration the signal-to-noise ratio is independent of the sideband amplitude.

6.2.5 Phase Noise

The frontal modulation technique was used to read-out the GW signal in all first-generation interferometers and it worked quite well. However, it is largely based on the goodness of the sideband fields: if these are for some reason noisy, if their transverse shape is bad, or if their amplitudes are largely unbalanced we might have some trouble. The typical examples are the *phase and amplitude noises*: the

Fig. 6.5 Coupling mechanism of phase noise into the interferometer output signal

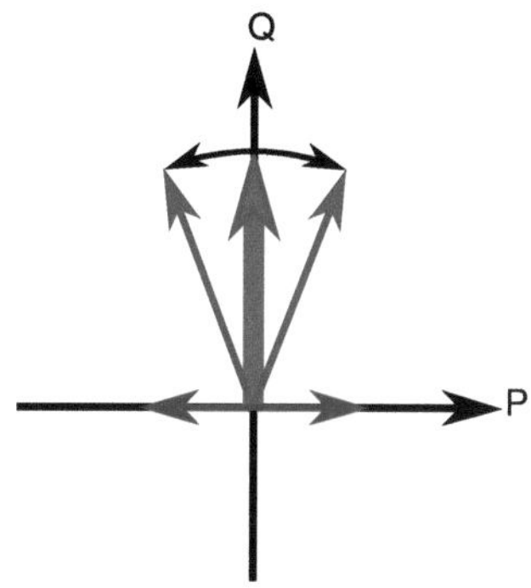

sinusoidal signal which is used both for the generation of the sidebands and for the demodulation typically contains some noise affecting both the amplitude and the phase. This noise can couple to the interferometer output signal. The mechanism is explained in Fig. 6.5. We can represent the output signal as a vector in a plane, where the two components are given by the in-phase and quadrature signals. If the demodulation phase changes, the vector rotates on this plane. The effect is clearly the same if the phase of the modulation signal changes. Let us assume that our signal of interest is in the in-phase (P) quadrature, which is typically zero because we are (magically) maintaining the interferometer close to the correct resonance condition. Let us assume however that some defects (typically unbalance of the two sideband amplitudes) create a large signal in the orthogonal (Q) quadrature. Phase noise will make the Q vector jitter around the vertical position, thus contributing with nonzero projection on the P signal:

$$\tilde{n}_P = \langle Q \rangle \, \tilde{n}_\phi \tag{6.36}$$

where $\tilde{n}_\phi$ is the phase noise of the modulation/demodulation sinusoid, $\langle Q \rangle$ is the average of the quadrature signal and $\tilde{n}_P$ is the noise induced in the main in-phase signal. A similar mechanism can couple amplitude noise in the in-phase quadrature, depending on its own mean value.

Another complication enters the game when the signal recycling cavity is added. This cavity can be detuned for the carrier, resulting in a detuning also for the sidebands. These could become very unbalanced at the exit of the SRC, resulting in a large coupling of phase noise in the main output signal. Even with the present state of the art oscillators, this would be a source of limiting noise in the sensitivity of Advanced Detectors.

6.2.6 Homodyne Detection

For the reasons explained in the previous section, among others, a different signal read-out strategy has been chosen for all the Advanced Detectors. The idea is quite

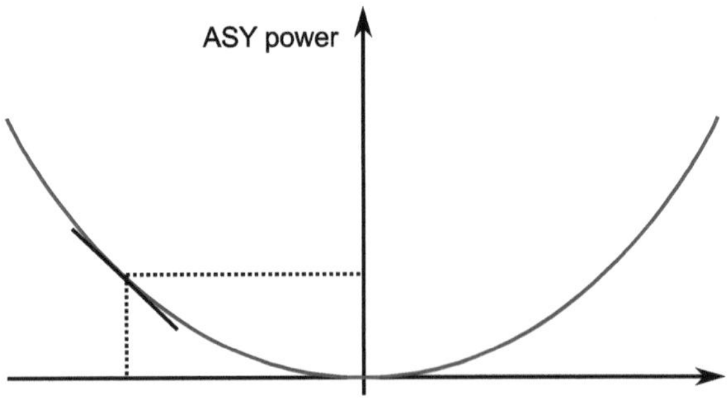

Fig. 6.6 Dependence of the antisymmetric port power on the differential tuning of the interferometer

simple: we add a static offset to the differential length of the arms, or in other words, we make one of the two cavities slightly shorter and the other slightly longer. The offset is very small in order to maintain the two cavities well inside their resonance width. This offset translates in a small difference in the static phase of the two beams recombining at the beam splitter: the perfect destructive interference is spoiled and a small fraction of the carrier field leaks to the antisymmetric port:

$$\Psi_{\mathrm{ASY}} = \Psi_{\mathrm{DC}} + x_0 \left[G(\omega) e^{i\omega t} + G(-\omega) e^{-i\omega t} \right] \tag{6.37}$$

Here, the first term is the static contribution coming from the offset and the second one is the time-varying one coming from the GW-induced differential motion x_0 multiplied by the interferometer gain $G(\omega)$. The photo-diode will as usual sense the power, which this time contains a term directly proportional to the GW signal times the static field:

$$P(t) = \cdots + \left[\Psi_{\mathrm{DC}}^{*} G(\omega) + \Psi_{\mathrm{DC}} G^{*}(-\omega) \right] x_0 e^{i\omega t} + \text{ c.c.} + \cdots \tag{6.38}$$

In this *homodyne read-out*, often called also *DC read-out*, scheme the static carrier field plays the role of the phase reference against whic the carrier GW signal beats to provide a readable output.

Another simple way to understand the working principle of the DC read-out approach is to look at how the carrier power reaching the antisymmetric port depends on the differential detuning of the arms. As shown in Fig. 6.6, for small offsets the dependence is quadratic. If the offset is zero, the slope of the power curve is also zero and therefore any fluctuation of the differential degree of freedom will not contribute to a variation of the power at the antisymmetric port. However, if we introduce a small offset, the slope gets different from zero and therefore we linearly convert differential signals into power signals. The gain is directly proportional to the offset or, which is the same thing, directly proportional to the square root of the static power at the antisymmetric port. As happened in the frontal modulation scheme, shot noise is also proportional to the square root of the power, so the signal-to-noise ratio is independent of the offset.

This approach to the extraction of the GW signal is pretty simple, but as usual in the real world it does not come without complications. First of all, any spurious field reaching the photo-diode will spoil the sensitivity, since it will increase the total power (and the shot noise) without increasing the optical gain and thus the signal. The sidebands are a first example of this spurious field. Indeed, it is not possible to switch them off since, as we will see in a moment, they are still needed to control all the other lengths of the interferometer. Therefore, we need a way to filter them out before the photo-diode. The only way is to add at the output port an additional short and high finesse cavity called the *output mode cleaner*. The line-width of this cavity must be significantly smaller than the lowest modulation frequency. In this way only the carrier field is transmitted.

Another difficulty introduced by the DC read-out scheme is the fact that arm cavities are working slightly out of resonance. As we learned in the previous lecture, this introduces strong radiation pressure effects. Since the two cavities are detuned in different directions, one will remain optomechanically stable, while the other will become unstable. The typical offsets foreseen for Advanced Detector is of the order of 10 pm. Using this number in the radiation pressure equations obtained in the previous chapter shows that the pendulum resonant frequency can be shifted as high as 40 Hz.

6.3 Longitudinal Control of the Interferometer

In a dual recycled interferometer the distances between the reflecting surfaces of the mirrors must be maintained fixed to the correct resonance conditions, which are typically expressed in terms of suitable multiples of the laser wavelength.

Referring to the definition shown in Fig. 6.7 the following physical degrees of freedom can be identified:

$$\text{DARM} = L_N - L_W \text{ (differential arm length)} \tag{6.39}$$

$$\text{CARM} = \frac{L_N + L_W}{2} \text{ (average arm length)} \tag{6.40}$$

$$\text{MICH} = l_N - l_W \text{ (differential Michelson short arm length)} \tag{6.41}$$

$$\text{PRCL} = l_P + \frac{l_N + l_W}{2} \text{ (power recycling cavity length)} \tag{6.42}$$

$$\text{SRCL} = l_S + \frac{l_N + l_W}{2} \text{ (signal recycling cavity length)} \tag{6.43}$$

The role of the longitudinal control system is to measure the variation of all these lengths with respect to the resonance position and to correct them. Let us first under-

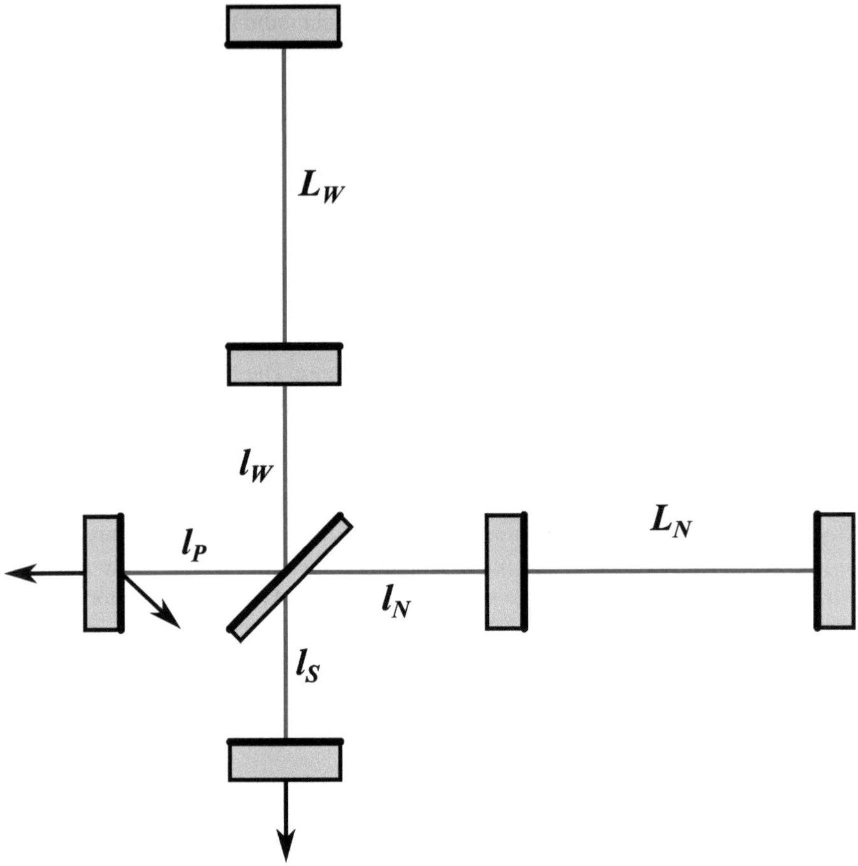

Fig. 6.7 Definition of lengths in a dual recycled interferometer. The *black arrows* indicate the probe beam that are sensed with photo-diode to extract the signals needed for the longitudinal control of the system

stand how we can measure the variation of these degrees of freedom. We already explained the strategy used for DARM, which corresponds to the GW channel. This degree of freedom is sensed using the power read by the antisymmetric port photodiode.

All the other degrees of freedom are typically called *auxiliary* and they are controlled using the frontal modulation technique. The antisymmetric port beam is not enough for all the degrees of freedom. For this reason other probe beams are extracted, as indicated by the black arrows in Fig. 6.7: the reflection of the interferometer and a pick-off of the beam circulating inside the power recycling cavity.

In a power recycled interferometer one modulation frequency is enough to control all the four degrees of freedom: CARM and MICH are sensed with the two quadrature of the PRC pick-off, while PRCL is sensed with one of the quadratures in reflection of the interferometer. To control a dual recycled interferometer an additional modulation is added, to be able to separate the signal and the power recycling cavities. The first modulation is chosen with a low frequency. This, together with a small Schnupp asymmetry (10–20 cm), makes the first sideband poorly transmitted to the antisymmetric port of the beam-splitter and thus almost insensitive to the SRC. The second modulation is instead chosen with higher frequency, in order to be significantly transmitted to the signal recycling cavity and therefore resonant therein. The interferometer reflected beam, together with the PRC pick-off, are demodulated at both frequencies, providing plenty of signals for the control of the interferometer.

6.3.1 The Problem of Lock Acquisition

All the optical signals that should be used for the interferometer control show behavior similar to the Pound-Drever-Hall signal in reflection of a single resonant cavity. This means that they are linearly proportional to a given d.o.f. displacement only in a small region around the working point. The width of this region is determined by the finesse of the corresponding cavity: DARM and CARM are linked with the arm cavities, PRCL and SRCL, with the power and signal recycling finesses.

When the interferometer is not controlled, the mirrors are free to move. Their typical low frequency peak to peak motion can span as much as one wavelength, since the seismic isolation system is designed to attenuate the high frequency residual motion but cannot do much for the very low frequency one. Here comes the problem of *lock acquisition*: how can we bring the system from a complete uncontrolled state to the final one, when all distances are tuned to the correct working point?

It is not possible to simply use the signals we identified before, since the linear range is very small and the probability that all distances are simultaneously close to the correct one is ridiculously small (something like 10^{-8}).

The *lock acquisition procedure* of a gravitational wave interferometer is a complex algorithm, which relies on different approaches depending on the detector. A detailed explanation would be too long for this chapter. The interested reader can refer to [6] for an introduction to the LIGO technique and to [7] for the Virgo one.

To briefly give an example, the Virgo lock acquisition procedure starts with a misaligned power recycling mirror. In this way the beam reflected by it cannot recombine with the main one. In this condition the power recycling cavity does not exist and the two arm cavities are optically separated. They can be *locked* independently using the PDH technique and looking at the peak in transmitted power at resonance as a trigger. Once the two cavities are robustly controlled, the entire interferometer behaves like a simple Michelson one and therefore the MICH degree of freedom can be controlled by looking at the amount of power exiting through the antisymmetric port. The interferometer is not locked at dark fringe yet, but instead at a gray fringe: the interference condition at the beam-splitter is such that half the power is exiting through the antisymmetric port and half is reflected back. At this stage the power recycling mirror is realigned. However, the Michelson interferometer reflectivity is rather small due to the choice of fringe. The field mixing due the presence of the PRC is not very large and the two arms can still be controlled independently. At this point also the PR mirror is controlled using a demodulated signal coming from the interferometer reflection. From this stage on all degrees of freedom are controlled, even if not yet at the final working point, which is adiabatically reached by moving the MICH degree of freedom to the dark fringe condition.

6.3.2 Laser Frequency Stabilization

So far we never considered the laser frequency ν as a variable in our system. Indeed, in all the computations we did in the previous chapter, the dephasing acquired by a field propagating over a distance L was written as

$$\phi = -kL = -\frac{2\pi}{\lambda}L = -2\pi\frac{\nu L}{c} \tag{6.44}$$

This equation shows that a variation of the dephasing can be obtained with a small change of the length but also with a small change in the laser frequency. Indeed

$$\delta\phi = -\frac{2\pi}{c}\left(\delta\nu L + \nu\delta L\right) \tag{6.45}$$

This equation is telling us something quite obvious: we are using our laser beam to measure length variations, therefore if the laser wavelength changes, our measurement also changes. The implication of this equation is that any variation of the laser frequency will be seen by our sensors as (almost [8]) equal to a length variation:

$$\frac{\delta\nu}{\nu} = \frac{\delta L}{L} \tag{6.46}$$

A free running laser like those used in GW detectors have a typical frequency noise

$$\delta\nu(f) = 100\left(\frac{100\ \text{Hz}}{f}\right)\ \text{Hz}/\sqrt{\text{Hz}} \tag{6.47}$$

with a total RMS of many kHz, larger than the arm cavity free spectral range. Since these frequency variation are quite fast, it is a good idea to stabilize them before the interferometer, otherwise it would be impossible to acquire the lock of the arm cavities. This is typically obtained by measuring the frequency variation with a Pound-Drever-Hall technique in reflection of a small rigid reference cavity made of a very low thermal expansion material.

In addition, when the entire interferometer is at resonance, the laser frequency variation is equivalent to a change in the mean arm length (the CARM degree of freedom). Therefore, the laser frequency is further stabilized on the mean length

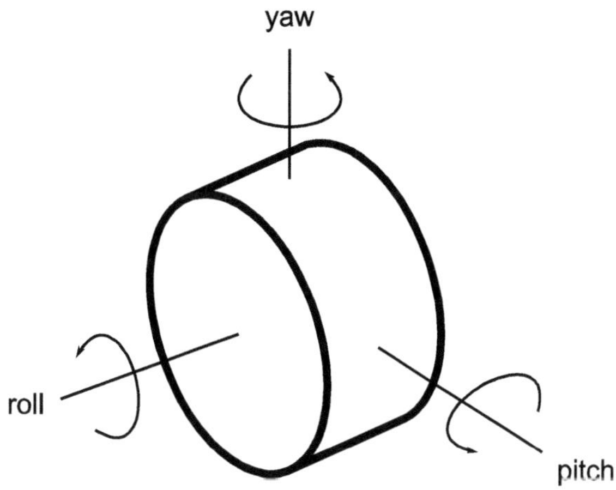

Fig. 6.8 Definition of pitch and yaw rotations

of the arms, obtaining a remarkable residual fluctuation of a few mHz on a second timescale, limited on longer times by the motion of the arm cavity lengths, which are no more mechanically controlled.

6.4 Angular Control

So far we considered only the longitudinal motion of the mirrors. However, they are rigid bodies and therefore there are five more degrees of freedom for each one. Two of them are rotations around axis orthogonal to the beam, often called *pitch* and *yaw*, see Fig. 6.8. The other three are not interesting for our applications: the rotation around the beam axis (roll) is irrelevant, since mirrors have spherical surfaces; the two translations orthogonal to the beam axis are equivalent to a combination of pitch or yaw with a longitudinal displacement, again because the mirrors have spherical surfaces.

Pitch and yaw degrees of freedom are instead very important and must be sensed and controlled as precisely as the longitudinal degree of freedom. There are two possible ways to measure the angular position of the mirrors. The first and simplest one is to use the ground as a reference. It is possible to send a small auxiliary laser to the surface of the mirror with a large angle and read out the reflected beam using a *position sensing device* (PSD): a sensor which returns the position of the laser spot in its surface. Using a proper combination of lenses at the output of this *optical lever* it is possible to reconstruct the angular motions in pitch and yaw of every single mirror. This measurement is however referred to the ground and therefore partially affected by seismic motion. It is moreover *local* in the sense that it measures the position of one single mirror, without any knowledge of the main laser beam position or of the other mirror positions.

This approach is not enough: this becomes clear when considering a Fabry–Perot resonant cavity. The optical axis of the cavity, the one that defines the resonant mode

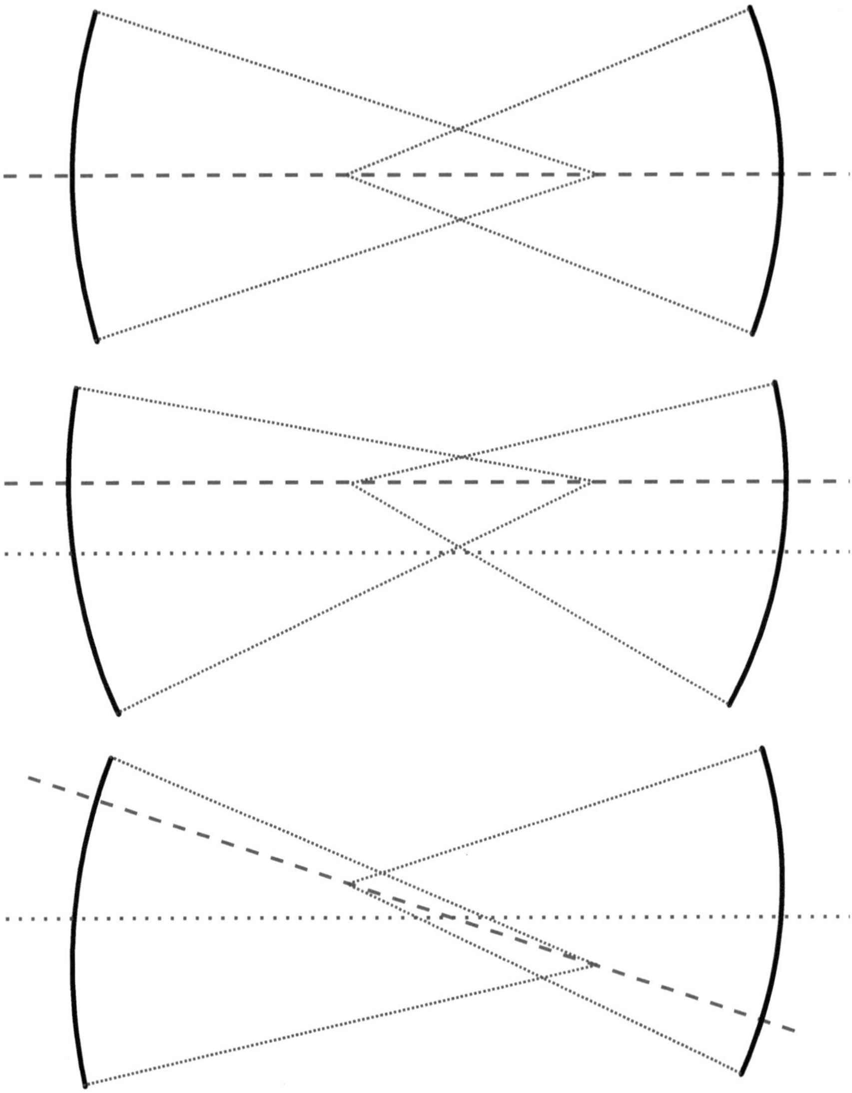

Fig. 6.9 Effect of tilting the mirrors on the cavity axis. The *blue dashed lines* are pairs of radii of the mirror surface. The *red dashed lines* show the position of the cavity axes, compared to the *green dotted lines*, which is the axis in the perfectly aligned case

we considered in the previous chapter, is defined as the only straight line which is perpendicular to both mirror surfaces at the impact points. Figure 6.9 shows the effect of mirror misalignment on the cavity axis, in the case of a stable bi-concave configuration. Simple geometrical considerations show that the cavity axis, which corresponds to the resonant mode axis, is the line passing through the two mirror centers. If we call α_1 and α_2 the tilts of the two mirrors (around their vertexes) with respect to the perfectly aligned case, the position of the two surface centers are written in coordinates with respect to the vertex of the first mirror as

$$C_1 = (R_1 \cos\alpha_1, R_1 \sin\alpha_1) \tag{6.48}$$

$$C_2 = (L - R_2 \cos\alpha_2, R_2 \sin\alpha_2) \tag{6.49}$$

Since we are considering small angular deviations, typically of the order of a few μrad, we can take only the leading order in the angles and write the equation of the line passing through the two points:

$$y = \frac{R_2\alpha_2 - R_1\alpha_1}{L - R_1 - R_2}(x - R_1) + R_1\alpha_1 \tag{6.50}$$

This axis is both tilted and translated with respect to the perfectly aligned one ($\alpha_1 = \alpha_2 = 0$). The tilt angle is already defined by the above equation, while to identify the translation we must decide a reference coordinate along the cavity axis. Since we are considering the resonant mode, the correct position to consider is where the waist of the resonant beam is

$$\alpha = -\alpha_1 \frac{R_1}{L - R_1 - R_2} + \alpha_2 \frac{R_2}{L - R_1 - R_2} \tag{6.51}$$

$$d = \alpha_1 \frac{R_1(R_2 - L)}{R_1 + R_2 - 2L} + \alpha_2 \frac{R_2(R_1 - L)}{R_1 + R_2 - 2L} \tag{6.52}$$

It is important to control the position of the beam axis. First of all, in our interferometers we have (at least) two cavities and if their axes does not coincide the two beams cannot properly recombine at the beam splitter and our precious interference conditions are spoiled. A second reason is that when the beam axis is not properly aligned with the geometrical axis of the interferometer, the spot is no more centered on the mirrors. This introduces a coupling of any residual angular motion of the mirror into longitudinal displacement, proportional to the mis-centering.

All these considerations make it clear that in GW interferometer it is necessary to implement a way to globally measure the position of thc bcam with respect to the mirrors and to maintain it close to the aligned one.

The standard technique foresees the use of photo-diodes with the sensitive surface divided into four quadrants. The sum of the power signal of each of the four quadrants will give the same result as a standard photo-diode. However, by taking the difference between the left and right halves and the top and bottom halves it is possible to derive a signal which gives a measurement of the spot position with respect to the sensor.

One drawback of this approach is that it is impossible to distinguish between motions of the beam and motions of the sensor. A solution for this problem is again based on the frontal modulation technique and it is usually called *wavefront sensing* or *Ward technique*. If we look at the beam reflected by a Fabry–Perot cavity with a quadrant-split photo-diode, we will have as in the longitudinal case two contributions:

the carrier field will be linked to the cavity axis and therefore it will be displaced on the sensor in accordance with it; the sideband field instead will be promptly reflected and therefore insensitive to what happens inside the cavity. If we take the difference between the sensor halves and demodulate it, we will use the sideband as a position reference. Any motion of the sensor will equally displace the sidebands and the carrier without any effect on the demodulated signal.

The angular control of a full dual recycled interferometer is clearly more complex than this. In particular, a simple counting shows that there are 18 degrees of freedom in total: the angular position in pitch and yaw of all seven mirrors and the pointing and translation of the input laser beam. A description of the Virgo angular control system can be found in [9].

Before leaving the topic, we can note that radiation pressure does play a role also in the angular control, even if the working point corresponds to a perfectly aligned system. As we saw, radiation pressure exerts a static force on the mirrors. If the beam is not perfectly centered, due for example to the residual angular motion of the mirrors, this static force will apply a nonzero torque to the mirror, therefore inducing a potential angular motion. It is easy to show that similarly to what happened in the longitudinal case, the effect is to introduce an additional optomechanical coupling that will change the stiffness of the system. Looking again at Fig. 6.9, one can see that for a misalignment like the one in the bottom row (*plus mode*) the torques applied to the mirror tends to move them back to the aligned configuration: the stiffness of this mode is increased as well as the corresponding resonance frequency. In the case of a misalignment like the middle row of Fig. 6.9 (*minus mode*), radiation pressure torques instead tend to move the mirrors further away from the aligned condition: the stiffness of the mode is therefore reduced and the mode can become unstable for high circulating powers. The design of the Advanced Detector arm cavities was carried out in such a way as to avoid this problem even at full power. A detailed analysis of the problem can be found in [10].

6.5 A Brief Introduction to Feedback Control Systems

So far we considered how to measure the deviation of the interferometer degrees of freedom from the desired one, but we did not consider how to actively and continuously act in order to maintain the right resonance conditions. There is nothing new to develop here, since the field of *feedback control systems* has been doing exactly this for a long time. We are mainly interested in the control of linear systems and for simplicity we will consider only one-dimensional systems.

A more detailed description of the basics of linear feedback systems can be found for example in [11, 12].

6.5.1 Linear Systems

The simplest model of a physical system with one single input and one output is a linear transformation:

$$y(t) = \int_{-\infty}^{t} h(t, \tau)\, x(\tau)\, d\tau$$

where the requirement of causality has been already taken into account. In general the integral kernel $h(t, \tau)$ is a function of two variables, since the behavior of the system can vary with time. If this is not the case, the system being time-invariant, the above expression can be simplified considering that it is possible to write $h(t, \tau) = h(t-\tau)$ and therefore

$$y(t) = \int_{-\infty}^{0} h(\tau)\, x(t-\tau)\, d\tau \qquad (6.53)$$

This is the most general expression of a *linear time-invariant system*. The vast majority of the systems involved in interferometric detectors are well described by these kinds of models.

A very convenient way to describe linear transformations, widely used in the theory of control systems for example, is the Laplace transform [13]. Given a signal $x(t)$, null when $t < 0$, its Laplace transform is defined as

$$\tilde{x}(s) = \int_{0}^{\infty} x(t)\, e^{-st}\, dt$$

It can be shown that this function of the variable s can be extended to a complex function with s varying in a defined domain of the complex plane. Applying the Laplace transform to Eq. 6.53 one obtains

$$\tilde{y}(s) = \tilde{h}(s)\, \tilde{x}(s) \qquad (6.54)$$

where the new function $\tilde{h}(s)$ is usually called the *(Laplace) transfer function* of the linear system. The Laplace transform is an extension of the Fourier transform, which can be recovered putting $s = i\omega$, if the imaginary axis is contained in the domain of existence of the transform.

In most cases the transfer function of a linear system can be described using rational functions

$$F(s) = \frac{P(s)}{Q(s)} = \frac{b_m s^m + b_{m-1} s^{m-1} + \cdots + b_1 s + b_0}{a_n s^n + a_{n-1} s^{n-1} + \cdots + a_1 s + a_0}$$

If both numerator and denominator are factorized, the function can be described as a list of poles and zeros. Moreover, if $F(s)$ is the transfer function of a physical system, all complex poles and zeros appear together with their complex conjugates, since in the time domain the integral kernel $h(\tau)$ must be real, being both $x(t)$ and $y(t)$ physical real signals.

A rational function is characterized by its poles and can therefore be written as

$$F(s) = \sum_i \frac{K_i}{(s - p_i)^{n_i}}$$

where n_i are the pole multiplicities and K_i the corresponding residuals. The anti-transform of each of these terms in time-domain is proportional to an exponential $e^{p_i t}$ (see Table 6.1).[3] This result is very useful when considering the response of a system to a step function or to an impulse (usually called *step response* and *impulse response*). In the latter case, being the Laplace transform of an impulse at $t = 0$ a constant, from Eq. 6.54 it follows that the impulse response of the system is given simply by the anti-transform of the transfer function. Therefore, it will contain oscillating terms at frequencies given by the imaginary part of each pole. If one pole has negative real part the oscillation is damped, otherwise it has constant amplitude if the pole is purely imaginary, or it grows up exponentially if the pole has positive real part. Therefore a system is stable, in the sense that its response to an impulse is asymptotically decreasing, if its transfer function has only poles with negative real part.

In developing feedback control system and in describing the properties of mechanical and optical systems, it is customary to describe transfer functions in terms of pole and zero frequencies and quality factors. A real (simple) zero at a frequency f_0 is given by

$$Z_1(s) = s - 2\pi f_0$$

while a real pole is given by the inverse of this expression. A complex (double) zero at frequency f_0 and with quality factor Q is given by

$$Z_2(s) = s^2 + \frac{2\pi f_0}{Q} s + (2\pi f_0)^2$$

The inverse of this formula gives the transfer function of a complex double pole (see Fig. 6.10): this corresponds to a system whose impulse response is a damped oscillation at frequency $f = f_0\sqrt{1 - 1/4Q^2}$, with time constant $\tau = \frac{Q}{\pi f_0}$: the higher the Q, the longer the oscillation will last. An example of such a system is a simple mechanical pendulum with friction. In electronic systems, an RC circuit provides an example of a simple pole, while a resonant RLC circuit of a complex pole.

6.5.2 Simple Feedback System

The simplest situation in a problem of control is a physical object (see Fig. 6.11), usually called *plant*, which can be described as a linear system with one single input

[3] This is true only if the pole multiplicity is 1. In case of multiple coincident poles, there is an additional multiplicative term proportional to a power of the time t.

Table 6.1 List of anti-transform of common second-order rational functions

Laplace transform	Time function (for $t > 0$)
$\frac{1}{s}$	1
$\frac{1}{s-s_0}$	$e^{s_0 t}$
$\frac{\omega_0^2}{s^2+\frac{\omega_0}{Q}s+\omega_0^2}$	$\frac{\omega_0}{\sqrt{1-\frac{1}{4Q^2}}} e^{-\frac{\omega_0 t}{2Q}} \sin\left(\omega_0 t \sqrt{1-\frac{1}{4Q^2}}\right)$

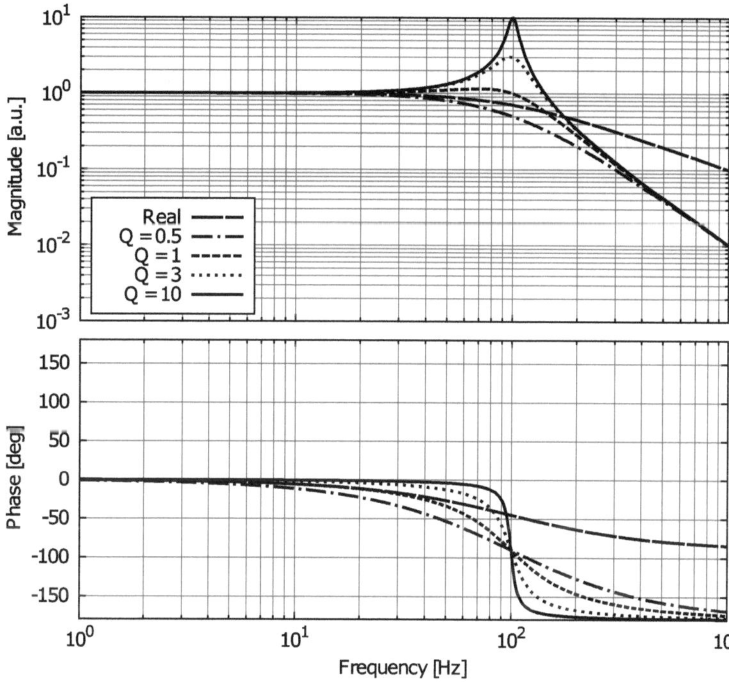

Fig. 6.10 Transfer function of systems described by a simple real pole or one complex pole with different quality factors Q

(for example a Fabry–Perot cavity and its length) and one single output, the *error signal* (a photo-diode demodulated output for example). This system is usually called in engineering books SISO (single-input single-output).

The goal of the control system is to keep the input degree of freedom as close as possible to a given function of time, the *reference*. Being the system linear, this can be taken as the value 0, and therefore controlling the input degree of freedom is equivalent to keeping the error signal close to 0. The response of the plant can in general be described using a linear transfer function $H(s)$ depending on the Laplace variable s. The error signal can be affected by some sensing noise which can be related to the sensor itself (for example shot noise for photo-diodes, ADC noise, etc.) or can be injected on purpose to measure the features of the system.

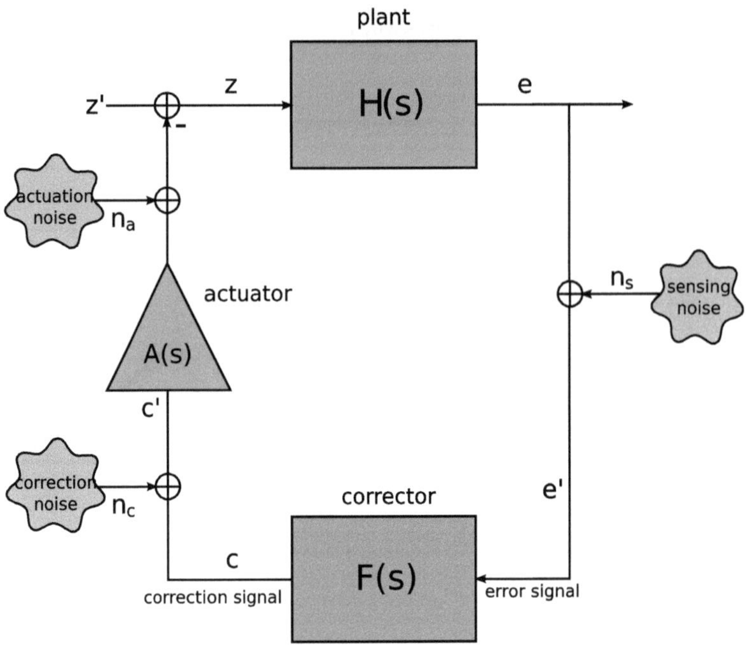

Fig. 6.11 Scheme of a simple feedback loop. See the text for details

The feedback is obtained passing the error signal through an ad hoc *corrector* $F(s)$ and feeding the result (*correction signal*) back to the plant by means of *actuators* whose response is modeled by $A(s)$. For example the length of a Fabry–Perot cavity is controlled by sending currents to coils attached to one of the mirrors. In general the actuators themselves can introduce some additional noise, called *actuation noise*, or as before an external excitation can be added on purpose to the correction signal. There is a general convention that puts a minus sign at the summation point of the correction and the system free motion. In this way the correction signal is in-phase with the motion that the feedback suppresses.

The behavior of the full linear system can be easily computed as

$$e(s) = H(s)\left[z'(s) - n_a(s) - A(s)\left(n_c + F(s)n_s + F(s)e(s)\right)\right]$$
$$z(s) = z'(s) - n_a(s) - A(s)\left(n_c + F(s)n_s + F(s)e(s)\right)$$

where $e(s)$ and $z(s)$ are the error and input signal respectively. These implicit equations can be solved to obtain

$$e(s) = \frac{H(s)}{1+H(s)F(s)A(s)} z'(s) - \frac{H(s)}{1+H(s)F(s)A(s)} (n_a(s) + A(s)n_c(s)) - \frac{H(s)F(s)A(s)}{1+H(s)F(s)A(s)} n_s(s) \tag{6.55}$$

$$z(s) = \frac{1}{1+H(s)F(s)A(s)} z'(s) - \frac{1}{1+H(s)F(s)A(s)} (n_a(s) + A(s)n_c(s)) - \frac{F(s)A(s)}{1+H(s)F(s)A(s)} n_s(s) \tag{6.56}$$

The fact that the output of the system is used also as part of the input justifies the name feedback and also explains why it is customary to refer to feedback control system as *loops* and to distinguish between the *open loop* configuration when the feedback signal is disconnected from the input and the *closed loop* configuration.

For simplicity the actuator transfer function can be considered as a part of the plant transfer function $H(s)$ and the correction and actuation noise can be treated on the same basis.

All signals that are measured inside a feedback loop are called *in-loop* (for example the error and correction signals) while those measured outside or added from outside are called *out-of-loop* (for example the free motion of the system or any external added perturbation). As an effect of the feedback all in-loop signals are reduced, with respect to the free system, by an amount given by the *suppression factor*

$$G_{\text{CLTF}}(s) = \frac{1}{1+H(s)F(s)}$$

also called *closed-loop gain* or *closed-loop transfer function* (CLTF) to differentiate from the *open-loop gain* or *open loop transfer function* (OLTF) given by

$$G_{\text{OLTF}}(s) = H(s)F(s)$$

Usually the value of the open loop transfer function at a given frequency is called the *gain* at that frequency. A change in the overall gain of the transfer function is simply a multiplication by a frequency-independent coefficient. The frequency (usually unique) where the gain is equal to 1 is called *unity gain frequency* UGF or *bandwidth* of the loop.

One widely used method to measure the open-loop transfer function of a system consists in adding a known source of noise $n(s)$ for example to the error point. If this noise dominates all the signals, in Eqs. 6.55 and 6.56 only terms proportional to $n(s)$ are relevant:

$$e(s) = -\frac{H(s)F(s)}{1 + H(s)F(s)} n(s)$$
$$e'(s) = \frac{1}{1 + H(s)F(s)} n(s)$$

where e' is the error signal measured after the noise injection point. Taking the ratio of these two equations, one can see that the OLTF is given by the opposite of the transfer function between the error signal measured after and before the addition of the noise.

Both the error and correction signals contain information about the residual motion of the system. At frequencies where the open-loop gain is higher than 1, the error signal gives a good estimate of the residual motion of the system, apart from a calibration factor given by the plant transfer function.

At the same frequencies on the other hand the correction signal is not a good estimate of the motion. For example consider a resonant Fabry–Perot cavity with its length controlled by applying a force to one of the two mirrors, using the demodulated transmission as error signal. If the input mirror is moving a lot, the correction signal must be large enough to induce a motion of the end mirror such that the cavity length is constant. Therefore, the correction can be quite large even if the error signal and the residual fluctuation of the cavity length are small.

6.5.3 Response to Variations of the Input Signal

One of the most interesting properties of a control system is the ability to track a varying input signal. This performance of a linear system, described by a transfer function $F(s)$, can be studied by analyzing the response to an input step:

$$x(t) = \begin{cases} 0 & \text{if } t < 0 \\ 1 & \text{if } t \geq 0 \end{cases}$$

The Laplace transform of this signal is $\tilde{x}(s) = \frac{1}{s}$ and therefore the system output is given by

$$\tilde{y}(s) = \frac{F(s)}{s}$$

The behavior of the anti-transform of this expression for large times can be obtained, assuming the system to be stable, by using the *final value theorem* [13]:

$$\lim_{t \to +\infty} y(t) = \lim_{s \to 0} s\,\tilde{y}(s)$$

In particular the response of the linear system to the input step for large times is given by

$$\lim_{t \to \infty} y(t) = F(s = 0)$$

Therefore, the output of the system goes to zero only if its transfer function includes a zero at $s = 0$, or in other words, if the global system gain is null at DC.

In terms of feedback systems, the input signal is an external disturbance added to the controlled d.o.f. and the output is for example the residual motion of the system. Therefore, the transfer function of the system is given by

$$F(s) = \frac{1}{1 + G_{\mathrm{OLTF}}(s)}$$

and it has a zero at the origin only if $G_{\mathrm{OLTF}}(s)$ includes a pole for $s = 0$ (a *pure integrator*). For this reason, real poles at zero frequency are usually added to the corrector filter.

The simplest case is that of a system with open loop transfer function given by a simple pure integrator with unity gain at ω_0:

$$G_{\mathrm{OLTF}}(s) = \frac{\omega_0}{s}$$

$$F(s) = \frac{1}{1 + G_{\mathrm{OLTF}}(s)} = \frac{s}{s + \omega_0}$$

Its response to an input unity step is given by the anti-transform of

$$\tilde{y}(s) = \frac{F(s)}{s} = \frac{1}{s + \omega_0}$$

which is exponentially decaying:

$$y(t) = e^{-\omega_0 t}$$

From this equation it is also possible to extract the *time constant* of the system, which is the typical timescale of its response to variations of the input signal

$$\tau = \frac{1}{\omega_0} = \frac{1}{2\pi f_0}$$

where f_0 is the unity gain frequency of the system.

In a similar way the behavior of a system with two real poles at the origin can be studied.

$$
\begin{aligned}
G_{\text{OLTF}}(s) &= \frac{\omega_0^2}{s^2} \\
F(s) &= \frac{1}{1 + G_{\text{OLTF}}(s)} = \frac{s^2}{s^2 + \omega_0^2} \\
\tilde{y}(s) &= \frac{F(s)}{s} = \frac{s}{s^2 + \omega_0^2}
\end{aligned}
$$

and the anti-transform of the last expression gives the time response to an input step:

$$
y(t) = \cos(\omega_0 t)
$$

which is not exponentially decreasing in time. Therefore the response of this system does not converge to zero when the input is perturbed, but rather a persistent oscillation at the unity gain frequency is present. This introduces the subject of stability of a feed-back system.

6.5.4 Stability

From the general point of view, a linear system described by a transfer function $H(s)$ is *proper* if $G(i\infty)$ is finite and *strictly proper* if $G(i\infty) = 0$. This is usually true for all physical system, since for very high frequency the response is always small.

Moreover, a system is *stable* if its transfer function is analytic in the closed right half-plane $s \geq 0$ or in other words if it has no poles in that region. It is *asymptotically stable* if there are also no poles on the imaginary axis.

As explained in Sect. 6.5.1, the anti-transform of the transfer function gives the response of the system to an impulse (usually called *impulse response* of the system). Any pole in the half right-plane will anti-transform to an oscillation with exponentially increasing envelope amplitude and any pole on the imaginary axis to a constant envelope amplitude oscillation. In the first case the response of the system to an impulse is therefore growing exponentially, while in the second case it remains constant with time. This explains the given definitions of stability: a stable system responds to an input impulse with a limited amplitude output, while an asymptotically stable system with an output which goes to zero when time tends to infinite. Since the system is linear, the behavior with any input is completely characterized by the impulse response.

Therefore any control system, in order to be able to maintain the controlled d.o.f. close to the desired value must be stable or better asymptotically stable. Otherwise any disturbance to the input will be amplified and will soon diverge exponentially. The control will fail to work as soon as the motion is larger than the linearity range of the system.

For a feedback system, the requirements of stability and asymptotic stability translate in requirements on the closed loop transfer function: a feedback system is

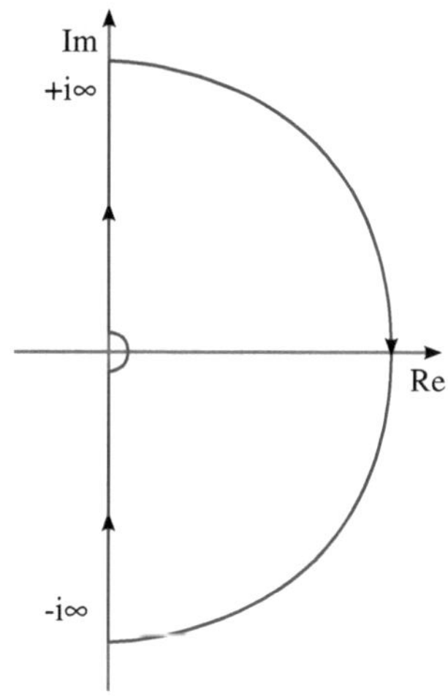

Fig. 6.12 Contour in the complex plane used for the derivation of the Nyquist stability criterion

asymptotically stable if and only if the closed loop transfer function

$$G_{\mathrm{CLTF}}(s) = \frac{1}{1 + F(s)H(s)}$$

has no pole for $\mathrm{Re}\, s \geq 0$.

6.5.4.1 Nyquist Stability Criterion

Usually the study of a system transfer function is carried out in the frequency domain rather than in the Laplace domain, mainly because only the values of the transfer function on the imaginary axis $s = i\omega$ correspond to physical frequencies and that can be directly measured. The link between the behavior of the system transfer function and the requirements of stability can be derived using standard complex analysis methods [12].

The starting point is the well-known *argument principle* [13]: the variation over a closed contour Γ of the phase of a function, analytic in the domain bounded by Γ except for a finite number of poles, is given by the following equation:

$$\frac{1}{2\pi i} \Delta \mathrm{Arg}\, F(s) = \oint_{\Gamma} \frac{F'(s)}{F(s)}\, \mathrm{d}s = n_z(\Gamma) - n_p(\Gamma) \tag{6.57}$$

where the contour is covered in counterclockwise sense, n_p and n_z denote the number of poles and zeros encircled by Γ. The integrand is called the *logarithmic derivative* of the function F.

The *Nyquist criterion* for the stability of a closed-loop feedback system can be derived starting from the open loop transfer function $G_{\mathrm{OLTF}}(s)$ and considering the contour in the Laplace s-plane shown in Fig. 6.12: it consists of the entire imaginary axis, with a small half-circle around the origin if $G_{\mathrm{OLTF}}(s)$ has poles there. The entire contour is closed by a half-circle at infinity and covered in clockwise direction from $-i\infty$ to $+i\infty$. The case of poles on the imaginary axis is not taken into account here. If the system is stable in the open-loop configuration, then the function $1+G_{\mathrm{OLTF}}(s)$ has no pole in the right half-plane. The closed-loop system is stable if and only if this function has also no zeros in the right half-plane. Using the argument principle, this is equivalent to the following integral being zero:

$$0 = \oint_{\Gamma} \frac{G'_{\mathrm{OLTF}}(s)}{1 + G_{\mathrm{OLTF}}(s)}\, ds = \frac{1}{2\pi i} \Delta \mathrm{Arg}\left(1 + G_{\mathrm{OLTF}}(s)\right) \qquad (6.58)$$

The *Nyquist plot* of the open loop transfer function is the image of the Γ contour, which is the curve obtained plotting the complex value of the transfer function while varying the frequency from $-\infty$ to $+\infty$. Since the time-domain response of the system must be real, the transfer function obeys the relation $G(-i\omega) = G^*(i\omega)$ and therefore it is enough to plot the transfer function for the positive physical frequencies and to complete it with its mirror image with respect to the real axis. The Nyquist plot must be covered in the direction corresponding to frequency going from 0 to $+\infty$ and then from $-\infty$ to 0. The right-hand term of Eq. 6.58 therefore is the total variation of the phase of the function $1 + G_{\mathrm{OLTF}}(s)$, which corresponds to the number of times the point -1 is encircled in clockwise sense by the *Nyquist plot* of the open-loop transfer function.

In conclusion the following criterion for the stability of closed-loop feed-back system is established:

Nyquist stability criterion. Given a system with stable open-loop transfer function, it is also stable in the closed-loop configuration if and only if its Nyquist plot does not encircle the *critical point* -1 in clockwise sense.

It is worth to better explain this result by means of some examples. If the system has no pole at zero frequency, the Nyquist plot has no point at infinity and the application of the criterion is straightforward. Figure 6.13 shows the Nyquist plots for the following, very simple, first and second order systems:

$$F_1(s) = \frac{1}{s - \omega_0}$$
$$F_2(s) = \frac{1}{s^2 - \frac{\omega_0}{Q}s - \omega_0^2}$$

Both systems are stable since the -1 point is never encircled. Both transfer functions go to zero when the frequency goes to $\pm\infty$ and are always finite, therefore there is no need to complete the Nyquist plot with the image of a circle at infinite frequency. In both cases if the sign of the transfer function is changed, which corresponds

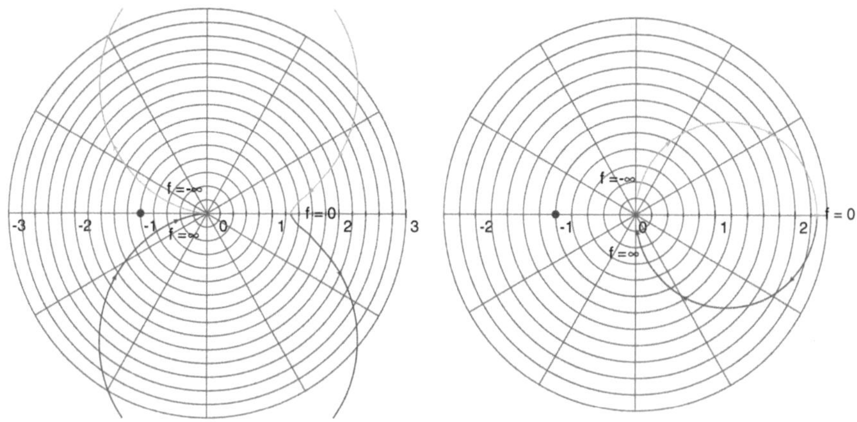

Fig. 6.13 Examples of Nyquist plot for a system of the first-order (*right*) and of the second-order (*left*). The plots in the *left panel* are closing out of the plot boundary

to changing the sign of the control filter, the system becomes unstable, since both Nyquist plots encircle the critical point.

As explained in the previous section, corrector filters often include one or more pure integrators. In this case the application of the Nyquist criterion is made slightly more complex by the presence of the singularity in the origin. The Nyquist contour Γ must include a semicircle around the origin, in the right half-plane. To understand the Nyquist plot at infinity, it is sufficient to consider the asymptotic behavior of the open loop transfer function, which is characterized only by the number of real poles at zero frequency. For a first-order system

$$H(s) \sim \frac{1}{s}$$

The small circle around the origin in the s-plane can be parameterized as

$$s = \varepsilon e^{i\phi}$$

with ϕ varying from $-\frac{\pi}{2}$ to $+\frac{\pi}{2}$. The corresponding section of the Nyquist plot describes a half-circle with very large radius in clockwise sense:

$$H(s) \sim \frac{1}{\varepsilon} e^{-i\phi}$$

starting from the point at $i\frac{1}{\varepsilon}$ and ending at $-i\frac{1}{\varepsilon}$. This circle moves to infinity when ε goes to zero. In any case it never encircles the critical point. If instead the sign of the open loop transfer function is changed, the portion at infinity of the Nyquist plot encircles the critical point and the system becomes unstable. This allows to state as a general rule that for a system including a pure integrator the phase of the open loop transfer function for frequencies going to zero must be $-90°$.

A similar result can be obtained for a double pole at the origin. It must be noted that a system whose open loop transfer function is described by only a double real pole at zero frequency cannot be stable, since whatever is the sign chosen, the Nyquist plot always passes through the critical point. Therefore, some more poles and zeros must be added to the corrector filter to stabilize the system. However, for what concerns the behavior at infinity of the Nyquist plot, the same trick as before can be used. If the open loop transfer function for small frequencies can be approximated by

$$H(s) \sim \frac{1}{s^2}$$

then the part of the Nyquist plot at infinity is a circle starting from a phase close to $-180°$ from above and ending again close to $-180°$ from below. This circle is covered in clockwise direction. As before, the system is stable with the chosen sign, while it is unstable with the opposite one. In other words, in the presence of a double pure integrator, the phase of the open loop transfer function in the limit of zero frequency must be $-180°$.

In general, in the presence of poles in the origin, the Nyquist plot must be completed with a contour at infinity consisting in one half-clockwise circle for each pole, starting from the point at positive infinite frequency and ending in the one at negative infinite frequency.

6.5.5 Bode and Nichols Plots

Together with the Nyquist plot, there are two other common ways to represent the open loop transfer function of a system.

The first one is the *Bode plot*: it simply represents in two different panels the absolute value of the transfer function and its phase as a function of frequency. This plot is useful to get an idea of where the unity gain frequency sits and of the gain of the transfer function at different frequencies. However, it is not straightforward to infer the stability of the closed-loop system from this kind of plot.

The *Nichols plot* of the system is obtained plotting the logarithm of the absolute value against the phase of the open loop transfer function (see Fig. 6.14) when the frequency varies from 0 to $+\infty$. The Nyquist stability criterion can be translated in terms of the Nichols plot. If the transfer function includes pure integrators, the plot must start for large amplitudes from the correct phase: $-90°$ for one pure integrator, $-180°$ for a double integrator and so on. The request that the critical point -1 is not encircled corresponds to the request that the Nichols plot crosses the unity gain axis between the two points corresponding to $-180°$ and $0°$ phases. If the unity gain line is crossed on the left of this segment, the plot must return again to gains greater than 1 remaining on the left of the critical point, otherwise this would correspond to

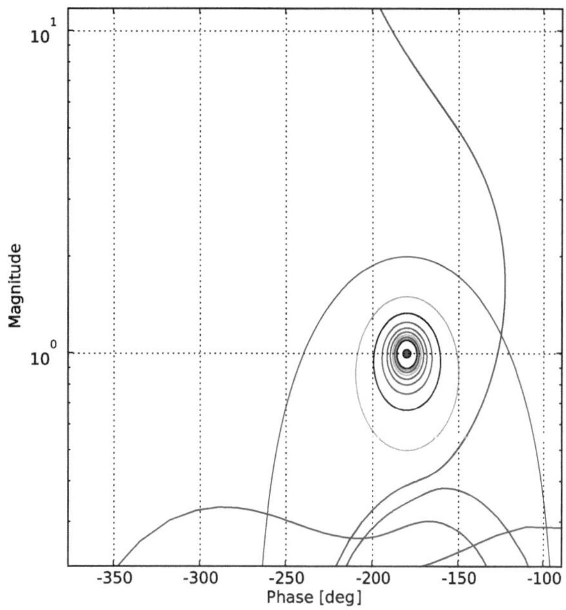

Fig. 6.14 Example of Nichols plot of the open loop transfer function of a stable system. The curves around the critical point show the minimum margin allowed for an overshoot going from 1 to larger values, in steps of one unity

a circle around -1 which would make the system unstable according to the Nyquist criterion.

The Nichols plot is particularly useful to easily obtain information on the behavior of the closed loop transfer function near the unity gain frequency. Indeed, if the open loop transfer function passes close to the critical point, the motion of the system can be enhanced instead of reduced at that frequency, since the closed loop transfer function:

$$G_{\mathrm{CLFT}}(s) = \frac{1}{1 + G_{\mathrm{OLTF}}(s)} \tag{6.59}$$

can become even much larger than 1. A limit on this *overshoot* ε can be imposed:

$$\left| \frac{1}{1 + G_{\mathrm{OLTF}}(s)} \right| \leq 1 + \varepsilon$$

Usually the system transfer function is much greater than 1 at low frequency and the closed loop transfer function is smaller than 1. It also decreases at high frequency and the closed loop transfer function is then very close to unity. Therefore to characterize the overshoot only the behavior around the unity gain frequency is needed. If the open loop transfer function is close to 1, it can be approximated as

$$G_{\mathrm{OLTF}}(s) \approx e^{M} e^{i\pi + i\phi} \approx -1 - M - i\phi$$

The request of Eq. 6.59 translates to

$$M^2 + \phi^2 \geq \frac{1}{(1+\varepsilon)^2}$$

which is the equation of a circle in the Nichols plot, with radius given by $\frac{1}{1+\varepsilon}$. The circles corresponding to overshoots from 1 to 9 are shown in Fig. 6.14.

Moreover, in the Nichols plot a change in the overall gain of the open loop transfer function corresponds to a vertical translation of the curve. If the translation is large enough to make the plot cross the critical point, the system becomes unstable. The distances along the vertical direction of the critical point from the Nichols plot are called *gain margins* since they give the maximum possible changes in the overall gain that leave the system stable. The two points of the Nichols plot that intersect the $-180°$ vertical line give also the typical frequencies of the instability oscillations that arise if the gain of the control loop changes too much.

Finally, the distance along the horizontal direction from the critical point to the Nichols plot gives the *phase margin* which is an indication of the robustness of the control against changes in the system transfer function.

6.5.6 An Example: Locking of a Fabry–Perot Cavity

The simplest application of the control theory to the GW interferometric world is the development of a feedback system to maintain a Fabry–Perot cavity at resonance. We already saw that the Pound-Drever-Hall signal in reflection is a good error signal to measure the cavity length. In the previous chapter we saw that the frequency response of the carrier reflected field is characterized by a simple pole. Since the sidebands are not affected by the cavity, the same pole is present in the response of the PDH signal. Without entering in all the details of the computation, the transfer function of the plant is given by the simple pole model

$$H(\omega_x) = -2J_0J_1 i \frac{(t_I t_E)^2}{1-(r_I r_E)^2} \frac{\mathscr{F}}{\pi} \frac{1}{1 - i\frac{\omega_x}{2\pi f_p}} \tag{6.60}$$

where $f_p = \frac{c}{4L\mathscr{F}}$ is the pole of the Fabry–Perot cavity. This transfer function is a low pass filter with a simple pole. We will consider here the initial Virgo arm cavities for which $f_p \approx 500\,\text{Hz}$.

The last missing point is the frequency response of the actuators: for the locking of a single cavity, acting only on the end mirror, this can be modeled by a simple pendulum with resonant frequency at about 600 mHz and very high quality factor.

The corrector filter designed to control the length of the cavity is a very simple one, see Fig. 6.15, consisting in a double integrator, compensated at 1 Hz with a simple zero and at 10 Hz with a complex zero and again at 800 Hz with a complex pole, to have a unity gain frequency around 50 Hz. In this way below the resonant frequency of the actuators the system is a double integrator, capable of maintaining

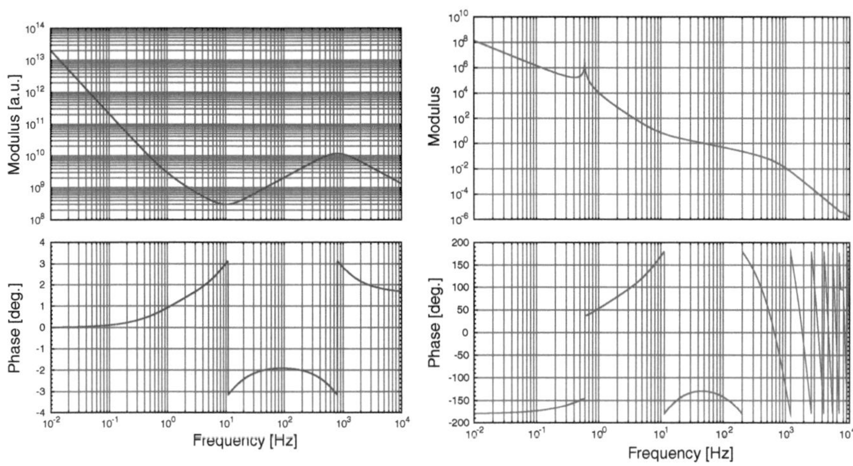

Fig. 6.15 *Left* Bode plot of the corrector filter implemented for the control of a single arm cavity. *Right* open loop transfer function for the control of a single cavity. The unity gain frequency is set at 50 Hz

fixed the length of the cavity even in presence of big motions of the input mirror. The pole and zero pair is a common method used to stabilize the system.

The Nichols plot of this system is shown in Fig. 6.16. The loop margins can be easily read from this plot: the overall gain can be increased or decreased by a factor of 5. The phase margin is 50°. This is a really robust feedback system, capable of maintaining the control of the cavity even with large changes in the optical transfer function. This robustness is indeed needed in the first moments of the lock acquisition, when the system response is highly nonlinear.

6.6 Actuation: How to Apply Forces to the Mirrors

So far we considered the problem of measuring the displacement of the interferometer degrees of freedom from the desired position and implementing a feedback control. The last missing point is how to apply the computed correction to the mirrors. It is obvious that we cannot touch in any way the mirror to apply the force, therefore we must use some kind of electromagnetic actuator. Two different strategies have been adopted in first-generation detectors: LIGO and Virgo used coil-magnet pairs, while GEO600 used electrostatic actuators. It is outside the scope of this chapter to describe the difference and the advantages of one solution with respect to the other. We will take into consideration only the coil-magnet case as an example.

The passive element (the magnet) is clearly attached to the mirror (Fig. 6.17). The active element (the coil) cannot be attached to the ground, otherwise it would rein-

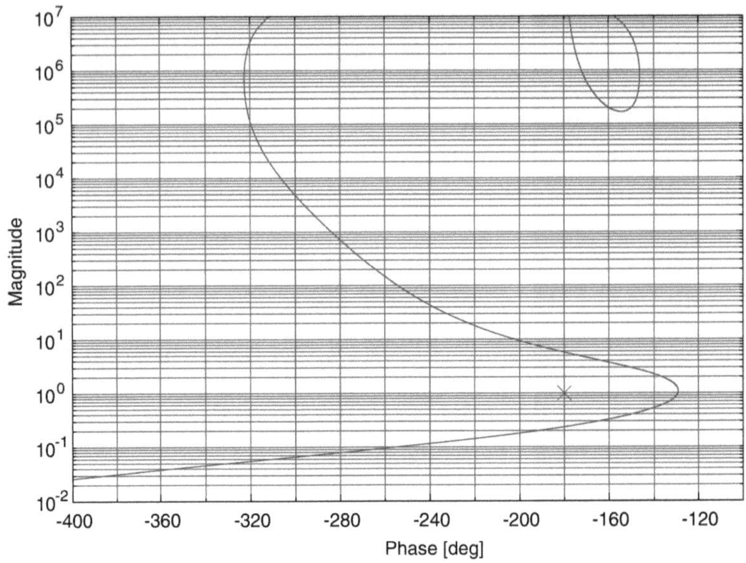

Fig. 6.16 Nichols plot of the control system for a single arm cavity with a finesse of 50

troduce seismic noise due to its relative motion with respect to the mirror. Therefore the coils must be attached to a *recoil mass* that must be also seismic isolated.

Moreover, since we need to control also the angular position of the mirror, one magnet is not enough. In Virgo and Advanced Virgo four magnets are used, placed in a plus configuration on the back surface of the mirror. By applying different corrections to each magnet it is possible to actuate in the longitudinal, pitch, and yaw directions.

6.6.1 Hierarchical Control

The maximum actuation force is determined by the needs of the lock acquisition procedure and by the maximum motion of the suspended mirror that must be compensated during the steady-state configuration. Both requirements typically bring to actuators with a relatively large peak force. For example, the largest contribution to the correction that we must apply to the cavity length comes from the strain due to luni-solar tidal forces. The typical measured peak to peak amplitude of the length variation over the 3 km baseline can be as large as 150 μm over 1 day. Considering a mirror of 40 kg attached to a pendulum system with a length of about 0.7 m, to compensate for this we would need a force of the order of 80 mN.

One key point on the actuation is the noise level of the correction. In the ideal world the force applied to the mirror is exactly equal to the correction computed by

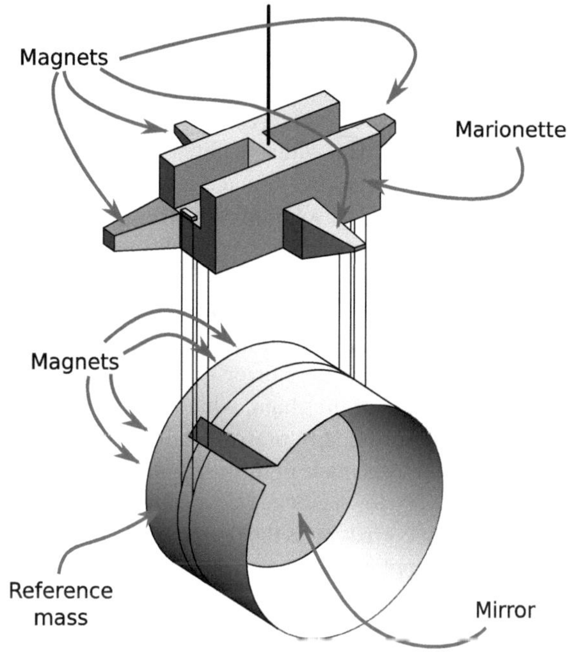

Fig. 6.17 Conceptual drawing of the mirror with the reference mass and the marionette as it was in Virgo. The concept is similar for Advanced Virgo, even if mechanical details will be largely different

the feedback system. In the real world, the force will contain several noise contributions coming from the electronic systems: arithmetic noise, digital-to-analog conversion noise, power amplifier noise, etc. All these translate to displacement noise directly applied to the mirror, which will therefore perfectly mimic a GW signal. All the amplitudes of these noise sources must remain well below the target science sensitivity.

Let us consider the implications of this requirement, assuming the maximum force needed is of the order of 80 mN as discussed above. If we consider a frequency-independent noise $\tilde{F}$ at the level of the force applied to the mirror, this will correspond to a displacement which is given for frequencies above the resonance of the last stage of the suspension:

$$\tilde{z}(f) = \frac{\tilde{F}}{4\pi^2 f^2 m} \tag{6.61}$$

This must be compared with the science sensitivity of, for example, the Advanced Virgo detector in the low frequency band. The most critical region is the one around 50 Hz, where the target in terms of strain is about 6×10^{-24} Hz$^{-1/2}$ which corresponds to 2×10^{-20} m Hz$^{-1/2}$. A flat force noise generates this displacement with an amplitude of about 7×10^{-13} N Hz$^{-1/2}$. We should require the force noise to be 10 times lower than this to avoid any contribution to the detector sensitivity: $\tilde{n}_F < 7 \times 10^{-14}$ N Hz$^{-1/2}$. This number must be compared with the maximum force needed:

$$\tilde{n}_F < 9 \times 10^{-12}\,\mathrm{Hz}^{-1/2}\,F_{\max} \tag{6.62}$$

These 11 orders of magnitude (220 db) are a huge amount. If we consider to generate the force with a electronic system with maximum output of 10 V, the noise level should be $10\,\mathrm{pV}/\sqrt{\mathrm{Hz}}$, which is unfeasible.

This conclusion seems to make noise-less actuation on an Advanced Detector a hopeless problem. The solution is the use of *hierarchical control*. The mirrors and its reference mass are suspended to complex isolation systems, that can be schematized, at least in the horizontal degrees of freedom, as a chain of simple pendulums with typical resonant frequencies below a few Hz. A simplified scheme is reported in Fig. 6.18. Three actuation points are used: at the level of the mirror, at the level of the mass from which the mirror is suspended (called in Virgo *marionette*), and at the topmost mass of the chain, which is clearly not rigidly connected to the ground but somehow seismically pre-isolated (in Virgo it is sitting on top of the *inverted pendulum*, which is made of three legs attached to the ground with flexible joints and providing a very low resonant frequency of a few tens of mHz).

The global correction is the output of our control system and gives the total force that must be applied to the mirror. Instead of sending everything to the actuators at the mirror level, we split the correction into two: the high frequency part (above a cut-off f_c here set to 8 Hz) is sent to the mirror actuators, while the low frequency part is sent to the intermediate mass (the marionette in Virgo).

The typical correction that must be applied to mirror is mostly concentrated at very low frequencies, below 1–2 Hz. The reason is linked to the performances of the suspension system: since all the pendulums have resonance frequencies around 500–1000 mHz, the filtering of seismic noise starts to be effective only above these values. This property of the correction helps us understand the first gain we get when using the described split strategy: the largest part of the correction is sent to the intermediate mass and only the residual high frequency part is still sent to the mirror. This strategy allows a reduction of many orders of magnitude of the correction that is sent to the mirror. In this way we can relax the requirement on the driving noise to a level which is still difficult, but feasible. However, it seems that the problem would appear again at the level of the intermediate mass, where now the correction is still quite high. We should consider that any force noise sent to the intermediate mass actuator must be transmitted by an additional pendulum before reaching the mirror. We gain therefore an additional $(f/f_0)^{-2}$ filtering above the resonant frequency which again relaxes enormously the driving noise requirement.

The final force reallocation is performed at very low frequency, using the global correction signal in place of displacement sensors used to control the top-most stage of the suspension. At very low frequency, where the gain of the locking feedback loop is very large, the correction is a direct measurement of the displacement of the mirror suspension point which, at frequencies below all the chain resonances, corresponds to the motion of the top stage. We can therefore substitute the displacement sensor (that is normally used to measure and control the motion of the top stage) with the global correction signal. This strategy has two advantages. The first one is to reduce the correction that must be sent to the intermediate mass, moving the very low frequency

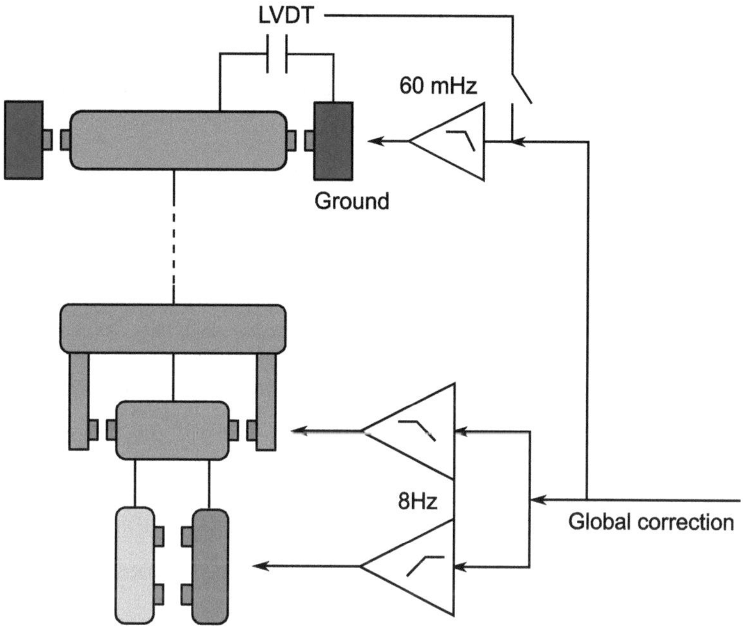

Fig. 6.18 Scheme of the hierarchical control in a multi-stage suspension system

($<$100 mHz) component to the top stage. Any actuation noise at the top level will not reach the mirror, due to the large number of seismic filters between them. The second advantage is that the top stage is no more controlled with signals measured with respect to the ground, but with global ones, which are measuring the relative displacement of mirrors. In other words we are only considering the relevant motions of the suspension system that affects the cavity length, neglecting all the others.

6.7 Exercises

6.7.1 Dither Control

The Pound-Drever-Hall technique is not the only one that can be used to generate an error signal to control a resonant cavity. Another possibility, which was used for example to control the Virgo and Virgo+ output mode cleaner cavity, is to modulate the length of the cavity itself, at a frequency which is still inside the cavity line-width.

1. Compute the response to a cavity length motion of the error signal obtained by demodulating the reflection of the cavity.
2. Compute the frequency response of the same signal.

3. We can use thermal actuation to change the cavity length. The response of such an actuator is a simple low-pass filter with a typically corner frequency of few tens of mHz, depending on the cavity material and mass. Develop a stable controller for the cavity length using this kind of actuator.

References

1. R.W.P. Drever, J.L. Hall, F.V. Kowalski, J. Hough, G.M. Ford, A.J. Munley, H. Ward, Laser phase and frequency stabilization using an optical resonator. Appl. Phys. B Lasers Opt. **31**(2), 97–105 (1983). doi:10.1007/BF00702605
2. E.D. Black, An introduction to Pound Drever Hall laser frequency stabilization. Am. J. Phys. **69**, 79–87 (2001)
3. R. Goldstein, *Electro-Optic Devices in Review* (Lasermetrics, Englewood, 1986)
4. Using the Jacobi-Anger expansion http://mathworld.wolfram.com/Jacobi-AngerExpansion.html
5. George B. Arfken, Hans J. Weber, *Mathematical Methods for Physicists*, 6th edn. (Harcourt, San Diego, 2005)
6. M. Evans et al., Lock acquisition of a gravitational-wave interferometer. Opt. Lett. **27**, 598 (2002)
7. F. Acernese et al., [Virgo collaboration], Lock acquisition of the Virgo gravitational wave detector. Astrop. Phys. **30**, 29–38 (2008)
8. M. Rakhmanov, R.L. Savage Jr, D.H. Reitze, D.B. Tanner, Dynamic resonance of light in Fabry Perot cavities. Phys. Lett. A **305**, 239–244 (2002)
9. F. Acernese et al., [Virgo collaboration], Automatic alignment for the first science run of the Virgo interferometer. Astropart. Phys. **33**, 131–139 (2010)
10. J.A. Sidles, D. Sigg, Optical torques in suspended Fabry Perot interferometers. Phys. Lett. A **354**, 167–172 (2006)
11. J. Doyle, B. Francis, A. Tannenbaum, *Feedback Control Theory* (Macmillan Publishing Co., New York, 1990)
12. E. Kappos, *Classical Control Theory: A Course in the Linear Mathematics of Systems and Control* (University of Sheffield, School of Mathematics and Statistics, Sheffield, 2002)
13. R. Courant, D. Hilbert, *Methods of Mathematical Physics* (Wiley-Interscience, New York, 1989)

Chapter 7
An Introduction to the Virgo Suspension System

Franco Frasconi and Piero Rapagnani

Abstract Suspending the mirrors is one of the most crucial tasks in gravitational wave interferometer technology. The performance of the suspensions must provide the required attenuation of seismic noise and reduction of thermal noise, two fundamental limits to the sensitivity of any gravitational wave detector. Moreover, the suspension system must be equipped with sensors and actuators which are used to actively control some relevant degrees of freedom, so to be able to keep the interferometer at its working point (i.e., "locked"). In the first part of this chapter, we deal with the basic principles behind the Superattenuator chains developed in Virgo to reduce the seismic noise. In the second part, we illustrate the techniques to reduce the thermal noise in the detection bandwidth, according to the theory illustrated in Chap. 8.

7.1 Introduction

The gravitational waves detection with a ground-based interferometer depends on the capability to include into the experimental apparatus free-falling masses, the test masses, well isolated from various noise sources, the influence of which has great relevance on the design sensitivity all over the detection bandwidth. Seismic noise gives the most important contribution at low frequency where gravitational waves emitted by pulsars and coalescing binaries are expected. For these reasons, the first generation of ground based and broadband interferometers have been equipped with an appropriate suspension system of the test masses, but also the upcoming

F. Frasconi
INFN, Pisa, Italy
e-mail: franco.frasconi@pi.infn.it

P. Rapagnani (✉)
Dipartimento di Fisica, Università La Sapienza, INFN, Roma, Italy
e-mail: piero.rapagnani@roma1.infn.it

M. Bassan (ed.), *Advanced Interferometers and the Search for Gravitational Waves*, Astrophysics and Space Science Library 404, DOI: 10.1007/978-3-319-03792-9_7,

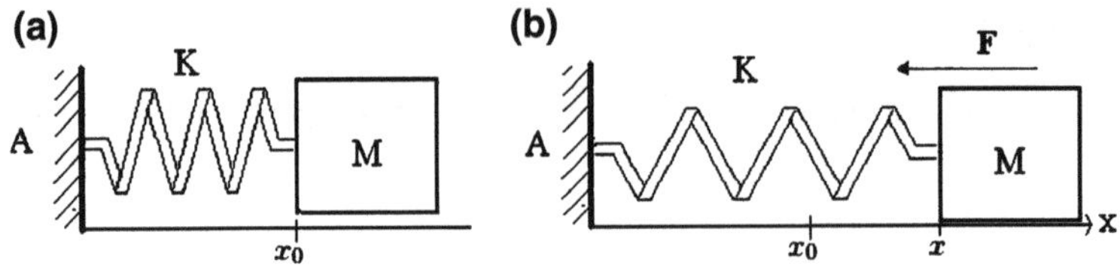

Fig. 7.1 One-dimensional harmonic oscillator at rest (**a**) and when an exciting force is applied (**b**)

second and future generations detectors will include complex mechanical structures studied to inhibit noise transmission to the suspended optics. To this purpose, the VIRGO Collaboration has developed and built a very sophisticated system to isolate from seismic noise and local disturbances the optical components of a Michelson interferometer having, along the arms, two Fabry-Perot cavities, each 3 km long. With this solution, it has been possible to extend the detection bandwidth in the low frequency region starting from a few Hertz.

7.2 The Harmonic Oscillator as a Mechanical Filter

Let us consider a one-dimensional harmonic oscillator formed by a mass m connected to a massless spring of elastic constant k, as represented in Fig. 7.1a. Applying a force in the right direction, the system will react and the equation of motion representing the status of the excited system (see Fig. 7.1b) can be written:

$$m\ddot{x} = -k(x - x_o) , \tag{7.1}$$

where x_o is the coordinate of the mass at rest, while x is the mass position when the force is applied with respect to the chosen reference frame. This is a second order differential equation which admits a solution $x(t) = xe^{i\omega t}$ in the frequency domain. Substituting it in Eq. 7.1, it turns out:

$$-m\omega^2 x = -k(x - x_o) , \tag{7.2}$$

and then defining ${\omega_o}^2 = k/m$ Eq. 7.2 becomes:

$$({\omega_o}^2 - \omega^2)x = {\omega_o}^2 x_o , \tag{7.3}$$

By introducing the concept of mechanical *Transfer Function* of the harmonic oscillator as the absolute value of the ratio between output and input signal of the system:

$$\mathrm{TF} = |\tilde{x}_{\mathrm{out}}/\tilde{x}_{\mathrm{in}}| , \tag{7.4}$$

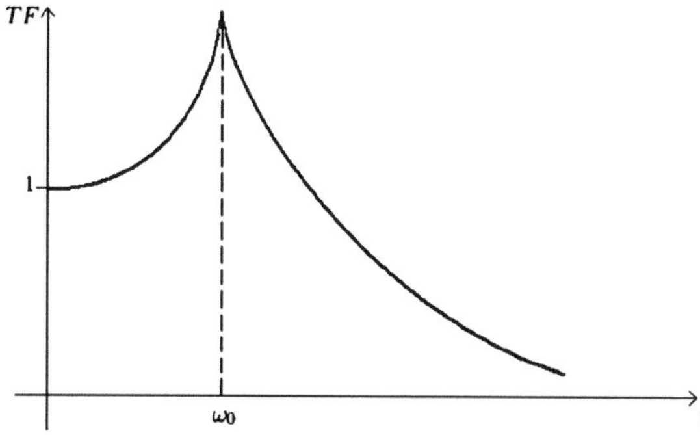

Fig. 7.2 The transfer function of a one dimensional harmonic oscillator

where $\tilde{x}$ represents the Fourier transform (frequency domain) of x and applying it to Eq. 7.3, it turns out:

$$\text{TF} = |\frac{\tilde{x}_{\text{out}}}{\tilde{x}_{\text{in}}}| = |\frac{{\omega_o}^2}{({\omega_o}^2 - \omega^2)}| \tag{7.5}$$

Analyzing the result obtained, the behavior of the Transfer Function can be reconstructed as follows, in accordance with the plot shown in Fig. 7.2:

$$\text{TF} = \begin{cases} 1/\omega^2 & \omega >> \omega_o \\ \infty & \omega = \omega_o \\ 1 & \omega << \omega_o \end{cases} \tag{7.6}$$

It can be argued that for frequencies ω higher than the harmonic oscillator fundamental mode $\omega_o (\omega >> \omega_o)$ the system behaves as a second order low pass mechanical filter.

It is evident that the harmonic oscillator represents a very good base element for the design of a suspension system for test masses of ground based interferometer developed for gravitational physics. With this solution it is possible to filter the transmission of seismic noise and local disturbances in all degrees of freedom.

The attenuation response of a mechanical structure based on the working principle of the harmonic oscillator can be enhanced by cascading a certain number of this base element. The comparison between the Transfer Function of a single pendulum with that one of a double pendulum having identical overall length is reported, as an example, in Fig. 7.3, where the improvement in the filtering performance at frequencies higher than the pendulum chain normal modes is showed.

Extending this approach to a pendulum chain with n-stages, it is evident that the horizontal motion of its suspension point at frequency ω higher than the frequencies of the chain normal modes ($\omega > \omega_0 > \omega_1 > \cdots > \omega_n$), is transmitted to the suspended mass with an attenuation proportional to ω^{-2n}. In particular, the ratio between the linear spectral density of the last mass horizontal displacement (the optical component) and the linear spectral density of the suspension point horizontal

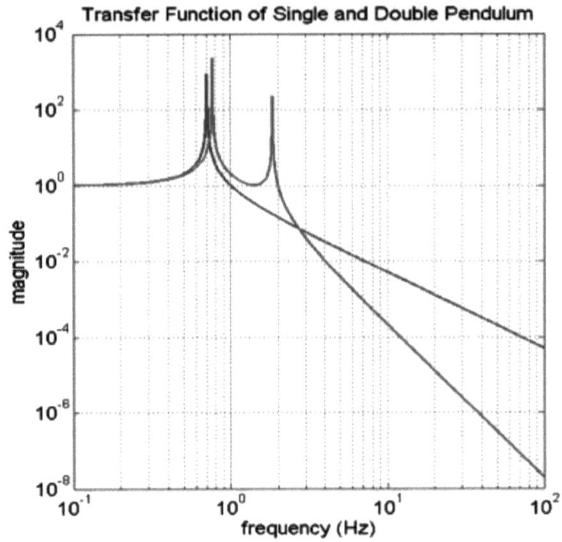

Fig. 7.3 Comparison of the mechanical transfer function between a single pendulum (*blue line*) with normal mode at 0.7 Hz and a double pendulum (*red line*) of the same overall length with normal mode at 0.8 Hz and 2 Hz respectively. The improvement of the attenuation performance is evident in the *red curve* (slope change) for frequency higher than 2 Hz (characteristic cut-off frequency of a double-stage pendulum chain)

displacement (where the excitation is applied) decreases as A/ω^{2n} where $A = \omega_0^2 \cdot \omega_1^2 \cdot \omega_2^2 \cdots \omega_n^2$. With this system a very large attenuation of the seismic noise horizontal component is reachable at frequencies above the highest pendulum mode. Therefore, the complete Transfer Function of this multistage pendulum for a frequency ω much higher than the normal modes of the chain ($\omega \gg \omega_0 \gg \omega_1 \gg \cdots \gg \omega_n$), is given by the product of the transfer functions of each harmonic oscillator forming the suspension chain:

$$TF = |\frac{\tilde{x}_{\text{out}}}{\tilde{x}_{\text{in}}}| = \frac{1}{\omega^{2n}} \prod_{i=1}^{n} \omega_1^2 \tag{7.7}$$

where $\tilde{x}_{\text{out}}$ is the linear spectral density of the test mass horizontal displacement while $\tilde{x}_{\text{in}}$ is that one of the suspension point and n is the number of stages.

7.3 The Real Suspension Systems in Gravitational Wave Interferometers

The first generation detectors [1] developed for the study of gravitational field have been equipped with a mechanical structure to isolate from seismic noise the free-falling masses. The solution adopted for the VIRGO interferometer is based on the idea to replicate a certain number of harmonic oscillators forming a filtering chain to enhance the natural capability of a single stage to attenuate the noise transmission at the test mass level. Seismic motion, always present on the ground, even in the absence of earthquakes, is a very powerful noise source for gravitational wave detectors. The typical linear spectral density of seismic displacement in all three space coordinates can be approximated, in the frequency range of VIRGO, by:

$$\tilde{x} \simeq \frac{\alpha}{f^2}, \tag{7.8}$$

measured, in $m \cdot H^{-1/2}$, as a function of the frequency f and of the parameter α, which ranges typically between 10^{-6} and 10^{-9} $\mathrm{m} \cdot \mathrm{Hz}^{3/2}$, depending on the geographical location. From the equation above, it turns out that the seismic motion reaching the test masses (the mirrors) has to be reduced by at least a factor 10^{12}, in order to have an acceptable sensitivity in the 10 Hz region. For these reasons, a high-performing device should be developed to suspend the free-falling masses, strongly reducing seismic induced motion.

The example of the previous paragraph has been focused on the attenuation performance of a generic harmonic oscillator in the horizontal degrees of freedom. A similar logic path should be followed for the noise transmission along the vertical direction. This is of great importance in the development of any mechanical system because the final goal is the construction of a supporting structure with comparable capability of seismic noise isolation in all degrees of freedom. With this intent a common detection threshold will be defined and a cut-off frequency, above which the system behaves as a chain of second order low pass filters, will represent the general characteristic of the chain. Adopting a multistage pendulum this goal can be achieved by replacing each suspension wire with a spring to form a cascade of vertical oscillators. The spring should support heavy load and, at the same time, it should be soft enough to exhibit a low resonant frequency. Paying particular attention during the design phase, it is possible to confine the vertical modes of the chain in a low frequency region obtaining a strong attenuation starting from a few Hz. While the pendulum mode (horizontal degrees of freedom) is a simple inverse function of the wire length l in accordance with the equation:

$$f = \frac{1}{2\pi}\sqrt{\frac{g}{l}}, \tag{7.9}$$

the construction of the suspension chain presents some difficulties in the design of the vertical harmonic oscillator which should have the same attenuation performance and similar cut-off frequency as the horizontal one. For the VIRGO interferometer [2], where the two end mirrors are suspended 3 km away, the situation is more complex because of the different directions of the vertical line on the curved Earth surface. This means that they are misaligned with respect to one another by about 3×10^{-4} rad. In addition any vertical vibration will be partially transmitted to the interferometer horizontal axis (laser beam direction) because of an unavoidable coupling among different degrees of freedom. With this intent the experimental apparatus will have a cut-off frequency above which the system behaves as a chain of second order low pass filters.

The horizontal displacements of the mirror along the laser beam direction can also be induced by rotations of the pendulum chain around the vertical axis (mechanical coupling between rotations and horizontal direction). To confine these rotational mode frequencies below the detection band, each pendulum mass has to be replaced by a structure having a high momentum of inertia. In addition, the diameter of the suspension wire connecting two consecutive stages, must be small enough to reduce its restoring torque (their diameter should be of the order of a few millimeters) which opposes the rotation of the chain and determines its rotational frequency. An interconnection of the stages at small distance and as close as possible to their centers of mass, guarantees low frequency tilt mode (i.e., rotational modes around the horizontal axes) and negligible coupling among different degrees of freedom, which can induce horizontal displacements of the suspended mass.

7.3.1 The Virgo Superattenuator

The Superattenuator (SA) is the mechanical structure developed by the INFN Pisa Group to isolate the test mass of the VIRGO interferometer [2] from seismic noise and local disturbances [3]. The system is a hybrid one formed by a chain of passive mechanical filters linked to each other by a thin metallic wire, a three leg structure based on the working principle of an Inverted Pendulum (see next paragraph) and a payload as last stage formed by a marionette, a recoil mass and the mirror (see Fig. 7.4). The design has been developed considering three important guide lines:

- The confinement of the multistage pendulum normal modes in the low frequency region;
- The introduction of a pre-isolation stage to be used as mechanical structure supporting the suspension point of the chain with the possibility to steer it by applying small forces;
- The introduction a soft suspension stage on top of the chain allowing Active Mode Damping (AMD) and seismic noise reduction by means of inertial sensors, position sensors and electromagnetic actuators. In addition, any noise introduced by acting on the top stage must be filtered out by the Superattenuator chain.

7.3.2 The Inverted Pendulum

After a preliminary version of the Superattenuator design, a three-leg mechanical structure based on the working principle of an Inverted Pendulum (IP) was added to the final design [4]. As shown in Fig. 7.4, each leg, an Aluminum cylinder about 7 m high, is connected to ground through a flexible joint while the tops of the three legs are interconnected to a mechanical ring (Top Ring) surrounding the first filter

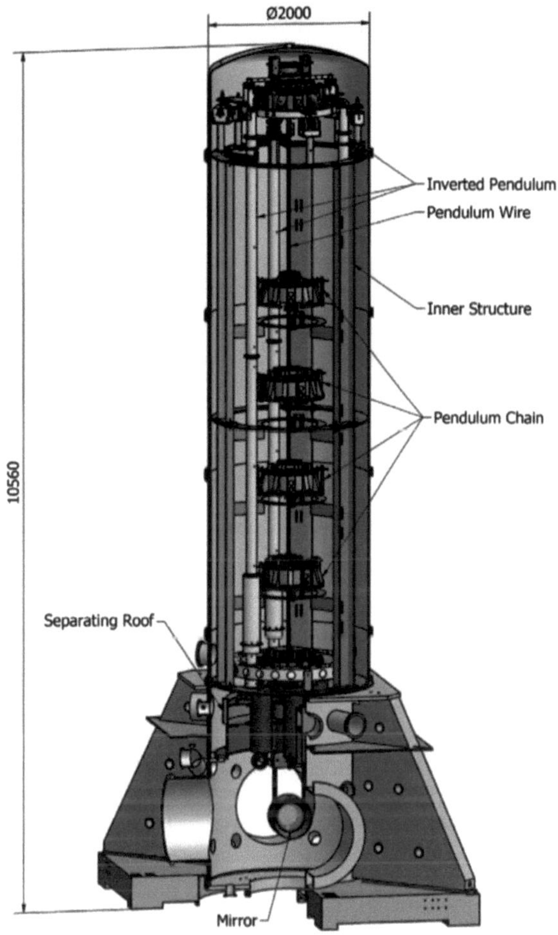

Fig. 7.4 A long Superattenuator of the VIRGO interferometer. All the elements of the mechanical structure are represented here within the vacuum tank divided in two sections: the *upper part*, where the suspension chain is accommodated, is maintained at a vacuum level of 10^{-7} mbar while the *bottom part* hosting the free-falling mass is at 10^{-9} mbar

of the chain (Filter 0). The flexible joints act on two horizontal degrees of freedom (displacements along x and z direction) as well as in the rotation of the structure around the vertical axis (θ_y d.o.f.). The Filter 0 is bolted on the Top Ring, and its action is limited to the vertical degree of freedom only.

An ideal inverted pendulum can be conceived as a mass-less vertical bar of length L, connected to ground by means of an elastic joint with stiffness k and supporting a mass m on its top (see Fig. 7.5). For a similar structure, the equation of the system can be written:

$$I\ddot{\theta} = -k\theta + mgL\sin\theta| \qquad (7.10)$$

where I is the cylinder moment of inertia, the first term of the equation right-hand side is the elastic torque due to the mechanical joint and the second one is the gravity torque.

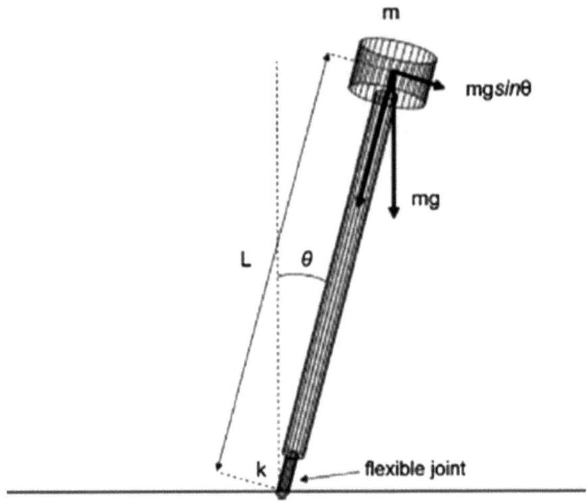

Fig. 7.5 A schematic view of the inverted pendulum working principle

Considering the system in the small oscillation regime, Eq. 7.10 can be written:

$$\ddot{\theta} + (\frac{k}{mL^2} - \frac{g}{L})\theta = 0 \tag{7.11}$$

having assumed $I = mL^2$ as the generic moment of inertia for a massless bar. The Inverted Pendulum resonant frequency will be:

$$f_{\mathrm{IP}} = \frac{1}{2\pi}\sqrt{\frac{k}{mL^2} - \frac{g}{L}} \tag{7.12}$$

From Eq. 7.12 it is evident that, with a fixed bar length, the resonant frequency depends on the mass m connected on its top. The frequency value can be lowered by increasing the mass m until the mechanical instability is reached while gravity acts as an anti-spring; such mechanical structure can operate at very low frequency (in Virgo around 50 mHz).

Attaching the suspension point of the multistage pendulum chain on the top of an inverted pendulum, we achieve an ultralow frequency pre-isolator stage that suppresses a large part of the input horizontal seismic noise. Moreover, due to its very low overall stiffness, the inverted pendulum can be moved applying very small forces.

7.3.3 The Top Ring and the Filter 0

As mentioned above, the upper parts of the three cylindrical legs of the Inverted Pendulum mechanical structure are connected by means of three thin steel wires (each 31 mm long) to a metallic ring, the so-called Top Ring. This last element

Fig. 7.6 The inverted pendulum Top Stage. Note the crowded Top Stage of a Superattenuator where the mechanical elements together with sensors and actuators are positioned

surrounding the Filter 0 forms the top stage of the Superattenuator (see. Fig. 7.6). Thanks to the triangular cantilever blades and the magnetic anti-springs installed on the Filter 0, which is bolted to the Top Ring (see next paragraph for the mechanical filter description), it has been possible to complete a seismic pre-attenuation stage in the three translational degrees of freedom. Seismic pre-isolation stage along the other two horizontal degrees of freedom has been fulfilled by using the flexible joints within the Inverted Pendulum structure [4].

The Superattenuator top stage represents a platform on top of which a set of sensors and actuators are installed for feedback control purpose [5]. In particular, three Linear Variable Differential Transformer (LVDT) displacement sensors, three horizontal accelerometers, two vertical accelerometers together with three coil-magnets actuators and three stepping motors are the devices used to perform a very low frequency control of the chain suspension point as well as the Superattenutator resonances Active Mode Damping.

Blending the LVDT signals, ground connected through the Inner Structure (see Fig. 7.4) and as a consequence non seismic noise free, with the accelerometer signals measuring the acceleration with respect to the fixed stars (inertial reference frame), a very accurate tidal strain and drift control can be performed up to about 200 mHz. This feedback, based on the acceleration measurements, is often called

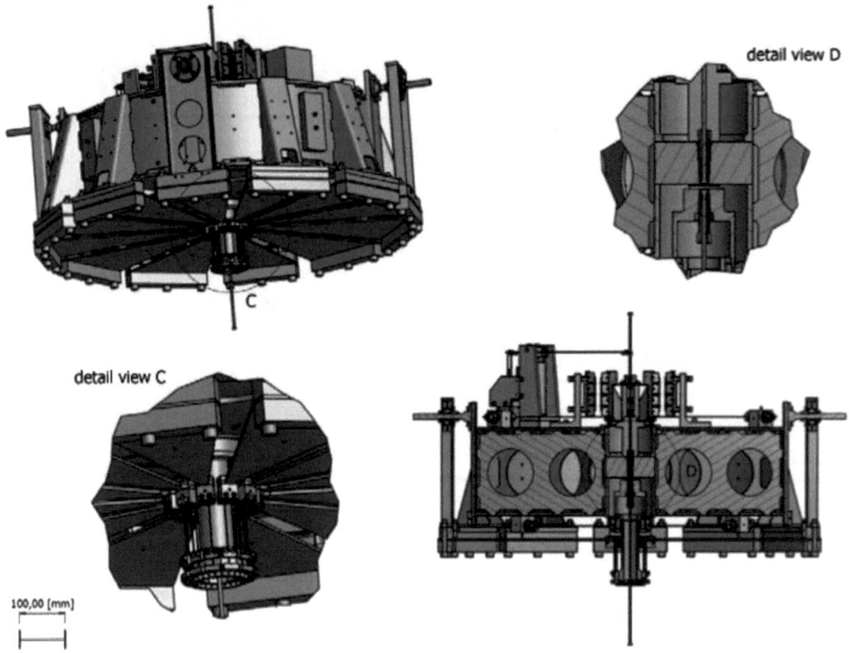

Fig. 7.7 Technical drawings of the mechanical filters for the VIRGO superattenuator. A complete mechanical filter is visible (*upper left*) where 12 cantilever triangular blades are clamped on the outer circumference of the filter body *bottom part*. A detail of the blade tip connection through a thin steel piano wire to the mobile central column, is shown in the *left bottom corner* (detail view C). The filter body cross section is shown on the *right bottom*. The *crossbar* (moving part of the mechanical filter) supporting the permanent magnets of the anti-spring system, is visible on *top* of the filter body. The fine alignment of these moving magnets with respect to those anchored to the filter body is obtained via the remotely controlled fishing rod, also shown. In the *right top corner*, a detail of the central column cross section (detail view D) shows the connection of two suspension wire sections in the mechanical filter center of mass

Inertial Damping in the VIRGO jargon. It plays a crucial role in the hierarchical control of the Superattenuator chain thanks to the Inverted Pendulum softness, conceived to allow a wide positioning of the suspension point in three degrees of freedom.

7.3.4 The Seismic Filters

In our system, each pendulum mass has been replaced by a rigid drum-shaped metallic structure also acting as an oscillator in the vertical direction (Fig. 7.7). A sequence of six mechanical filters guarantees good seismic noise isolation of the optical components according to the working principle of a multistage pendulum.

Each mechanical filter (70 cm diameter, 18.5 cm high for a total weight of about 100 kg) is suspended as close as possible to its center of mass. On the outer circumference of the filter body bottom part, a set of triangular cantilever spring blades is clamped. Each blade (3.5 mm thick and 385.5 mm long) is bent at a constant curvature radius and with a different base width according to the load to be supported. These differently curved blades become flat and horizontal under loads ranging between 48 and 96 kg. Each blade tip is connected by a 1 mm diameter steel piano wire to a mobile central column (see detail view C in Fig. 7.7), coaxial to the filter body. Any movement of the central column, apart from the vertical direction, is prevented by two sets of four centering wires stretched on the top and on the bottom of the filter body. A crossbar, bolted on the upper part of the central column, is used as a mechanical support for the magnetic anti-spring system described in the next section. The central column and the crossbar represent the moving part of the mechanical filter from which the load of the lower stages is suspended by a steel suspension wire. By connecting each filter to the next one, a chain of mechanical oscillators in the vertical direction is obtained.

The set of triangular steel blades mounted under each filter (see Fig. 7.7), sustains the weight of the lower part of the chain. The support capability of each spring set is tuned varying the number of blades from 12 to 4, the base width of each blade between 180 mm and 110 mm and the curvature radius without load

Once properly loaded, the main vertical resonance frequency of each filter is about 1.5 Hz. In order to reduce the peak amplitude of the blade first flexural mode (about 100 Hz), a rubber/metal damper is mounted in the middle of each blade. Another special damper, mounted on the crossbar, suppresses a mode at about 50 Hz, due to the suspension wire connecting the crossbar to the next filter and acting as a stiff spring.

Maraging steel has been used in place of standard steel for blade construction [6, 7], in order to minimize micro-creep effects[1] due to the high load applied. The same material has been also used to machine the suspension wires with nail-heads at both ends [9]. Since the suspended load decreases as we move from the top to the bottom of the chain, the wire diameter changes accordingly, from 4 mm to 1.85 mm. In this way, it has been possible to confine the wire violin modes (see [10]) within the high frequency band and to reduce their angular stiffness, which determines the rotational frequencies around the vertical axis. The two nail-heads of the wires connecting a filter to that above and below are into the central part of its body at a relative distance of 5 mm (see detail view D in Fig. 7.7), very close to the filter center of mass. As mentioned above, this guarantees a small restoring torque to rotations of the filter around a horizontal axis and thus a low tilt frequency.

---

[1] *A micro-creep event is a sudden release of the stress accumulated in the lattice (microstructure of the material), which can propagate all the way to the bottom of the chain, to the suspended mirror, thus simulating a Gravitational Wave burst event. Though the micro-creep is a noise of thermal origin, it comes from a non linear phenomenon and it has a non-Gaussian spectrum* [6–8].

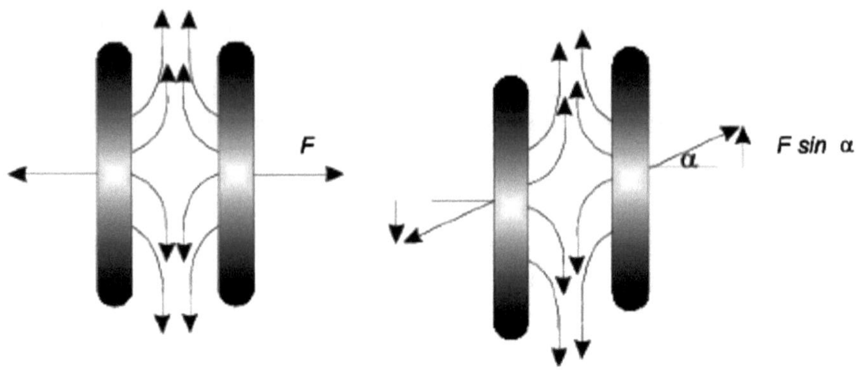

Fig. 7.8 The working principle of the magnetic anti-spring system: when the magnets are displaced in vertical direction (*right side*) a vertical component of the repulsive force appears

7.3.5 The Magnetic Anti-Spring

The chosen suspension wire length of 1.15 m sets the pendulum resonant frequency of each stage at about 0.5 Hz. In the vertical direction, the stiffness of the triangular blade springs fixes the natural resonant frequency at about 1.5 Hz.

In order to reduce the vertical stiffness of the blades, and then to confine the main vertical resonant frequency of each filter below that of the pendulum, a system of magnetic anti-springs was conceived [11, 12]. This consists of four sets of permanent magnets (two assembled on the crossbar and two on the filter body) facing each other with opposite horizontal magnetic moment (namely in a repulsive configuration). The two matrices on the crossbar are constrained to move in vertical direction only. The working principle of this system is sketched in Fig. 7.8. When the magnets are perfectly faced, the repulsive force has a null vertical component, but as soon as the crossbar (and its magnets) moves in the vertical direction, a vertical component of the magnetic force appears. Considering a small relative displacement (Δy), the vertical component of the repulsive force (F_y) is proportional to Δy:

$$F_y \simeq F_0 \cdot \Delta y/d \tag{7.13}$$

where F_0 is the module of the repulsive force and d is the magnet separation (see Fig. 7.8). Such a device is equivalent to a vertical spring with a negative elastic constant (anti-spring) whose module is $k_{AS} = F_0/d$. Its rest position is thus the position for which the two couples of matrices are perfectly faced. On a seismic filter the magnetic anti-springs act on the crossbar in parallel with the blade mechanical springs, so that the overall elastic constant is reduced and the vertical mode frequencies of the chain can be confined below the highest frequency of the horizontal modes. To this purpose writing down the equation of the mechanical system with the combined action of the blade springs and the magnetic anti-spring, it turns out:

$$m\ddot{y} = -k_b \Delta y + k_{\text{AS}} \Delta y \tag{7.14}$$

where k_b is the stiffness of the blade springs and and k_{AS}, as defined above, is that one obtained by the fine tuning of the magnets distance mounted on the crossbar and on the filter body respectively. From Eq. 7.14, defining:

$$\omega_o{}^2 = \frac{k_b - F_0/d}{m} \tag{7.15}$$

it is evident that the elastic constant of the system and then the vertical mode frequency ω_0 (see Eq. 7.15) can be changed modifying the distance of the permanent magnets pairs bolted on the mechanical filter in accordance with Eq. 7.13.

The device working principle requires that in the rest position all the magnetic anti-springs are perfectly aligned, in order to null the vertical component of the magnetic force. To this end, a small remotely adjustable cantilever blade (the so-called fishing rod, visible in Fig. 7.7) has been added on top of each filter, to bring the anti-springs to the correct working point.

7.3.6 The Filter 7

The last mechanical filter of the chain is the so-called Filter 7 [13]. Since it has been designed to steer the payload, it has been equipped with four aluminum legs (about 900 mm-long and 250 mm in diameter) bolted on the filter body bottom part (see Fig. 7.9). These legs are used as a mechanical support for the coils mounted in front of the permanent magnets on the marionette wings (see next paragraphs for the marionette description). The coil-magnet pairs are used to apply forces and torques to the marionette, controlling its position in three degrees of freedom: translation along the beam direction (z), rotation around the vertical (θ_y), and transverse axes (θ_x).

In addition, the Filter 7 is equipped with different vacuum-compatible stepping motors. Two of them are dedicated to the filter alignment around its vertical axis. The first one is accommodated on top of the filter body and is used to rotate the filter with respect to the upper part of the Superattenuator together with the whole payload. The second is mounted under the filter and rotates the marionette and the rest of the payload with respect to Filter 7, adjusting the relative positions of coils and magnets.

A second pair of vacuum-compatible stepping motors moves two small masses attached to mechanical trolleys on top of the filter body, performing a fine adjustment of the tilts (rotations around the two horizontal axes).

The set of sensors and actuators is completed with three position sensors with permanent magnets bolted on the cylindrical mantel of the filter body (not visible in Fig. 7.9 because their installation is performed after the Filter 7 integration within the

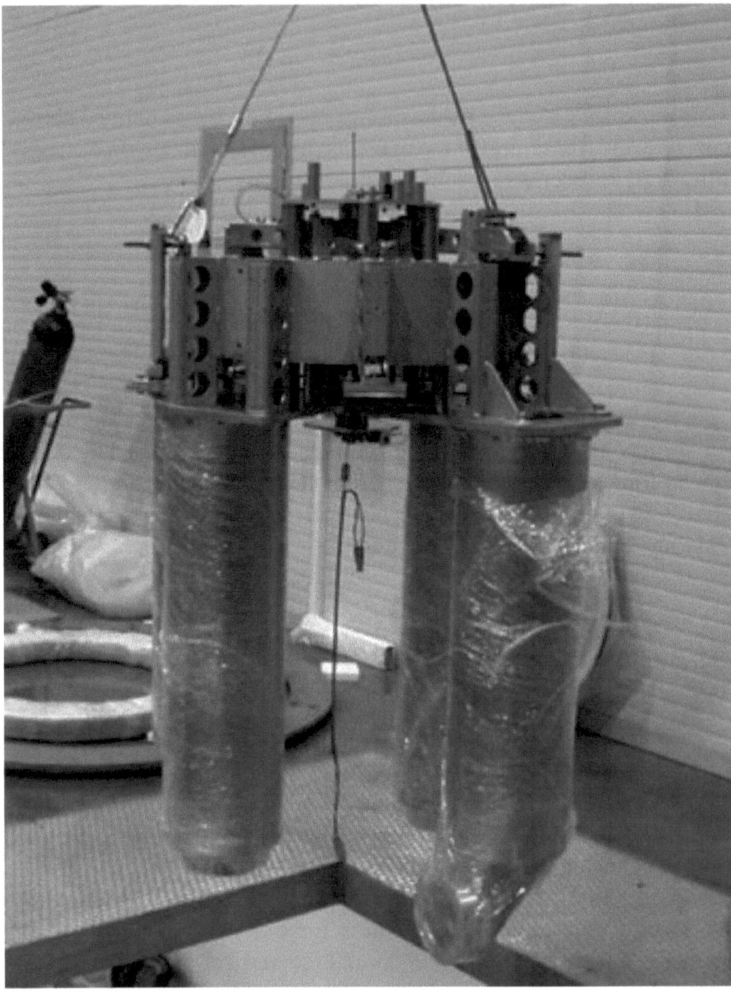

Fig. 7.9 The Filter 7 during the integration phase within a suspension chain of the Superattenuator. The four aluminum legs used as mechanical support for the actuation coils are visible. The first stepping motor used to rotate the system Filter 7-payload with respect to the *upper part* of the suspension chain is visible (platform for its installation) on *top* of the mechanical body. The second one, devoted to the payload rotation with respect to the Filter 7, is also visible on the *bottom part* filter body

vacuum tower). These are installed reproducing the geometry of the IP top stage (pin-wheel configuration at 120° from one other) and, together with a set of three 110 mm diameter actuation coils, mounted on the ground connected bottom ring, represent the system used to feedback control the Filter 7 position during interferometer alignment. The large size of the coils, having a field gradient uniformity of better than 2 % over a 30 mm distance, ensures that seismic noise reintroduction at this stage is negligible and the coupling among different degrees of freedom is minimized.

7.4 The Last Stage Suspension

7.4.1 Overview of the Last Stage Suspension

The main role of the last stage of the suspension is to steer the optical components to maintain the interferometer at its working point [14]. Passive attenuation is obtained by suspending the mirror with wires so to have a pendulum, which provides a further mechanical filtering at frequencies higher than the pendulum resonances. However, such a scheme must be integrated with the capability to apply a feedback control to the position of the optical components. For this reason, a series of actuators acts directly on the mirror and on the mass from which it is suspended, the so-called *marionette*. Indeed, for these actuators to be effective, the mirror must be hung with two wires, so that, changing the inclination of the plane of the suspension, also the inclination of the mirror changes.

The marionette is also suspended from the last filter of the Superattenuator chain, the above described Filter 7. In this way, the suspended marionette is also a pendulum acting as any other mechanical filter and providing further passive attenuation for seismic noise.

By construction, the marionette orientation and inclination define orientation and inclination of the mirror. These degrees of freedom of the marionette are controlled using coils which are rigidly connected to the Filter 7, and act on magnets on the marionette arms. Injecting a current in the coils we can, for instance, push the marionette and the mirror along the z axis (which in Virgo reference frame is horizontal and coincides with the optical axis) and control the length of the Fabry Perot cavity. However, there is a strong geometrical constraint that must be followed if we want this action to be effective: the forces acting on the magnets of the marionette in the z direction should result, as much as possible, in a motion of the marionette along the z direction only. This can be achieved if the suspension point of the marionette, its center of mass, and the magnets dipoles lie all on the same geometrical plane. This is not pursuable with infinite precision and therefore a force along the z axis will always give also a motion (a coupling) along other degrees of freedom, e.g., a torsion along the vertical axis (the y axis in Virgo reference frame). In any case, we want this coupling to be as small as possible: acting on z (our input) must result in a displacement of the mirror along the z direction only: this is sometimes referred to saying that our controls must be *orthogonal*. Currently the coupling reduction that has been achieved between different degrees of freedom in Virgo is $\simeq 10^{-3}$.

A further difficulty comes from the fact that the mirror is connected to the marionette by the suspension wires: this connection is not a problem if we want to change the static position of the mirror, for instance with a dc current in the coils, but, of course, acts as a mechanical filter also when we want to move the mirror at a frequency much higher than the pendulum frequency of the mirror suspension, for instance to feedback the motion due to the seismic noise at, say, 5 Hz. In Virgo, the typical pendulum frequency of the mirror suspension is 0.6 Hz, so that the motion imposed on the marionette at 5 Hz is attenuated by a factor $\simeq (0.6/5)^2 \simeq 10^{-2}$ when it reaches the mirror. To overcome this problem, we must be able to act directly on the mirror, and not only via the action on the marionette: to this purpose, we glue some small magnets on the mirror and position some coils in front. In this way, if we feed a current to the coils, we exert a force directly on the mirror, without any filtering effect from the marionette.

However, "putting some coils in front of the mirror" requires a mechanical support, which again must be connected somewhere. Of course, we cannot just bolt it on the ground, because in that way the seismic noise would move directly the coils and be transmitted to the mirror via the e.m. coupling with the magnets, spoiling the Superattenuator efficiency. The safest thing to do is suspend the support in the same way as the mirror: in this way, we are sure that any seismic noise reaching it is attenuated at the same level as for the mirror itself. So we suspend our support from the marionette, with a two wire suspension, like the mirror: the motion of the marionette will move the mirror and the coil support (which is called *recoil mass* in Virgo) in the same way. Of course, this is true if the center of mass of the mirror and of this recoil mass are coincident. This puts some strong constraints on the shape of the recoil mass: its typical geometry is that of a hollow cylinder surrounding the mirror, with coils placed on one face. As a by-product, this set-up proved to be quite useful also to protect the mirror from accidents like wire breaking and excessive motion of the Superattenuator chain due to a failure in feedback control, and has been used both in initial Virgo and in Virgo+. A simplified view of this apparatus can be seen in Fig. 7.10a.

The Marionette, the Mirror and the Recoil Mass are what we call a branched triple pendulum, according to the scheme shown in Fig. 7.10b.

The frequencies of the coupled modes of these pendula can be evaluated solving a simple system of equations:

$$\begin{cases} M_1\ddot{x}_1 = -k_1x_1 - \beta_1\dot{x_1} + k_2(x_2 - x_1) + \beta_2(\dot{x}_2 - \dot{x}_1) + k_3(x_3 - x_1) + \beta_3(\dot{x}_3 - \dot{x}_1) \\ M_2\ddot{x}_2 = -k_2(x_2 - x_1) - \beta_2(\dot{x}_2 - \dot{x}_1) - \beta_{2,3}(\dot{x}_2 - \dot{x}_3) \\ M_3\ddot{x}_3 = -k_3(x_3 - x_1) - \beta_3(\dot{x}_3 - \dot{x}_1) - \beta_{2,3}(\dot{x}_3 - \dot{x}_2) \end{cases} \tag{7.16}$$

which gives the frequencies of the pendulum modes of the branched system.

7.4.2 *The Last Stage Suspension and the Local Control of the Mirror*

The motion of the mirror at these frequencies, as well as the motion due to the other low frequency modes of the attenuation chain, must be damped in order to keep the interferometer locked on its working point: this damping effect is provided by a system, the *Local Control*, that monitors the payload position, and damps the internal angular and longitudinal modes involving the mirror, that the inertial damping of the Superattenuator chain is not able to control. In that way, it is possible to recover the reference position of each mirror separately and to approach the interferometer working point.

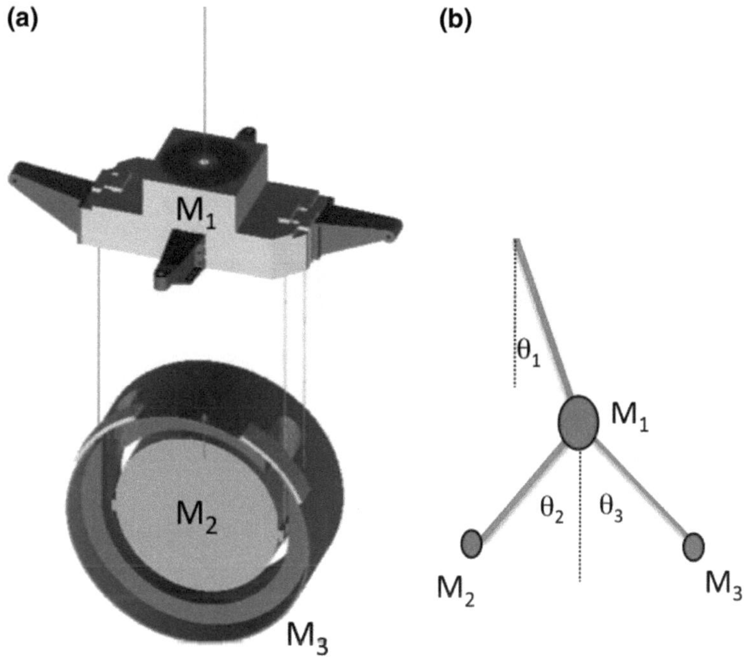

Fig. 7.10 **a** Scheme of the marionette—recoil mass— mirror system. **b** The equivalent system of three pendula

Longitudinal and angular mirror positions are reconstructed by means of two optical levers based on two low-power red laser diodes: one laser beam points directly the mirror, while the second one is reflected by an auxiliary mirror installed below the marionette. The reflected beams are detected by position sensing devices (PSD), as shown in Fig. 7.11.

When the oscillation amplitude is larger than the dynamical range of the PSD sensors, the coarse position is reconstructed using a camera CCD which monitors the relative position of the diffusive markers attached to the mirror itself. The local control system also receives the global interferometric signals for the longitudinal and angular sensing and control, transmitted to the central and terminal buildings by digital optical links (DOL). In order to reduce as much as possible the required dynamics of the actuators, part of the control force is re-allocated from the mirror-reaction mass actuators to higher stages: the very low frequency component is sent to the inertial damping control system (in particular to the filter 0); the intermediate frequency component is actuated on the marionette and only the high frequency part of the signal is sent to the reaction mass coils.

Indeed, the local control of the mirror position has an accuracy of $\simeq 1\ \mu$rad and it is not sufficient to fulfill the interferometer requirements for keeping the detector on the resonance condition. Furthermore, this control system is referred to the ground,

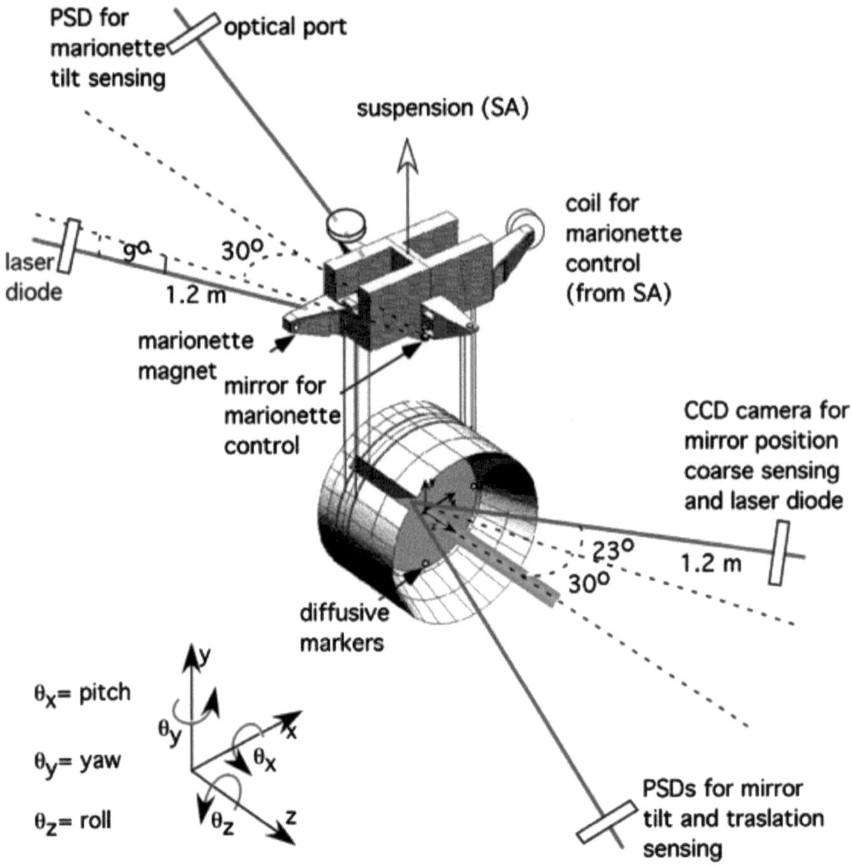

Fig. 7.11 Scheme of the local control sensing system, including the marionette and mirror optical levers

and as a consequence, limited by the seismic motion. Moreover, it is also affected by temperature drifts, and so it is not stable for long-time periods. For these reasons, a *global control* is necessary, as it has been discussed in Chap. 6.

The specification for mirror swing over a long period of time is $<1\ \mu\text{m}$, with velocity $<1\ \mu\text{m/s}$. The reached values are $0.7\ \mu\text{m}$, $0.25\ \mu\text{m/s}$, respectively. The specifications for angular control are to reduce the angular swings from a few dozens of μrad to 1 μrad. The attained value is $\simeq 1\ \mu$rad for both pitch and yaw (rotations defined in Fig. 7.11).

A more detailed description of this fundamental system can be found in [15, 16].

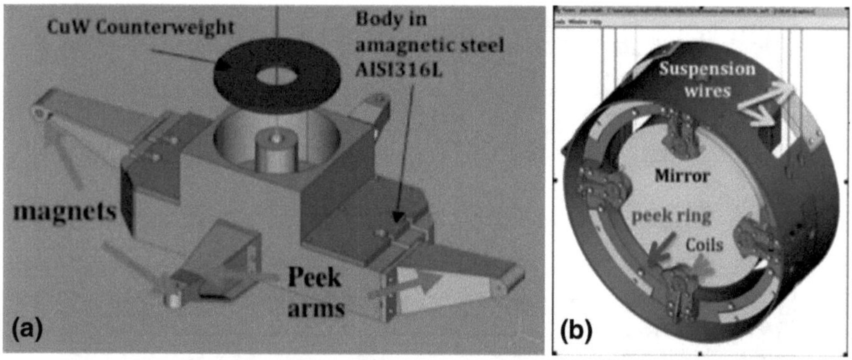

Fig. 7.12 **a** Details of marionette design. **b** Details of recoil mass design

7.4.3 The Real Design of Last Stage Suspension Elements

The construction details, both for marionette and recoil mass, must be compliant with strict requirements of ultra-high vacuum compatibility and cleanliness. Moreover, these parts must be made of materials which give a negligible contribution to the dissipation of energy of the pendulum motion of the mirror by means of eddy currents. Such additional complications are reflected in the real design of a marionette and of a recoil mass, that can be seen in Fig. 7.12a, b.

The counterweight is a piece of high density alloy of Copper and Tungsten which is used to set the center of mass of the marionette on the suspension point. The arms holding the magnets are in peek, a UHV compatible plastic. In this way, any effect due to coupling of the magnetic field of the magnets with the arms is avoided. For the same reason, the body of the marionette is also made of a special nonmagnetic stainless steel.

The recoil mass has the same constraints as the marionette: for instance, its center of mass coincides with that of the mirror, and amagnetic steel and UHV PVC (peek) are used for many pieces. Also, not shown in the figure is a series of stops to prevent the mirror from falling down if its wires break. For safety, the wire suspending the recoil mass are dimensioned so to be able to withstand the entire weight of both mirror and recoil mass. In Fig. 7.13 is the beam splitter payload of Virgo and Virgo+: this is how a real payload looks like.

7.4.4 Last Stage Suspension and Thermal Noise

If we look at the curve shown in Fig. 7.14 we can now recognize the contribution due to the seismic noise (green line) and the peaks in the transfer function due to the resonances of the Superattenuator chain. We can see that, because of the high efficiency of Virgo mechanical filters, the seismic noise becomes negligible already

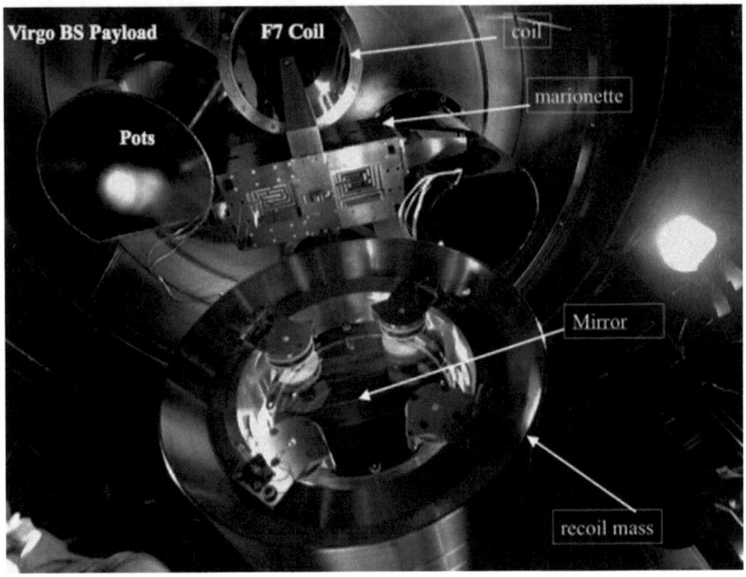

Fig. 7.13 The Beam Splitter mirror payload in Virgo

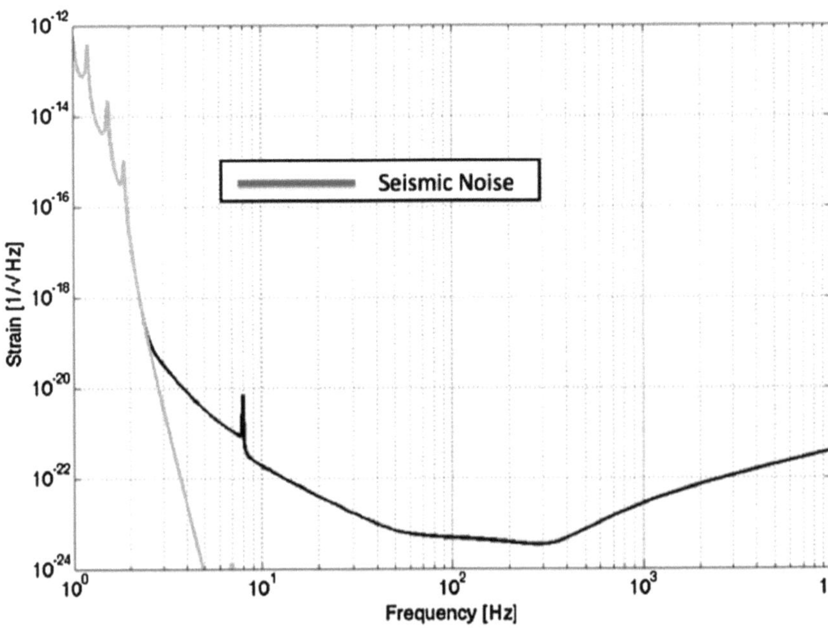

Fig. 7.14 The Virgo sensitivity *curve* for the advanced stage: the limiting contributions of seismic noise at low frequency is shown

at frequencies of the order of 4 Hz, and another kind of noise becomes dominant: the thermal noise. To reduce such noise is one of the most important functions of the last stage.

The general theory of the thermal noise is extensively described in Chap. 8: here, in the following, we shall mainly deal with the methods that are being devised to reduce it, referring again principally to the Virgo case.

In any case, literature on this subject is large and rapidly growing. An extensive list of references, as well as a more detailed discussion of the topics that we shall treat in the following, can be found in [17].

7.4.4.1 The Thermal Noise Contribution of the Mirror and of Its Suspension

In order to evaluate how the sensitivity of the interferometer is limited by thermal noise, we must apply the theoretical model analyzed in Chap. 8 to the real case of a mirror suspended to a marionette, as described above. It is evident that in such a configuration, the motion of the coated face of the mirror, which is detected by the laser beam resonating into the F-P cavity, is due also to the thermal motion of the entire bulk of the mirror, with a spatial shape and frequency that depends on its normal modes of resonance. Typical values for the resonance frequencies in a mirror are of the order of a few kHz. So, in the frequency spectrum at the output of the interferometer we shall see a series of peaks, due to the thermal motion of the mirror at the resonance of each normal mode. The behavior of the thermal noise at frequencies lower and higher than the peak value is given by the fluctuation-dissipation theorem, and depend on the dissipation processes going on inside and on the surface of the mirror (see Chap. 8). Moreover, the suspended mirror behaves like a simple pendulum, with a frequency of about 1 Hz. Also this oscillator will have its thermal motion, and this noise will be detected by the interferometer. However, in this case, we have the advantage of the so-called *dilution factor* (see for instance [18]). Indeed, the motion of a pendulum is mainly due to a restoring force which is gravity, and gravity is intrinsically non dissipative, while only a small contribution is given by the elasticity of the wire, whose dissipations can be described by the fluctuation-dissipation theorem. In particular, if $k_g = mg/l$ is the elastic constant due to gravity, and $k_{el}(1+i\phi_{struct})$ is the complex elastic constant taking into account the structural dissipative effects of the wire material, described by ϕ_{struct}, the equation of motion of the pendulum can be written as:

$$m\ddot{x} = k_g x - k_{el}(1 + i\phi_{struct})x \tag{7.17}$$

which can also be written as:

$$m\ddot{x} = -(k_g + k_{el})(1 + i\frac{k_{el}}{k_g + k_{el}}\phi_{struct})x \tag{7.18}$$

This equation shows that it is possible to define an effective loss angle ϕ_{eff} given by:

$$\phi_{\text{eff}} = \frac{k_{\text{el}}}{k_g + k_{\text{el}}}\phi_{\text{struct}} \simeq \frac{k_{\text{el}}}{k_g}\phi_{\text{struct}} = D_p\phi_{\text{struct}} \tag{7.19}$$

since usually $k_{\text{el}} << k_g$. The factor $D_p = \frac{k_{\text{el}}}{k_g} << 1$ is the pendulum dilution factor. If we insert in this formula, the explicit expressions for k_{el} and k_g, we find:

$$D_p = \frac{k_{\text{el}}}{k_g} = \sqrt{\frac{EI}{TL^2}} \tag{7.20}$$

where E is the Youngs modulus, I is the moment of the cross section, T is the tension applied to the wire of length L. Thanks to the dilution factor, the effective measured loss angle ϕ_{eff} is much lower than the value which is due only to the properties of the wire material.

The value of ϕ_{eff} includes all the dissipative effects that play a role in the pendulum motion. For instance, another important dissipation mechanism is due to the heat transfer between zones of the wire having different stresses: the temperature in a compressed zone is higher than in an extended zone, and as heat moves to even out the temperature difference, energy is lost. This mechanism is more efficient when the motion of the wire is at a certain frequency, depending on the characteristics of the materials: if we plot the losses of the material as a function of frequency, this effect appears as a characteristic peak in dissipation, and is often referred to as the thermal peak. Another important effect contributing to the overall losses is due to the fact that, in a material, surfaces usually have more defects than the bulk. So the losses of the material due to the status of the surface are usually higher and taken into account separately. Finally, further dissipations of the pendulum motion are due simply to the recoil of the supporting structure. These dissipations can be limited by careful design, but are not due to a part of the pendulum, thus, are not mitigated, like the others, by the dilution factor.

In summary, we must account for all these losses when we write down explicitly ϕ_{eff}:

$$\phi_{\text{eff}} = D_p(\phi_{\text{struct}} + \phi_{\text{therm}} + \phi_{\text{surf}}) + \phi_{\text{recoil}} \tag{7.21}$$

Using ϕ_{eff} we can write down the thermal noise estimation for the pendulum mode:

$$S_X^{\text{pend}}(\omega) = \frac{4k_BT}{\omega m}\frac{\phi_{\text{eff}}\omega_p^2}{(\omega^2 - \omega_p^2)^2 + \omega_o^4\phi_{\text{eff}}^2} \tag{7.22}$$

where $\omega_p = \sqrt{\frac{g}{l}}$ is the simple pendulum resonant frequency and k_B is Boltzmann's constant.

In a typical interferometer, the thermal noise due to the pendulum of the mirror suspension is dominant from a few Hz to about 100 Hz, while the thermal noise due

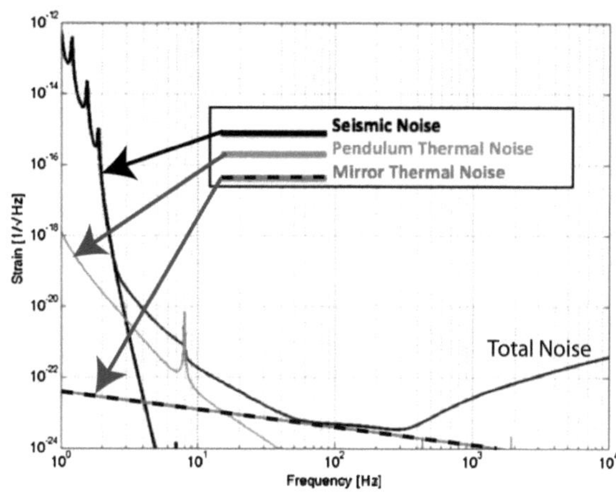

Fig. 7.15 Sensitivity of Advanced Virgo with both seismic and thermal noise contributions put in evidence

to bulk of the mirror is dominant from 100 Hz to a few kHz. In Fig. 7.15 we show the same sensitivity curve of Virgo of Fig. 7.7 (the black line), but now we highlight also the thermal noise contributions due to the pendulum (grey line) and to the mirror (dashed line).

7.4.5 The Monolithic Suspensions

The thermal noise sources which we described in the preceding section depend mainly on the intrinsic characteristics of the mirror (substrate and coating) and of the suspension wires. Moreover, an important contribution is also given by the friction on the clamping points of the suspension. The best way devised so far to reduce these sources of noise, has been to use fused silica (i.e., glass) wires, attached to the mirror by welding or using a procedure called silica bonding, which can reproduce the connection between materials at the molecular level [19]. For this reason, they are often referred to as *monolithic suspensions*.

Indeed, fused silica has a very low internal loss angle $\phi_{FS} \simeq 10^{-9}$ [20, 21]. For a comparison, the value for steel is $\phi_{STEEL} \simeq 10^{-4}$. As an additional vantage, fused silica wires have a tensile strength of 4 GPa (for a comparison, it is 2.9 GPa for C85 steel).

However, these good properties are balanced by some important drawbacks, that make the implementation of fused silica fibers for mirror suspension rather complex and challenging: in particular, the tensile strength is strongly dependent on cracks and defects present on the fiber surface. Moreover, such cracks can also be induced by a long exposure to environment pollution. Therefore, it is crucial to avoid contaminants and impurities during the fibers production and also it is necessary to store them in a high performance clean room (class 100 or better) avoiding contact with virtually anything.

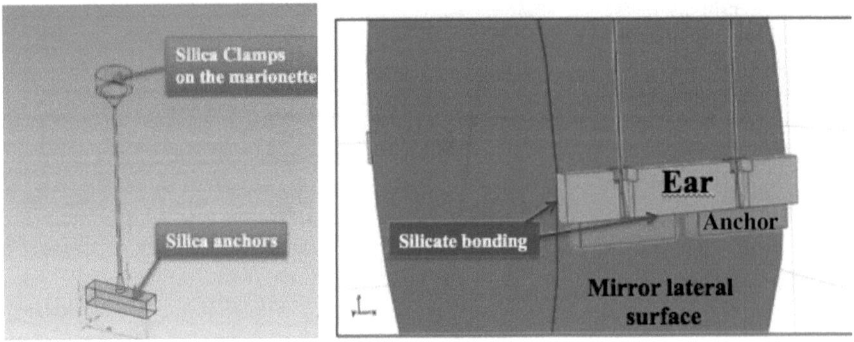

Fig. 7.16 Scheme of monolithic suspension in Virgo+

As a consequence, design, production, and integration of monolithic suspension is a difficult task. The first step is to identify the optimal geometry to minimize the pendulum thermal noise. This choice must be compatible also with the requirements for an optimal control of the test mass, and must, at the same time, ensure safety and reliability.

In Virgo, each fiber is welded to two fused silica blocks (see Fig. 7.16) which are then connected to the marionette on one side and to the mirror on the other side. From the mirror side, the fused silica piece welded to the fiber (we call this piece the *anchor*) is glued with a silica bonding technique to a section of fused silica protruding from the mirror (*the ear*). Machining the ear out of the mirror is rather difficult, so the ear itself is silica bonded to the mirror before the fiber is assembled. If silica bonding is correctly applied, fiber, anchor, ear, and mirror are a continuous body, and we have a monolithic suspension. The other end of the fiber, welded to another fused silica piece, is connected to the marionette with a stainless steel interface (*the upper clamp* assembly).

Fiber length, fiber stiffness, and position of bending points are constrained by the available space and the choice of resonant frequency of the various modes of the suspension. This choice depends strongly on control issues. For instance, the vertical bouncing frequency (for a description of this and other typical modes of a string, see for instance [10]) of the last stage represents the lower limit of the detection band because, although the vertical to horizontal coupling is small (a value of 10^{-3} is assumed, as said above), the vertical oscillation does not have any dilution factor. The result is that the vertical bouncing mode is clearly visible in the thermal noise spectrum and so, in order not to spoil the sensitivity, this frequency must be kept below 10 Hz, i.e., below the low frequency limit of the detection bandwidth. This can be achieved with a careful choice of the dimensions and shape of the fiber. Also, in order to have the detection band with the smallest possible number of resonant modes, its first violin mode [10] should be as high as possible. *Violin modes* are resonances of a stretched string clamped at its ends, like in a guitar or in a violin. The frequency of the first violin mode and of its harmonics depends on the fiber cross

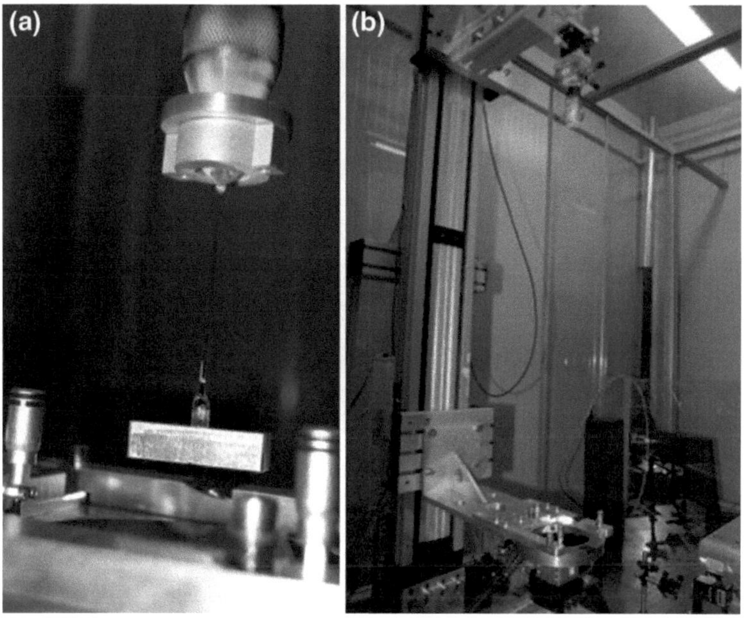

Fig. 7.17 **a** A fused silica fiber being ready to be stretched by the facility in Virgo. **b** The fiber has been stretched to its final length

section and then ultimately on its breaking stress. The first violin mode of our fibers is at about 430 Hz, about 100 Hz higher than what can be obtained with steel wires.

In Virgo, each fused silica fiber is produced starting from high purity fused silica cylindrical bars, commercially available (suitable materials are Herasil or Suprasil), 10 cm long and 1.5 mm thick. This small cylindrical bar is clamped at both ends and heated in the central region using a 100 W CO_2 commercial laser with 10.6 μm wavelength. Subsequently, the two ends are pulled apart, extending the fiber at the desired length and shape. In fact, once melted, the fused silica material becomes highly viscous and it can be plastically deformed. This process is performed by a dedicated machine developed at the University of Glasgow for GEO gravitational interferometer and duplicated in Virgo in a dedicated laboratory (Fig. 7.17).

After being produced, the fibers are tested to a load at least double the operation load. Then, if the fiber survives, its bending length is measured. The bending length λ of the suspension is the distance of the fiber bending point from the clamped end. This is an important information, because tells us where is the point where the fiber bends during the pendulum motion of the mirror. That is also the point that must be contained in the plane passing by the suspension point of the marionette and the actuation axis of the coil-magnet actuators. Positioning the bending point on the center of mass plane of both the marionette and the mirror allows minimum coupling between the different degrees of freedom; it is also possible, through a simple model of the fiber based on the knowledge of λ, to study its dynamical behavior. The tapered

Fig. 7.18 Test of a fiber after production

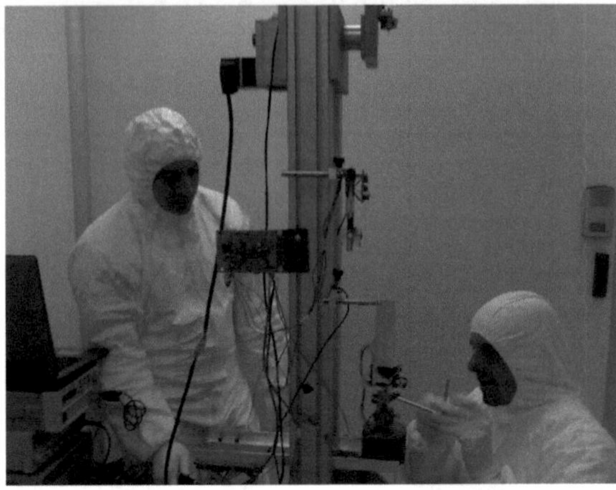

profiles of the fibers near the heads require this quantity to be directly measured. For doing this, a specific instrument was realized (Fig. 7.18).

The fiber is clamped to a rotary stage, suspending a load as similar as possible to the working conditions. If the bending point is placed exactly on the rotation axis of the stage, when the top head of the fiber is rotated, the bottom part of it actually does not move. Acting on a micrometric translator moving along a diameter of the rotary stage it is possible to find a position where the fiber movements get their minimum amplitude, thus having a measurement of the bending length. It is found that a repeatibility of some micrometers is achieved. This device also provides an easy way to measure the first violin and bouncing frequencies.

After being produced and validated, the fibers are then placed in position, clamping the upper part to the marionette and inserting the lower anchor below the lateral supports bonded to the mirror. In the end, the anchor and the supports are bonded together through silicate bonding.

In Fig. 7.19, we can see the assembly of the fused silica fibers into the payload.

The complete payload is then inserted in a container with controlled humidity and cleanliness and transported under the tower for integration, with a continuous monitor of acceleration during transport (Fig. 7.20).

The insertion of the payload into the tower and its integration with the Super-attenuator is a complex procedure: after several hours of stressful work the mirror is suspended at the end of the Super Attenuator chain by the monolithic fibers (see Fig. 7.21), and the weeks long procedure to activate the local control and subsequently the global control can start.

In 2009, the solution we have described has been implemented in Virgo+. That has been the first time that such a technology was used on a large interferometer: the four mirrors of the 3 km Fabry-Perot arms were in operation for about 2 years, suspended to glass fibers of 0.4 mm in diameter, before being disassembled to start the upgrade of Virgo to the Advanced state. While this new set up was a success as a technical achievement, the actual values of Q measured for violin modes were in

Fig. 7.19 The four fibers silica bonded to the mirror

Fig. 7.20 The transport of the monolithic payload for integration with its Superattenuator

general a factor 10 lower than expected. The cause of these extra dissipations was probably in the design of the upper clamp of the fibers. This design is being revised now in order to implement monolithic suspension in Advanced Virgo.

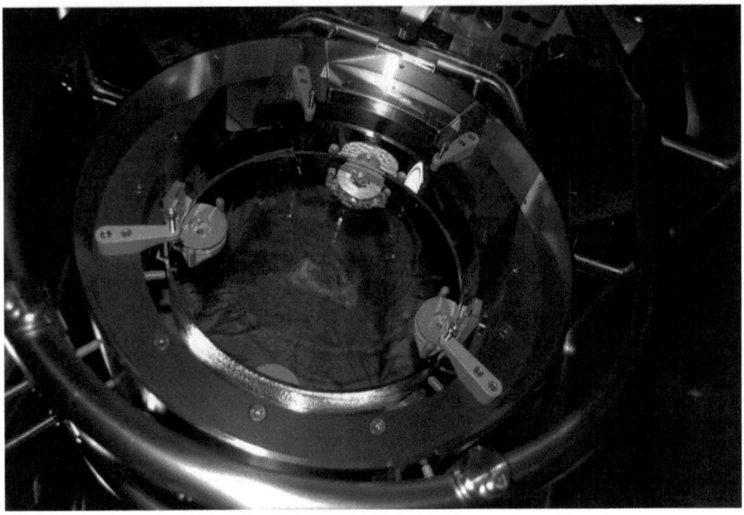

Fig. 7.21 The NI mirror monolithic payload of Virgo+ integrated with its Superattenuator suspension

7.4.6 The Last Stage Suspension in Advanced Virgo

A revised design of the clamps of the monolithic suspensions is not the only difference that will be implemented in the payloads for Advanced Virgo. In this new configuration of the detector, the power circulating into the interferometer and stored in the cavities will be increased by at least an order of magnitude, to reduce the shot noise contribution to the sensitivity. This increased power will produce a deformation of the mirrors that must be compensated by installing large optical pieces (the so-called "compensation plates") near the mirrors at the input of the Fabry-Perot cavities, and some heaters near the lateral surfaces of all mirrors (the "ring heaters"). Also, having more power inside the interferometer means having more light of the main beam scattered around the mirrors, toward the wall of the vacuum chamber. Scattered light has been found to be one of the main limits in sensitivity in previous versions of the interferometer and this problem will probably be more severe in Advanced Virgo. Hence large baffles (about 1 m in diameter) to intercept and damp any scattered light are being designed and produced, and will be installed either ground connected inside the vacuum chambers or suspended around all mirrors.

As a consequence of the requirements of this new set up, we had to make a deep change of the basic scheme of the mirror suspension. In brief, we found that it is very difficult to design and implement a suspended recoil mass supporting all needed items (baffles, compensation plates, ring heaters...), so we decided to take out the recoil mass suspension and to rigidly connect it to the body of the Filter 7. In this way, we can support easily large pieces very near to the mirror, andhave the additional

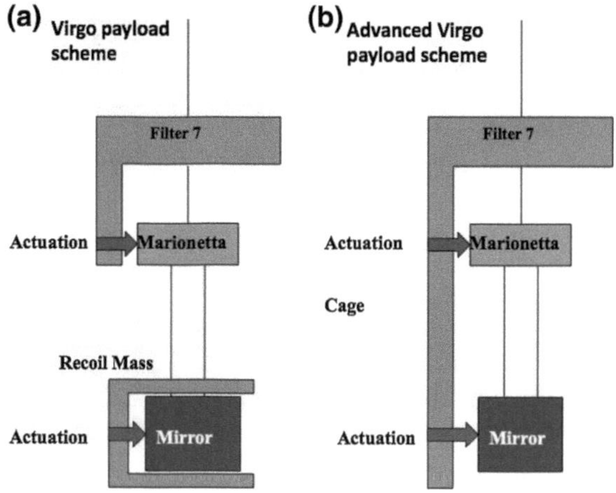

Fig. 7.22 Scheme of last stage suspension for mirrors in Virgo (**a**) and Advanced Virgo (**b**)

advantage to take out some low frequency resonances of the suspended recoil mass (e.g., the bouncing frequency of the wires of the recoil mass suspension) that were difficult to control. In Fig. 7.22, we can see a comparison between the old and the new scheme of the payload.

Of course, this new scheme has some disadvantages as well: the control of the Filter 7 must be improved, because now also the positioning of the coils facing the magnets on the mirror depends on it. Moreover, the metal wall that in Virgo separated the upper vacuum of the tower from the lower one, where the mirror is located, must be moved over the Filter 7. This change requires also that the Filter 7 must now meet the strict requirements of cleanliness and UHV compatibility of all other parts of the payload. These constraints drove a thorough redesign of the Filter 7 and of some elements of the vacuum chamber of the suspension, that is now underway.

The new payloads for Advanced Virgo are currently (october 2013) being designed and tested at the prototype level. In Fig. 7.23 we can see a first implementation, designed and built for the Beam Splitter Mirror.

We hope that with Advanced Virgo we shall be able to improve the sensitivity by an order of magnitude with respect to the past, as shown in Fig. 7.24. In such a way, we shall increase the volume in the detector range, and consequently the detection rate, by a factor of one thousand.

After the advanced detectors, it will be difficult to reach a further increase in sensitivity by improving the dissipations of suspensions beyond the already challenging limits imposed by the monolithic technique. In future, a further reduction of the thermal noise in gravitational wave interferometers will be probably given by the use of cryogenic technologies. The perspectives that such a development will open are dealt with in Chap. 14.

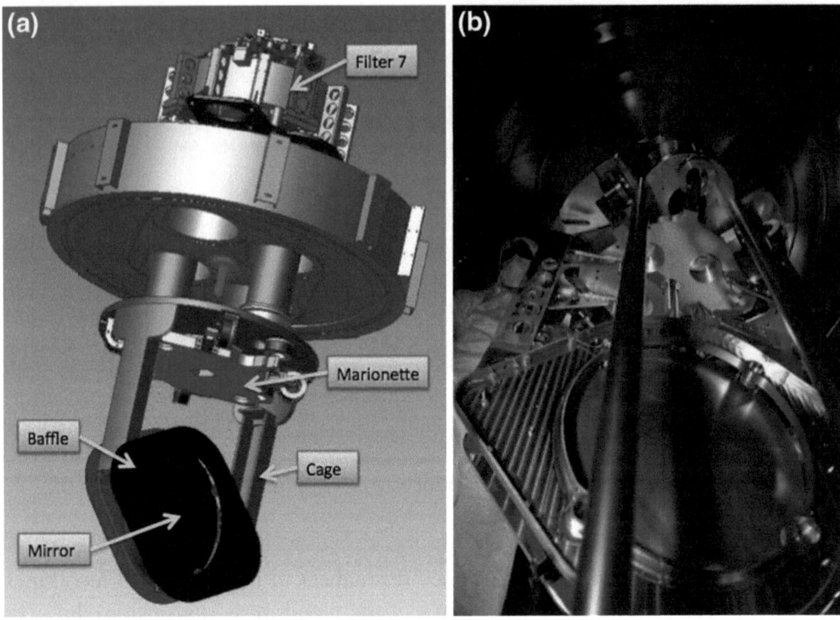

Fig. 7.23 **a** Design of the new Beam Splitter payload. **b** The prototype of the Beam Splitter payload ready to be integrated with a Superattenuator for testing

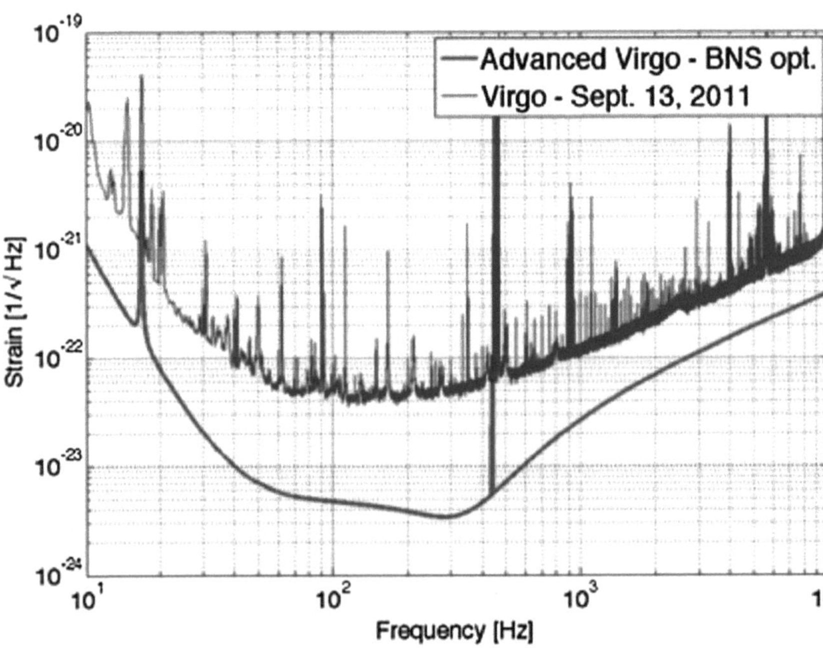

Fig. 7.24 Sensitivity of Virgo compared to the sensitivity expected for Advanced Virgo

References

1. T. Accadia et al., Virgo: a laser interferometer to detect gravitational waves. J. Instrum. **7** P03012, (2012)
2. T. Accadia et al., VIRGO collaboration, Virgo: a laser interferometer to detect gravitational waves, J. Instrum. **7** P03012 (2012). doi:10.1088/1748-0221/7/03/P03012
3. G. Ballardin et al., Measurement of the VIRGO superattenuator performance for seismic noise suppression. Rev. Sci. Instrum. **72**, 3643 (2001)
4. G. Losurdo et al., An inverted pendulum pre-isolator stage for the Virgo suspension system. Rev. Sci. Instrum. **70**, 2507 (1999)
5. G. Losurdo, G. Calamai, E. Cuoco, L. Fabbroni, G. Guidi et al., Inertial control of the mirror suspensions of the VIRGO interferometer for gravitational wave detection. Rev. Sci. Instrum. **72**, 3653 (2001) [gr-qc/0105110]
6. M. Beccaria et al., The creep problem in the Virgo suspensions: a possible solution using maraging steel. Nucl. Instrum. Meth. Phys. Res. A **404**, 455 (1998)
7. S. Braccini et al., The maraging-steel blades of the Virgo super attenuator. Measur. Sci. Tech. **11**, 467 (2000) [Erratum Measur. Sci. Tech. **15**, 599 (2004)].
8. G. Cagnoli et al., Mechanical shot noise induced by creep in suspension devices. Phys. Lett. A **237**, 21 (1997)
9. A. Delapierre, F. Frasconi, Stress distribution on the suspension wires of the Virgo superattenuator, Virgo Internal Report VIR-TRE-PIS-4600-137, (1997)
10. M. Bassan, F. De Marchi, L. Marconi, G. Pucacco, R. Stanga, M. Visco, Torsion pendulum revisited. Phys. Lett. A **377**, 1555 (2013)
11. M. Beccaria et al., Extending the Virgo gravitational wave detection band down to a few Hz: metal blade springs and magnetic antisprings. Nucl. Instrum. Meth. Phys. Res. A **394**, 397 (1997)
12. S. Braccini et al., An improvement in the Virgo super attenuator for interferometric detection of gravitational waves: the use of a magnetic antispring. Rev. Sci. Instrum. **64**, 310 (1993)
13. G. Ballardin et al., Measurement of the transfer function of the steering filter of the Virgo super attenuator suspension. Rev. Sci. Instrum. **72**, 3635 (2001) [Erratum Rev. Sci. Instrum. **73**, 233 (2002)]
14. A. Bernardini, E. Majorana, P. Puppo, P. Rapagnani, F. Ricci, G. Testi, Suspension last stages for the mirrors of the Virgo interferometric gravitational wave antenna. Rev. Sci. Instrum. **70**, 3463 (1999)
15. T. Accadia et al., Virgo: a laser interferometer to detect gravitational waves. J. Instrum. **7** P03012 (2012)
16. F. Acernese et al., A local control system for the test masses of the Virgo gravitational wave detector. Class. Quantum Gravity **20**, 617 (2004)
17. M. Colombini, Thermal noise issue in the monolithic suspensions of the Virgo+ gravitational wave interferometer, Ph.D. Thesis, Rome La Sapienza University, 2012
18. G. Cagnoli et al., Damping diluition factor for a pendulum in an interferometric gravitational waves detector. Phys. Lett. A **272**, 39 (2000)
19. G. Cagnoli et al., Suspension losses in low-frequency mechanical pendulums. Phys. Lett. A **213**, 245 (1996)
20. G. Cagnoli et al., Low-frequency internal friction in clamped-free thin wires. Phys. Lett. A **255**, 230 (1999)
21. A.M. Gretarsson, G.M. Harry, Dissipation of mechanical energy in fused silica fibers. Rev. Sci. Instrum. **70**, 4081 (1999)

Chapter 8
Thermal Noise in Laser Interferometer Gravitational Wave Detectors

Raffaele Flaminio

Abstract Thermal noise is one of the major limitations to the sensitivity of present and future laser interferometers devoted to gravitational wave detection. According to the fluctuation-dissipation theorem any mechanical oscillator is affected by a motion of thermal origin directly related to its thermodynamic temperature. The mirrors and their suspensions that are used in gravitational wave detectors such as Virgo or LIGO are examples of such mechanical oscillators. As a consequence their position is affected by this thermal vibration and the sensitivity of the gravitational wave detector is thermal noise limited over a wide range of frequencies. After recalling briefly the fluctuation-dissipation theorem and its origins, this chapter describes the main types of thermal noise affecting gravitational wave detectors. In the last part of the chapter a special emphasis is given to the thermal noise due to dissipation in the mirrors optical coatings.

8.1 Introduction

Gravitational wave detectors such as Virgo [1] or LIGO [2] use long baseline laser Michelson interferometers to measure the vibration of space–time associated with a gravitational wave. Gravitational waves act as quadrupolar tidal forces on the suspended mirrors of the interferometer (the so-called test masses) thus inducing a change in the interferometer arms lengths difference. Present gravitational wave detectors have reached a strain sensitivity of a few times $10^{-23}/\sqrt{\mathrm{Hz}}$. Such a strain sensitivity corresponds to mirror displacements of the order of $10^{-19}\mathrm{m}/\sqrt{\mathrm{Hz}}$ or equivalently to a displacement of about $10^{-18}\,\mathrm{m}$ over a time of $10\,\mathrm{ms}$. Future

R. Flaminio (✉)
Laboratoire des Materiaux Avances, CNRS/IN2P3, Universite de Lyon, 7 Avenue Pierre de Coubertin, 69622 Villeurbanne, Lyon, France
e-mail: flaminio@lma.in2p3.fr; raffaele.flaminio@nao.ac.jp

National Astronomical Observatory of Japan, Mitaka, Tokyo181-8588, Japan

M. Bassan (ed.), *Advanced Interferometers and the Search for Gravitational Waves*, Astrophysics and Space Science Library 404, DOI: 10.1007/978-3-319-03792-9_8,

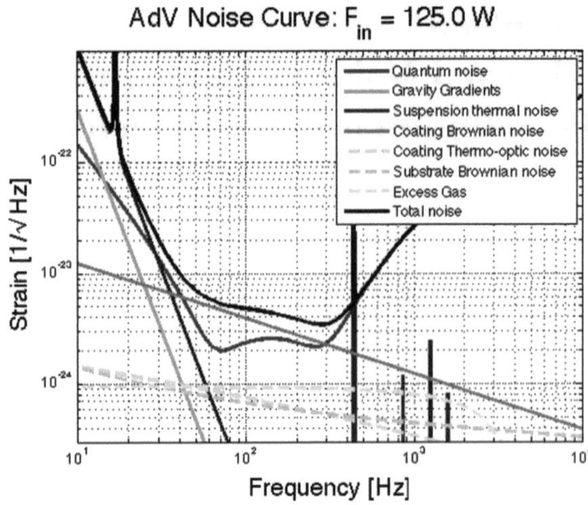

Fig. 8.1 Main noises limiting the strain sensitivity of Advanced Virgo as a function of the frequency

advanced detectors such as Advanced Virgo [3], Advanced LIGO [4], or KAGRA [5] aim at improving this sensitivity by about a factor of 10 thus decreasing the residual mirror displacements to the level of $10^{-20}\text{m}/\sqrt{\text{Hz}}$. Figure 8.1 shows the planned strain sensitivity of Advanced Virgo reaching a level of about $5 \times 10^{-24}/\sqrt{\text{Hz}}$ for frequencies around 100 Hz. At such a tiny level of displacement thermal noise becomes an issue. According to the fluctuation-dissipation theorem any mechanical system subject to some form of dissipation will be affected by a fluctuation of its position directly related to the system temperature. The equipartition theorem establishes that the total thermal energy of each degree of freedom of a mechanical system should be equal to $\frac{1}{2}kT$. A simple application of this theorem to the case of the main mechanical mode of a 20 kg mirror shows that the total vibration amplitude will be in the range of 10^{-15} m, much larger than the effect of the gravitational waves one wishes to detect. Luckily, most of the thermal vibration is concentrated at the resonance of the mirror mechanical mode, the spectral displacement off-resonance being considerable smaller. The distribution of the thermal noise off resonance is known from the fluctuation-dissipation theorem [6, 7] once the dissipation is known. As a consequence, it is of primordial importance to have a detailed knowledge of the dissipation mechanisms in order to foresee and possibly reduce the thermal noise amplitude.

Later in this chapter, the fluctuation-dissipation theorem is briefly recalled while in Sect. 8.3 its application to the case of a harmonic oscillator is discussed. Different cases of dissipation are discussed. In Sect. 8.4 the case of the pendulum thermal noise is discussed since this is one of the main limitations to the sensitivity of gravitational wave detectors at low frequencies, i.e., below 100 Hz. The discussion of thermal noise in continuous system is presented and discussed in Sect. 8.5. In the following section, we present the case of the mirror substrate thermal noise. Both the Brownian noise and the thermoelastic noise are dealt with. Section 8.7 is devoted to a discussion of thermal noise due to the losses in the optical coatings that are deposited on the substrate to provide the required reflectivity. This is at present the main limitation to the performance of gravitational wave detectors around the 100 Hz region, i.e., where interferometers reach their best strain sensitivity. This section includes a brief summary of the ongoing R&D to have a better understanding of this limitation and of the paths envisaged to reduce it. Finally, some conclusions are given in Sect. 8.8.

8.2 The Fluctuation-Dissipation Theorem

According to the Fluctuation-Dissipation theorem [6, 7], any linear physical system affected by some form of dissipation is subject to a fluctuation directly related to its thermodynamic temperature. The first evidence of such an effect was noted by Einstein while studying the so-called Brownian motion of pollen grains suspended inside water. Einstein noticed that the random motion of the pollen grains was originated by its collision with the water molecules, the latter being the same as causing the friction of a body moving inside water [8]. He deduced a relationship among the amplitude of the fluctuation of the random motion, the liquid viscosity, and the liquid temperature. This is the first example of an application of the Fluctuation-Dissipation theorem.

More than 20 years later in 1926, B. Jonhson [9] at the Bell Laboratories noticed that it was possible to measure a fluctuating voltage across any conductor. He empirically found a relationship between the resistance of the conductor and the amplitude of the voltage fluctuation measured across it. The square of the voltage fluctuation measured over a frequency bandwidth Δf is equal to:

$$< V^2 > = 4k_B TR\Delta f \tag{8.1}$$

with k_B the Boltzmann constant, T the conductor temperature and R its resistance. Once again a relation is empirically found between the amplitude of the dissipation (the resistance) and the amplitude of the fluctuation (the voltage). The relation was theoretically explained by H. Nyquist [10]. It took more than 20 years to arrive to the general form of the so-called Fluctuation-Dissipation theorem [6, 7]. The theorem applies to any linear physical system. It provides a relation between the power spectrum of the fluctuating thermal force applied on a generalized coordinate $x(\omega)$ of the system and the system impedance $Z(\omega)$, the latter being defined as the ratio between the system speed $v(\omega) = i\omega x(\omega)$ and the generalized force $F(\omega)$ applied to the same generalized coordinate. The theorem states that:

$$F_{th}^2(\omega) = 4k_B TR(\omega) \tag{8.2}$$

$R(\omega)$ being the real part of the system impendence $Z(\omega)$. The analogy with the expression of the voltage fluctuation across a resistor is clear, the force replacing the voltage and the real part of the impedance being the resistance. From the previous expression one can deduce the fluctuation of the generalized coordinate $< x_{th}^2(\omega) >$ using the relation $x(\omega) = \frac{1}{\omega}F(\omega)/Z(\omega)$:

$$x_{th}^2(\omega) = \frac{1}{\omega^2}4k_BTRe(Y(\omega)) \tag{8.3}$$

$Y(\omega)$ being the system admittance $Y(\omega) = 1/Z(\omega)$. The conclusion is that the real part of the system admittance (or impendence) is the parameters which defines the distribution of the fluctuations as a function of frequency.

8.3 Application to the Harmonic Oscillator

Let us consider the case of a damped harmonic oscillator whose equation of motion is given by:

$$m\ddot{x} + \beta\dot{x} + kx = F \tag{8.4}$$

m being the oscillator mass, k the elastic constant, and β the viscous damping factor. The impedance of such a system is given by:

$$Z(\omega) = im\omega + \beta + k/i\omega \tag{8.5}$$

And the real part of the impedance is the viscous damping factor β. The admittance of the system is

$$Y(\omega) = \frac{-im\omega^3 + \beta\omega^2 + ik\omega}{(k - m\omega^2)^2 + \beta^2\omega^2} \tag{8.6}$$

and as a consequence the thermal noise affecting the position of the harmonic oscillator is:

$$x_{th}^2(\omega) = \frac{4k_BT\beta}{(k - m\omega^2)^2 + \beta^2\omega^2} \tag{8.7}$$

Assuming a pure viscous damping, i.e., β constant as a function of frequency, the spectrum of the harmonic oscillator position due to thermal noise is shown in Fig. 8.2.The spectrum exhibits a sharp resonance at the oscillator resonance frequency. The resonance sharpness, defined as the ratio of the resonance frequency f_0 and the resonance width δf, is also known as the oscillator quality factor $Q = f_0/\delta f$. Its value depends on the damping factor, the quality factor increasing as the damping factor decreases. If τ is the oscillator damping time and T its period the quality factor is also equal to the number of oscillations before the oscillation amplitude decreases by a factor e, i.e., $QT = \tau$. The thermal noise spectrum is constant at low frequencies, i.e., below the resonance. In this region of frequencies the noise decreases as the quality factor increases or equivalently the damping decreases. Something similar occurs at high frequency (above the resonance) but in this case the noise decreases as a function of frequency. On the contrary, the amount of noise at the resonance

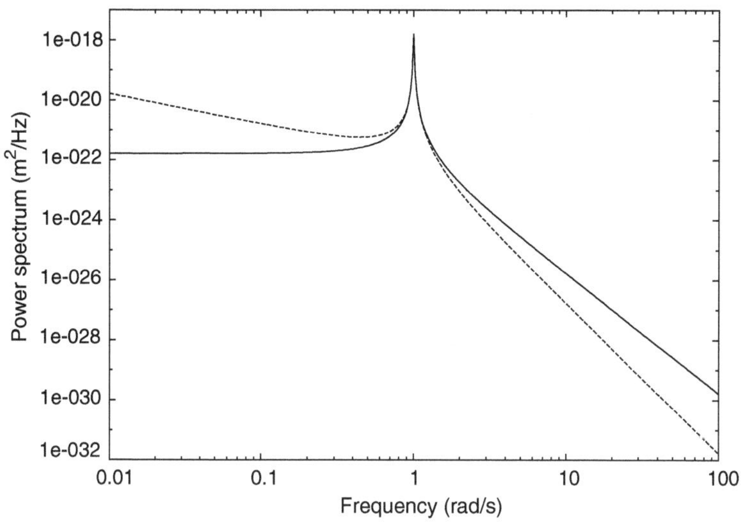

Fig. 8.2 Thermal noise power spectrum of an harmonic oscillator. The *continuous line* refers to the case of viscous damping. The *dashed line* refers to the case of structural damping. In both cases the quality factor is 100

increases when the damping decreases. In other words when the damping decreases the thermal noise is more and more concentrated at the resonance and the power spectrum outside the resonance decreases. If one looks for signals with frequencies different from the mechanical resonances of the mirrors and of the suspensions, as is the case in laser interferometer gravitational wave detectors, it is possible to reduce the thermal noise limitation by improving the mechanical quality factors of the system.

A classic case of viscous damping might be due to the residual gas around the mirror. This is usually negligible given the very low pressure maintained inside the vacuum chambers hosting the interferometer (10^{-7} Pa in the case of Advanced Virgo). Some effects may appear if a very small gap is left between the back surface of the mirror and the front face of the so-called actuation mass (or reference mass), the mass used to hold the electrostatic actuators to act on the mirror position. In this case the same gas molecules bouncing back and forth between the two surfaces hit several times the mirror surface increasing the overall noise. Equivalently, a small gap increases the damping due to the residual gas contained in the volume of the gap. A detail study of this situation can be found in [11].

Another classic case of viscous damping is the one of the damping due to eddy currents. This can be due to the magnets attached on the mirror producing stray magnetic fields on metal parts placed around the mirror. For this reason it is important to avoid placing conducting parts near the magnets. As an example the Virgo+ reference mass were partly made of Tecapeek CF30 to avoid this problem. Ideally the solution is to use masses made of glass around the mirror.

Another kind of mechanical losses could originate from dissipation inside the material composing the harmonic oscillator. These are the so-called structural losses. In their simpler form they are described by a model where the elastic constant k includes an imaginary part $k\phi$, ϕ being called the loss angle. Physically, this translates into a delay between the spring length and the restoring force. Analogously in a continuous body this would translate into an imaginary part of the Young modulus of the material composing the system. A delay between the displacement and the force implies a dissipation and as a consequence a fluctuation of the oscillator position. The equation of motion of such an oscillator is the following:

$$m\ddot{x} + k(1 + i\phi)x = F \tag{8.8}$$

From the equation of motion one can deduce the oscillator impedance, its admittance, and finally the oscillator fluctuation due to thermal noise:

$$x_{th}^2(\omega) = \frac{1}{\omega} \frac{4k_B T k\phi}{(k - m\omega^2)^2 + k^2\phi^2} \tag{8.9}$$

Figure 8.2 shows the thermal noise power spectrum of such an oscillator assuming the structural losses ϕ to be constant as a function of frequency. Similarly to the case of viscous damping, the smaller the losses are (the value of ϕ) the more the fluctuations are concentrated in the region of the resonance. Compared to the case of viscous damping the power spectrum increases inverse proportionally to the frequency as the frequency decreases. To avoid divergence and to satisfy the Kramers-Kronig relations there should be a cut-off at some very low frequency where ϕ starts decreasing to become equal to zero at DC. A similar difference is visible above the resonance; there the power spectrum decreases as f^{-5}, i.e., more rapidly that in the case of viscous damping.

In many cases the loss factor $\phi(\omega)$ turns out to be approximately constant with frequency. There is no precise theory explaining this general behavior. As mentioned above there should be a frequency cut-off below which ϕ decreases reaching zero at zero frequency. The Kramers-Kronig relations require $\phi(\omega)$ to be an odd function. In some cases the loss angle $\phi(\omega)$ exhibits a broad maximum at a characteristic frequency $f_c = 1/\tau$. These are the so-called anelastic solids.

Such a behavior can be modeled as an oscillator with a classical spring k_1 in parallel with a second spring k_2 in series with a dashpot. The latter is a pure viscous damping element (see Fig. 8.3) characterized by a damping factor β. At low frequency the dashpot does not oppose any resistance and the overall elastic constant is given by k_1. At very high frequency the dashpot becomes rigid and the overall elastic constant will be the sum $k_1 + k_2$. In between, the elastic constant increases with frequency. According to Bode's theorem if the gain of a linear system (in this case the elastic constant) increases with ω^n a phase lag equal to $n\pi/2$ should appear in the system response. Similarly, in the case of an anelastic solid modeled as in Fig. 8.3, a phase lag will appear between the force and the displacement with a broad maximum at an intermediate frequency between the two regimes corresponding to $k = k_1$ and

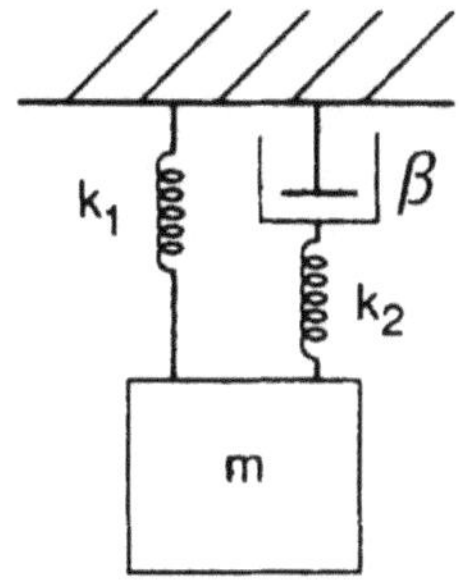

Fig. 8.3 Model of an anelastic oscillator. Figure is from [12]

$k = k_1 + k_2$. The amplitude of the phase lag at its maximum will depend on the difference between the high frequency and the low- frequency elastic constants, i.e., on k_2. Using the model given in Fig. 8.3, it is possible to deduce the behavior of the elastic constant as a function of frequency. One finds that:

$$F = k_1 \left[1 + \frac{k_2}{k_1} \frac{\omega^2 \tau^2}{1 + \omega^2 \tau^2} + i \frac{k_2}{k_1} \frac{\omega \tau}{1 + \omega^2 \tau^2} \right] \tag{8.10}$$

where $\tau = \beta / k_2$ is a relaxation time determined by the viscous damping of the dashpot and the spring elastic constant k_2. The phase lag is given by the imaginary part. The latter exhibits a broad maximum for frequencies around $f_c = 1/\tau$. The ratio k_2/k_1 is the relaxation strength and determines the amplitude of the dissipation. The real term is also frequency-dependent and shows the increase of the elastic constant with frequency which follow the behavior described above.

An example of losses due to anelasticity is the so-called thermoelastic damping. This is particularly relevant in small flexures like wires or thin cantilevers. Thermoelastic damping is related to the coefficient of thermal expansion of the material. As a wire is bent, one side of the wire gets hotter and the other gets colder. As a consequence heat flows from one side of the wire to the other, thus reducing the restoring force from its original value. The process being irreversible it ends up damping the wire oscillation. The model shows that such a process is well described by an anelastic oscillator as the one described above. In this case the relaxation strength is given by the following expression:

$$\delta = \frac{Ea^2 T}{C} \tag{8.11}$$

where E is the Young modulus, a the thermal expansion coefficient, T the temperature, and C the specific heat per unit volume. The relaxation frequency (or time) is given by:

$$f_c = \frac{1}{2\pi \tau} = \frac{2.16k}{Cd^2} \tag{8.12}$$

where k is the thermal conductivity and d the wire diameter. One notes that the relaxation strength, i.e., the maximum of the damping at the relaxation frequency, only depends on the thermal and mechanical properties of the wire material. Thus this is a fundamental limit inherent to each material according to its thermomechanical properties. On the other hand, the wire geometry (i.e., its diameter) determines the relaxation time. If the relaxation frequency is larger than the frequency of interest, it is possible to reduce the effect of the thermoelastic damping at such frequency by pushing the thermoelastic peak toward higher frequencies. This can be achieved by reducing the wire diameter.

8.4 Pendulum Thermal Noise

All the mirrors of laser interferometer gravitational wave detectors are suspended as pendulum having wires length L in the range between 0.5–1 m. Pendulums are classical examples of harmonic oscillators subject to losses. Actually the main restoring force is provided by gravity which, of course, is lossless. But a small fraction of the restoring force comes from the elasticity of the wires that are used to suspend the mirrors. This part of the restoring force is affected by losses. The total equivalent elastic constant for a pendulum of length L suspended with N_w wires of diameter d is given by:

$$k = \frac{Mg}{L} + N_w \frac{\sqrt{T_w EI}}{2L^2}(1 + i\phi_w) = k_g + k_{el}(1 + i\phi_w) \tag{8.13}$$

where M is the mirror mass, g the gravitational acceleration, E the wire Young modulus, I the wire cross-section moment of inertia, and ϕ_w the losses in the wires. k_g and k_{el} represent respectively the restoring forces due to gravity and to the elasticity of the wires. This can be rewritten as:

$$k = k_g(1 + \frac{k_{el}}{k_g} + i\frac{k_{el}}{k_g}\phi_w) \approx k_g(1 + i\frac{k_{el}}{k_g}\phi_w) \tag{8.14}$$

assuming that the restoring force due to gravity will be always much larger than the one due to the wires. From the last equation one sees that the losses due to the wires are diluted by the ratio k_{el}/k_g so that the pendulum loss angle is $\phi_p = (k_{el}/k_g)\phi_w$. In the case of the initial Virgo mirrors which had a weight of 20 kg and were suspended with four steel wires 0.7 m long and 300 microns in diameter, the dilution factor was about 10^{-3}. The intrinsic losses of steel wires being around 10^{-3} the final pendulum losses ϕ_p were around 10^{-6}, i.e., the pendulum quality factor was of the order of 10^6. This was slightly reduced by additional friction between the wires and the mirrors at the point of contact between the wire and its standoff.

Using the result of the previous section one can evaluate the pendulum thermal noise. For frequencies above the pendulum resonance (which is typically of the order of 1 Hz or less) the thermal noise spectrum is well approximated by the relationship:

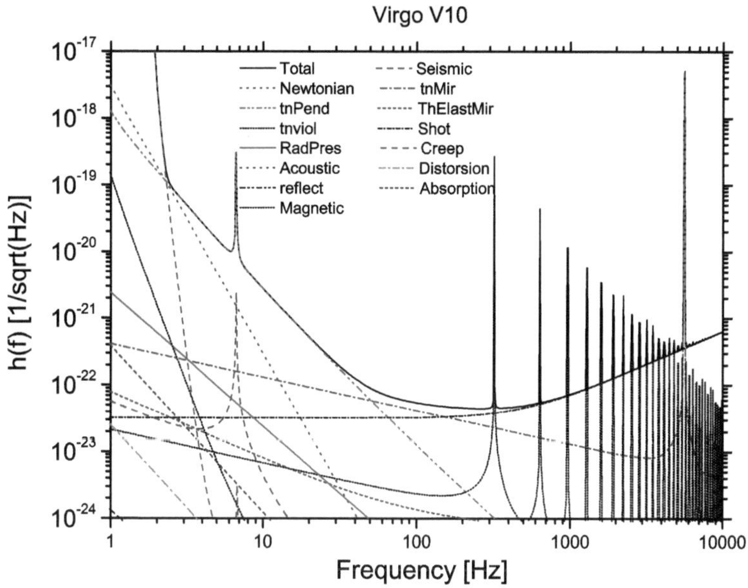

Fig. 8.4 Main noises limiting the strain sensitivity of Virgo as a function of the frequency

$$x_{th}^2(\omega) = \frac{4k_B T}{\omega_0^2 M \omega^4} \frac{\phi_p}{\omega} \tag{8.15}$$

where ω_0 is the pendulum angular frequency at resonance. One may note that the thermal noise power spectrum decreases with the inverse of the square root of the mirror mass. The reason is that the pendulum losses ϕ_p increase with the square root of the mirror mass. In fact increasing the mirrors mass requires to increase the diameter of the wires and the wire tension. In the end, there is still a gain in increasing the mirror mass and this is one of the reasons one tries to have mirrors as massive as possible.

Figure 8.4 shows the Virgo sensitivity and the various noises limiting its initial sensitivity. The pendulum thermal noise is the main limitation in the low-frequency region. This consideration has pushed the development of the so-called monolithic suspensions. In this case the mirrors are suspended by silica fibers. The silica fibers are welded to small silica blocks that are silicate bonded to the sides of the mirrors. This solution avoids loose contacts between the suspension wires and the mirrors. Moreover, it takes advantage of the excellent quality factor of the silica fibers whose losses are in the range of 10^{-7}. Due to larger uncertainties on the fiber strength a larger security factor is usually required and as a consequence the dilution factor is in the range of a few hundreds. Still, the overall pendulum quality factor exceeds 10^9. This change in the mirror suspension should allow improving considerably the sensitivity of laser interferometer gravitational wave detectors. The improvement of the

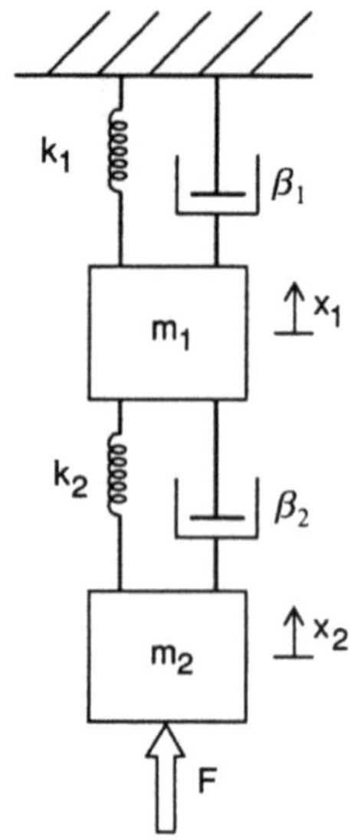

Fig. 8.5 Model of a double oscillator. Figure is from [12]

Advanced Virgo planned sensitivity at 10 Hz (Fig. 8.1) compared to Virgo (Fig. 8.4) is largely due to this change in the design.

In laser interferometer gravitational wave detectors the mirrors are suspended to pendulums which are themselves suspended to a chain of pendulums (or to other kinds of oscillators) providing the required seismic isolation. The overall Q of the suspension will depend on the losses of all the stages. One can then expect that the thermal noise affecting the mirror will depend on all these losses. The fluctuation-dissipation theorem helps answering this question. For simplicity let us consider the case of a double oscillator (or equivalently of a double pendulum) like the one schematized in Fig. 8.5.

The equations of motions can easily be written as a function of the masses, elastic constant, and viscous damping. To calculate the system impedance one should apply a force on the last masses (the mirror in our case) and calculate its speed. A detailed treatment of this example can be found in [12]. The result is that while both resonances of the system have quality factors which depend on both damping factors, the last stage thermal noise for frequencies larger than the resonant frequencies is only affected by the losses in the last stage. This is of course true only if the latter does not become negligible. The loss in the upper stage is a source of noise as well, but this noise is filtered by the second oscillator. Figure 8.6 shows the thermal noise affecting the mirror in the case where $\beta_1/m_1 = 10^{-2}$, $\beta_2/m_2 = 10^{-6}$, $m_1 = m_2 = 1$kg, and $k_1/m_1 = k_2/m_2 = 1\,\text{Hz}^2$. The dotted line shows the level of the noise if only the second oscillator was present. One sees that above the resonance the thermal noise is equal to the one of the lower oscillator as if it was alone. In case the second oscillator has zero losses then the thermal noise is due to the losses in the first oscillator, but in this case the noise goes as ω^{-8} at high frequency (instead

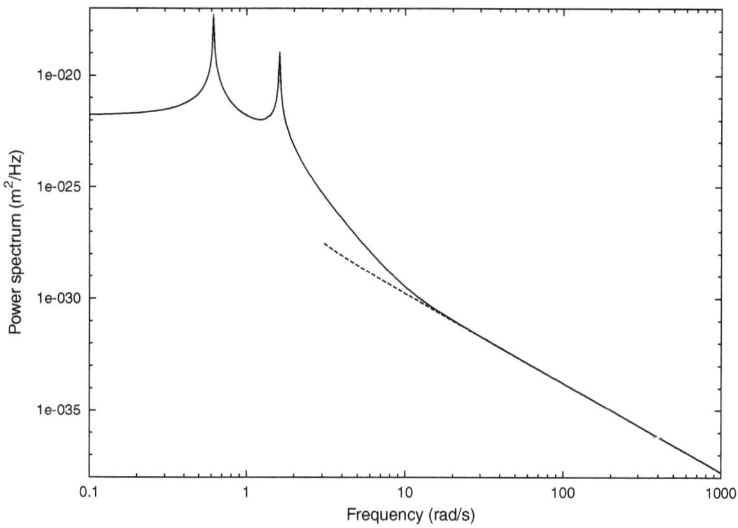

Fig. 8.6 Thermal noise power spectrum of a double oscillator. The *dotted line* shows the thermal noise if only the lower oscillator was present, i.e., the case where the upper ocillator has no losses

of ω^{-4}). As the frequency decreases and approaches the resonances the effect of the upper oscillator starts to be visible. If the lower stage has losses much lower than the upper stage, the latter can start playing a role. This is in fact the case for the planned Advanced Virgo sensitivity around 10 Hz which starts to be limited by the losses of the penultimate pendulum.

8.5 Thermal Noise in Continuous Systems

A continuous mechanical system can be thought of as made of an infinite number of harmonic oscillators corresponding to the normal modes of the system. Each mode will have its own resonant frequency, effective mass, and damping time. When thinking of the thermal noise affecting the system one should consider which degree of freedom to look at. This will depend on the sensor used to measure the system. An example could be that of a mirror suspended to a wire. The wire itself is a continuous system and the coordinate of interest is the displacement of the mirror (considered in this case as point-like) suspended to the wire. Another example could be the mirror itself and its internal modes. In this case the coordinate one is sensing the displacement of the mirror surface as seen by the Gaussian laser beam. In both cases one could think the system as made by an infinite number of eigenmodes $\mathbf{u}_n(\mathbf{r})$ where $\mathbf{r}$ is a vector giving the position in the system. Each eigenmode has a resonant frequency ω_n. The eigenmode functions $\mathbf{u}_n(\mathbf{r})$ are normalized with the density function $\rho(\mathbf{r})$ such that:

$$\int \rho(\mathbf{r})\mathbf{u}_n(\mathbf{r})\mathbf{u}_m(\mathbf{r})d\mathbf{r} = \delta_{nm} \tag{8.16}$$

The system position $\mathbf{u}(\mathbf{r}, t)$ will be a specific superposition of eigenmodes:

$$\mathbf{u}(\mathbf{r}, t) = \sum q_n(t)\mathbf{u}_n(x) \tag{8.17}$$

where the $q_n(t)$ are the generalized coordinates and should obey the dynamic law:

$$\ddot{q}_n(t) + \omega_n^2(1 + i\phi_n)q_n(t) = Q_n(t) \tag{8.18}$$

$Q_n(t)$ is the generalized force, i.e., the projection of the force $\mathbf{f}(\mathbf{r}, t)$, applied on the system, on the eigenmode $\mathbf{u}_n(\mathbf{r})$. Using the equation above one can calculate the impedance (or admittance) of each mode and deduce the thermal noise affecting each mode. The result is:

$$q_{n,th}^2 = \frac{4k_B T}{\omega} \frac{\omega_n^2 \phi_n}{(\omega_n^2 - \omega^2)^2 + \omega_n^4 \phi_n^2} \tag{8.19}$$

In general, none of the eigenmodes will coincide with the observable $x(t)$ one is looking at. The latter will be a specific combination of the system coordinates $\mathbf{u}(\mathbf{r}, t)$ and can be written as an arbitrary superposition of the eigenmodes:

$$x(t) = \int d\mathbf{r}\boldsymbol{\alpha}(\mathbf{r})\mathbf{u}(\mathbf{r}, t) = \sum_n q_n(t) \int d\mathbf{r}\boldsymbol{\alpha}(\mathbf{r})\mathbf{u}_n(\mathbf{r}) \tag{8.20}$$

From the equation above one can deduce the noise affecting the observable $x(t)$ as

$$x_{th}^2(\omega) = \frac{4k_B T}{\omega} \sum_n \frac{\omega_n^2 \phi_n}{(\omega_n^2 - \omega^2)^2 + \omega_n^4 \phi_n^2} \left[\int d\mathbf{r}\boldsymbol{\alpha}(\mathbf{r})\mathbf{u}_n(\mathbf{r}) \right]^2 \tag{8.21}$$

This formalism can be applied to the case of the thermal noise due to the losses in the pendulum wires and showing up at the resonances of the wire vibration modes. The thermal noise affecting a mass m suspended to a wire of length L and linear density ρ is:

$$x_{th}^2(\omega) = \frac{4k_B T}{\omega} \sum_n \frac{\omega_n^2 \phi_n}{(\omega_n^2 - \omega^2)^2 + \omega_n^4 \phi_n^2} \frac{2\rho L}{\pi^2 m^2 n^2} \tag{8.22}$$

where ω_n and ϕ_n are the angular frequency and the loss angle of the n-th wire's mode (usually known as violin modes). The result in the case of the Virgo sensitivity is shown in Fig. 8.4 (see the curve labeled 'tnviol'). The peak at 300 Hz and its harmonics results from the thermal noise at the wires vibration modes (known as violin modes). This same method can be used to calculate the mirror thermal noise.

The main limitation of this formalism is that it assumes the losses to be uniform across the system. The case of the double pendulum discussed previously is the best demonstration of this limitation. At each of the resonances the thermal noise of the suspended mass depends on the losses of both stages while well above the resonance frequencies only the losses at the second stage play a role. A model describing the thermal noise as a simple composition of the two modes each with its own quality factor would not work in this case. The other limitation is that in some cases many modes need to be considered. This may become computationally expensive. For instance in the case of the mirror thermal noise all the modes with spatial wavelength equal to a non-negligible fraction of the beam size need to be included. More details can be found in references [13, 14].

For these reasons, a more effective method was introduced by Levin [15] to calculate the thermal noise of continuous systems and applied to the case of the mirror thermal noise. In this case the observable $x(t)$ is the displacement of the mirror surface $u(\mathbf{r}, t)$ as monitored by the laser beam, i.e.,

$$x(t) = \int I(\mathbf{r})u(\mathbf{r}, t)dS \tag{8.23}$$

where I($\mathbf{r}$) is the laser beam intensity as a function of the position $\mathbf{r}$ on the mirror surface and u($\mathbf{r}$,t) is the displacement of the mirror surface as a function of time and of the position on the surface. In the case of a Gaussian laser beam $I(\mathbf{r})$ will be $Nexp(-2r^2/w^2)$, w being the beam radius and N a normalization constant. To calculate the mirror mechanical impedance a force having the same spatial distribution of the laser beam is applied on the mirror surface, i.e., $F(\mathbf{r}, t) = I(\mathbf{r})F(t)$. The mirror admittance will thus be:

$$Y(\omega) = \frac{i\omega x(\omega)}{F(\omega)} \tag{8.24}$$

$x(\omega)$ being the Fourier transform of the observable $x(t)$ when the system is subject to the force $F(\omega)$. The real part of the admittance is proportional to the dissipated energy. In fact:

$$Re\left[Y(\omega)\right] = Re\left[\frac{v(\omega)F^*(\omega)}{|F(\omega)|^2}\right] = \frac{W_{\rm diss}(\omega)}{|F(\omega)|^2} \tag{8.25}$$

where $v(\omega)$ is the speed Fourier transform and $W_{\rm diss}(\omega)$ is the power spectrum of the dissipated energy. From the real part of the admittance one deduces the thermal noise affecting the observable x using the fluctuation-dissipation theorem:

$$x^2_{th}(\omega) = \frac{4k_BT}{\omega^2}Re\left[Y(\omega)\right] = \frac{8k_BT}{\omega^2}\frac{W_{\rm diss}(\omega)}{|F(\omega)|^2} \tag{8.26}$$

The calculation of $W_{\rm diss}$ allows evaluating the thermal noise amplitude without using the modal decomposition previously described. Let us consider the simple case of a mirror with homogenous structural losses, i.e., whose Young modulus E can be written as $E(\omega) = E_0(1 + i\phi(\omega))$, $\phi(\omega)$ being the loss angle. Since the loss angle is the phase shift between the force and the displacement, the energy dissipated is given by:

$$W_{diss}(\omega) = \omega\phi(\omega)U_{max}(\omega) \tag{8.27}$$

where $U_{max}(\omega)$ is the maximum energy stored in the mirror during the forced oscillation at frequency ω. Since the frequency of the gravitational waves one is looking for is in general lower than the resonance frequencies of the mirror modes, the calculation of U_{max} at $\omega = 0$ is sufficient. As a consequence to evaluate the thermal noise it is sufficient to calculate the energy stored in the mirror when it is subject to a constant force having the same spatial distribution of the laser beam used to sense its surface. The exact calculation gives [15]:

$$U_{max} = \frac{1-\sigma^2}{2\sqrt{\pi}E_0 w}|F(\omega)|^2 \tag{8.28}$$

where σ is the Poisson ratio, E_0 the mirror Young modulus, and w the laser beam radius. From the expression above one can deduce the amplitude of the mirror thermal noise as:

$$x_{th}^2(\omega) = \frac{4k_B T}{\omega^2}\frac{1-\sigma^2}{\sqrt{\pi}E_0 w}\phi(\omega) \tag{8.29}$$

As expected the thermal noise increases with the loss angle $\phi(\omega)$. Moreover, one notes that the thermal noise decreases with the stiffness of the material composing the mirror and with the laser beam size. Using larger beam sizes allows reducing the thermal noise, the limit being the loss of light out of the mirror edges.

8.6 Substrate Thermal Noise, Thermo-Elastic Noise, and Thermo-Refractive Noise

From the equation of the mirror thermal noise developed in the previous section it is clear that the structural losses in the material composing the mirror substrate are of primordial importance. The choice of the substrate material consists in finding the best optical materials with the lowest possible mechanical losses. The material that has been used so far is fused silica since it combines very low optical absorption, tiny birefringence, and excellent homogeneity with very low mechanical losses. An extensive study of fused silica mechanical losses [16] has shown that losses depend on the material shape and that they are a combination of intrinsic losses and surface losses. For cylindrical shapes having an aspect ratio (radius to thickness ratio) of the order of one as the substrates commonly used in gravitational wave detectors the loss angle can be as low as 10^{-9}. Other materials that have been considered are sapphire, calcium fluoride, and silicon. These are crystalline materials with rather different thermal and mechanical properties compared to silica [17]. Even if their structural losses are quite low as well, thermoelastic losses need to be investigated in these cases.

As in the case of structural losses, the fluctuation dissipation theorem can be used to evaluate thermoelastic noise in the mirror substrate. The method is the same previously described and developed by Levin. As a force having the Gaussian shape of the laser beam is applied on the substrate surface, the material gets compressed and heats up. The heat transfer toward the colder regions produces a reduction of the restoring forces and as consequence a loss of energy. The calculation of the energy dissipated and the fluctuation dissipation theorem allow extracting the equation of the thermoelastic noise [18]:

$$x_{th}^2(\omega) = \frac{8k_B T}{\omega^2} \frac{2\alpha^2(1+\sigma)^2 kT}{\sqrt{\pi} C^2 \rho^2 w^3} \tag{8.30}$$

where α is the material thermal expansion coefficient, k is the thermal conductivity, and C the specific heat per unit volume. As in the case of structural losses the noise decreases when the temperature is decreased and when the beam size is increased. The rest of the expression depends on the intrinsic thermal and mechanical properties of the material. In particular one notes that the noise increases when the thermal expansion coefficient and the thermal conductivity increase. On the other hand, an increase of the thermal capacity induces a reduction of the thermoelastic effect. An alternative way to deduce this equation [18] is to consider the temperature fluctuations affecting any material of volume V as the source of the noise. Using the thermomechanical equation of the material under consideration one finds the same expression. Given some temperature fluctuation one can expect the noise to be larger when the thermal expansion coefficient gets larger.

In general, the thermal properties of materials depend on the temperature considered. At room temperature crystalline materials such as sapphire or silicon exhibit a thermal expansion coefficient and a thermal conductivity considerably higher than fused silica. As a consequence their thermoelastic noise is larger. Figure 8.7a shows the expected substrate thermal noise (or Brownian noise) and thermoelastic noise for silica, sapphire, and silicon. The structural losses are respectively 4×10^{-10}, 3.8×10^{-9}, and 10^{-8} [16, 19, 20]. While the difference in the substrate Brownian noise is relatively small and in any case, as we will see, small compared to coating thermal noise (see Sect. 8.7), below 100 Hz the effect of thermoelastic noise becomes relevant both for sapphire and silicon. This result excludes the possibility to use these materials for gravitational wave detectors operating at room temperature. It should be mentioned that thermoelastic noise in sapphire mirrors has been directly measured using a test interferometer (known as Thermal Noise Interferometer). This interferometer used a configuration where the laser beam size on the mirrors was of only a few hundred microns in order to increase the sensed thermal noise and make it dominate the other concurrent noises [21].

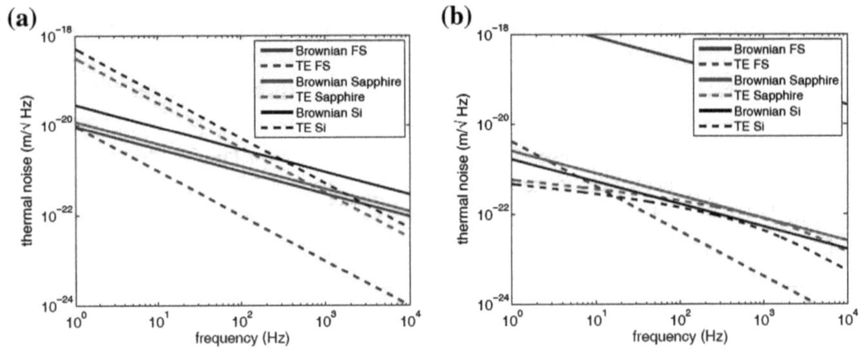

Fig. 8.7 Brownian noise and thermoelastic noise of silica, sapphire, and silicon substrates at 300 K (**a**) and at 10 K (**b**)

At cryogenic temperatures the results are quite different. First of all, the structural losses in silica increase dramatically as the temperature decreases. This prevents the use of silica as substrate material at low temperatures. On the other hand, the evolution of the thermal properties of crystalline materials at low temperatures, and in particular the decrease of the thermal expansion coefficient, reduces considerably the thermoelastic noise making it negligible compared to the effect of structural losses. The case of silicon is particularly interesting since it has been observed that the thermal expansion coefficient goes through zero at 20 K and at 120 K thus nulling the thermoelastic noise at these temperatures [22]. At the same time the structural losses of both sapphire and silicon decrease considerably at low temperature reaching values as low as a few parts in 10^{-9} as the temperature drops to the 10–20 K region [17]. For all these reasons, crystalline materials are considered for laser interferometers planned to operate at cryogenic temperatures. The Einstein telescope project [17] is considering silicon while the KAGRA interferometer [5] is planned to use sapphire. Figure 8.7b shows the Brownian noise and the thermoelastic noise in the case of a mirror operating at 10 K and made, respectively of silica, silicon, or sapphire [17].

The same dissipative process which is at the origin of the thermoelastic-noise also causes the so-called thermo-refractive noise. In this case the process which is at the origin of the effect is the variation of the material index of refraction due to a change in its temperature. The amplitude of the effect depends on the so-called thermo-refractive coefficient of the material. This causes a noise on the laser beam phase if transmissive optics are used as, for instance, in the case of the Fabry–Perot cavity input mirrors or in the case of the beam-splitter. The exact calculation of the effect can be found in [23–25]. It is found that the thermo-refractive noise is equivalent to a displacement of the mirror surface equal to:

$$x_{th}^2(\omega) = \frac{4k_BT}{\omega^2}\frac{4kTl}{\pi(\rho C)^2w^4}\left(\frac{dn}{dT}\right)^2 \tag{8.31}$$

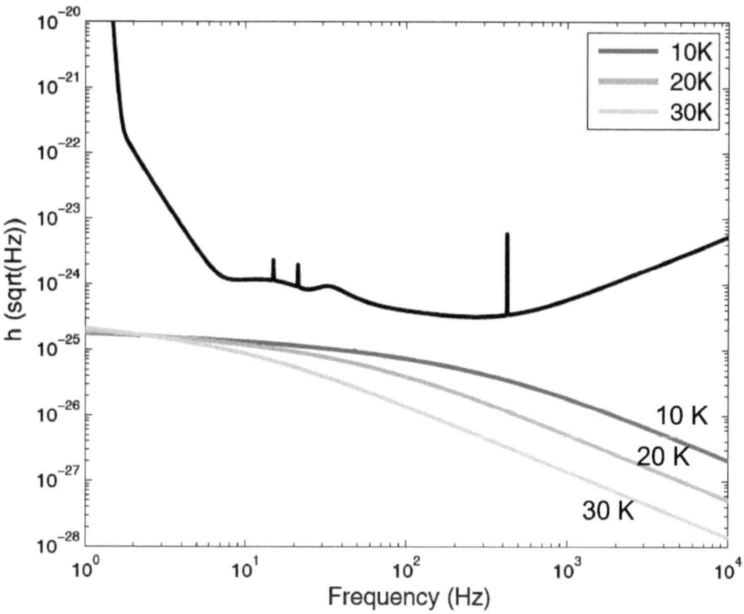

Fig. 8.8 The effect of thermo-refractive noise in the Fabry–Perot input mirrors compared to the ET planned sensitivity. The mirrors substrates are made of silicon and the temperatures of operation are those shown. $\frac{dn}{dT}$ is supposed constant and equal to $10^{-6}K^{-1}$

where l is the optics length and dn/dT is the thermo-refractive coefficient. As mentioned above, the effect increases when the thermo-refractive coefficient increases. Of course, the noise also increases with the optics length and, as usual, it decreases when the laser beam size gets larger. In most cases this effect is found to be negligible compared to the other noise souorces considered so far. Figure 8.8 shows an evaluation of the thermo-refractive noise in ET assuming the input mirrors of the Fabry–Perot cavities to be made of silicon (see [17] for the complete set of parameters used for the ET design).

8.7 Coating Thermal Noise

Surprising as it may be, the main contribution to the mechanical losses in a mirror weighing several tens of kilograms comes from the few microns of coating deposited on its surface in order to obtain the required reflectivity. The coating deposited on the substrate surface is typically made of a multilayer of alternating low refraction index and high refraction index layers. Usually, the low refraction index material is silica (refraction index around 1.45) and the high index material is tantalum pentoxide (refraction index slightly above 2) also known as tantala. In a standard design each layer will have an optical thickness of $\lambda/4$, λ being the operating wavelength (usually 1064 nm for current gravitational wave laser interferometers). The number of layers depends on the required reflectivity and typically is in a range between 15 and 40.

Several experiments have shown that when a multilayer coating is deposited on a substrate of silica the quality factor drops considerably. From these measurements one finds that the coating loss angle is in the range of 10^{-4} [26]. A more detailed analysis [27–29] has shown that the loss angle is dominated by the losses in the tanatalum pentoxide layers (found to be around 3–4$\times 10^{-4}$) while the loss angle of the silica layers is considerably lower, usually around 5×10^{-5}.

Coatings with low optical absorption, as those required for gravitational wave detectors, are made by ion beam sputtering. An ion beam source combined with an electron source is used to produce a beam made of an inert gas (typically Argon) having a typical energy of 1 keV. The atomic beam is used to sputter a target made of the material one wants to deposit. The sputtered atoms having an energy of tens of eV reach the substrate surface and condensate on it thus forming the thin film. The substrate is moved around typically in a simple rotational motion in order to make the coating more uniform.

Despite the fact that the coating represents only a very small fraction of the mirror total mass its thermal noise dominates the mirror total thermal noise. This is due to a combination of two effects. The first is that the losses in the coating are considerably higher than the losses in the silica substrate. The second is due to the fact that the coating losses are concentrated on the surface where the beam is reflected. The calculation of the energy $U_{\max}$ stored in the system when a gaussian force is applied on the mirror surface (see Sect. 8.4), shows that the largest part of this energy is stored at the surface itself, i.e., where the coating is located. As a consequence the losses in the coating play a larger role compared to losses located elsewhere. The exact calculation can be found in [27, 30]. The coating thermal noise is given by the following expression:

$$x_{th}^2(\omega) = \frac{4k_BT}{\omega}\frac{1-\sigma^2}{\sqrt{\pi}E_0 w}\phi_{eff,c}(\omega) \tag{8.32}$$

where $\phi_{eff,c}(\omega)$ are the effective coating losses:

$$\phi_{eff,c}(\omega) = \frac{t}{\sqrt{\pi}w}\left(\frac{E_0}{E_\perp}\phi_\perp(\omega) + \frac{E_{||}}{E_0}\phi_{||}(\omega)\right) \tag{8.33}$$

In the expression above t is the coating total thickness, w is the laser beam radius, E_0 is the Young modulus of the substrate, $E_\perp$ is the Young modulus of the coating along a direction normal to the layers, and $E_{||}$ is the Young modulus of the coating along a direction parallel to the layers. Similarly, $\phi_\perp$ and $\phi_{||}$ are the loss angles of the coating respectively along a direction perpendicular and parallel to the layers. The Young moduli and the losses $\phi_\perp$ and $\phi_{||}$ are given by:

$$E_{\perp} = \frac{t_1 + t_2}{t_1/E_1 + t_2/E_2} \tag{8.34}$$

$$E_{||} = \frac{E_1 t_1 + E_2 t_2}{t_1 + t_2} \tag{8.35}$$

$$\phi_{\perp} = \frac{E_{\perp}}{t_1 + t_2}\left(\frac{t_1}{E_1}\phi_1 + \frac{t_2}{E_2}\phi_2\right) \tag{8.36}$$

$$\phi_{||} = \frac{E_1 t_1 \phi_1 + E_2 t_2 \phi_2}{E_{||}(t_1 + t_2)} \tag{8.37}$$

As expected the thermal noise increases with the coating thickness and the coating losses. As usual it decreases when the laser beam size increases.

Coating thermal noise has been measured directly with the thermal noise interferometer [31]. It has been found to be also the limitation to the frequency stabilization of lasers to high finesse cavities [32]. In both cases the result are well in agreement with expectations. The conclusion is that coating thermal noise is expected to be the main limitation to the Advanced Virgo sensitivity around 100 Hz, i.e., where the detector will reach its best strain spectral sensitivity (see Fig. 8.1).

A considerable R&D program was pursued over the last years to reduce the losses of the high index materials. Other high index materials such as ZrO_2, Nb_2O_5, TiO_2, and others have been tested. None has shown better performances than the tantalum pentoxide. The measured properties are summarized in the ET design studies and are reproduced in Table 13 of [17]. The only material, ZrO_2, having slightly lower mechanical losses has considerably larger optical absorption. Niobium pentoxide has slightly higher mechanical losses and absorption. The higher optical index of refraction of Nb_2O_5 helps compensating part of these lower performance but does not allow reaching lower thermal noise.

Another approach consisted in doping the tantalum pentoxide with other materials. Cobalt, tungsten, and titanium were tested at the Laboratoire des Materiaux avances (LMA) [33]. The best performances were obtained by doping the tantalum pentoxide with titanium [27]. This mixture, made of a ratio titanium/tantalum of about one-third, allows reaching loss angles as low as 2×10^{-4}. On top of this result, this material also shows a lower optical absorption and a higher refraction index (see Fig. 8.9). An higher index of refraction allows reducing the amount of material to be deposited in order to obtain a given reflectivity, thus reducing the final thermal noise of the coating.

Even when using titanium-doped tantala, the high index material remains the material with larger losses. An approach to reduce the overall coating losses consists in revisiting the coating design and try to reduce the amount of high index material. Typically, a high reflectivity mirror is made of a stack of quarter wavelength layers. By properly changing the stack design it is possible to decrease the amount of tantala and increase the amount of silica while keeping the same optical response. The total deposited thickness is increased but the ratio tantala over silica is reduced and the

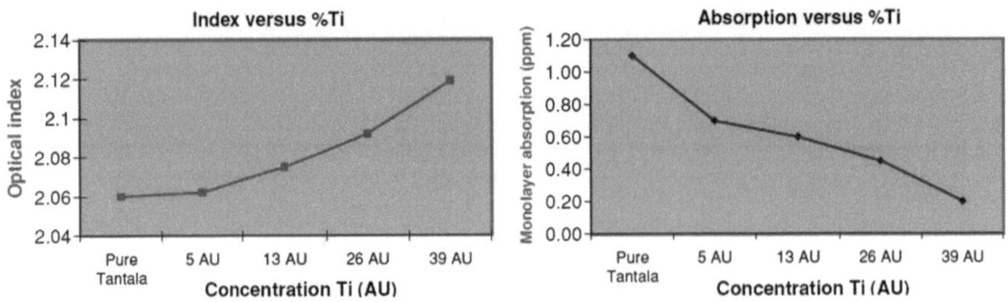

Fig. 8.9 Refraction index (*left*) and absorption of titania doped tantala as function of the Ti concentration in arbitrary units

Fig. 8.10 Dissipation in vitreous silica as a function of temperature. The different curves correspond to different resonant frequencies. A dissipation peak is visible on all the curves. The peak temperature increases as the mode frequency increases

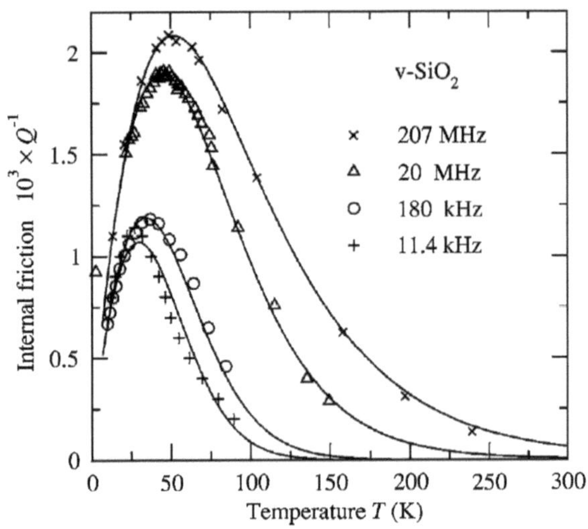

total mechanical losses are decreased [34]. These so-called optimized coatings have been realized at LMA and tested on the Thermal Noise Interferometer at Caltech [35]. These types of coatings will be used in Advanced LIGO. They should allow to reduce the spectral density of the coating thermal noise by about 10–15 %.

The need to find coating materials with lower mechanical losses has triggered a R&D program to better understand the origin of mechanical losses in amorphous materials. The study of the mechanical losses as a function of temperature has allowed having some insight into the origin of the losses. From the literature, it is known that mechanical losses in silica glass increase considerably at cryogenic temperature. A dissipation peak is found around a few tens of degrees Kelvin, the exact temperature depending on the frequency of the resonant mode being studied. For a given sample, it is usual to find the dissipation peak temperature increasing with the resonance frequency of the vibration mode considered (see Fig. 8.10 [36]).

This behavior is consistent with a model where the atoms in the material undergo a dissipative motion inside double well potentials. These double well potentials

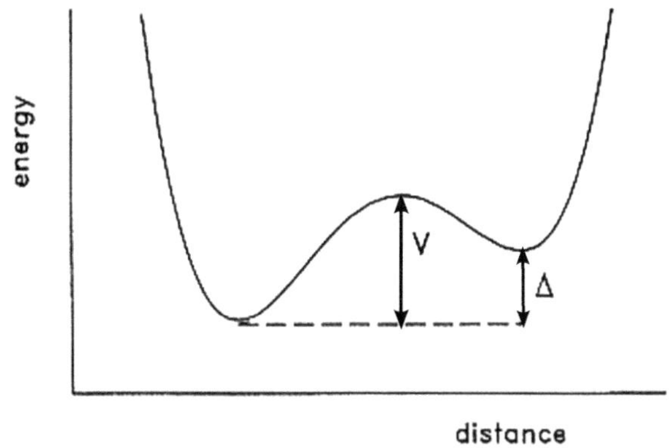

Fig. 8.11 The asymmetric double well potential used to describe dissipation inside amorphous materials such as silica

(Fig. 8.11) are characterized by the barrier height V and the potential asymmetry Δ, the latter being related to the material deformation [37]. It can be shown that each single relaxation mechanism having an asymmetric double well potential produces an effect on the elastic response of the material that can be modeled with the standard anelastic solid (see Sect. 8.3) with some relaxation time τ. The relaxation time is a function of the barrier parameters, an intrinsic time constant τ_0, and the temperature [37]:

$$\tau = \tau_0 sinh\left(\frac{\Delta}{k_b T}\right) exp\left(\frac{V}{k_b T}\right) \tag{8.38}$$

To each double well potential thus corresponds a dissipation peak (known as Debye peak) having the same shape seen in the case of an anelastic solid at angular frequency $\omega = 1/\tau$. Each double well potential contributes to the imaginary part of the material Young modulus which depends on the potential asymmetry. The total imaginary part of the Young modulus will be the sum of all these contributions:

$$E_I(\omega) = \int\int \Delta E(\Delta) \frac{\omega\tau(\Delta, V)}{1 + \omega^2\tau^2(\Delta, V)} g(\Delta) f(V) \mathrm{d}\Delta \mathrm{d}V \tag{8.39}$$

with $g(\Delta)$ and $f(V)$ being the distribution of the double well potential parameters. It can be demonstrated that if $g(\Delta)$ is a Gaussian distribution and $f(V)$ is an exponential distribution, than $E_I(\omega)$ has to have a maximum for a given temperature. This is what is found is silica glass and is thought to be linked to dissipative motion of the oxygen atoms in a double well potential whose asymmetry depends on the material stress.

A similar behavior of the losses versus the temperature is seen in dielectric coatings (see Fig. 8.12). Both the coatings made of tantalum pentoxide and of silica exhibit a dissipation peak at cryogenic temperatures typically in the range between 10 and 30 K [17, 38, 39]. In the case of tantala the height and the width of the dissipation peak vary with the coating annealing temperature. This is thought to be due to change in the material structure as the crystallization temperature is approached.

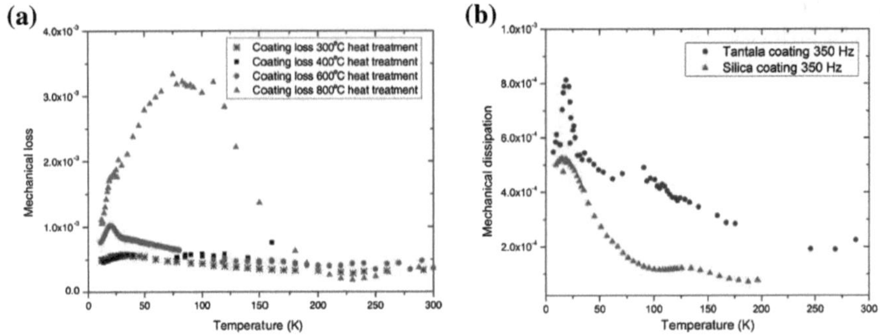

Fig. 8.12 **a** Measured values of the coating loss of tantala annealed at different temperatures and **b** Comparison of 600 °C heat treated tantala and silica coatings at low temperatures and 350 Hz [17]

To study the coating material structure transmission electrons microscopy has been used. The reduced density function (RDF) of the electron diffraction pattern has been used to study the structure of the amorphous coatings and to look for correlation with mechanical losses [40, 41]. The reduced density function exhibits peaks which are representative of the average distances between the atoms. The first peak is believed to provide information concerning the distance between the metal atoms (Ta) and the oxygen atoms. The second peak is related to the metal–metal average distance. The peak widths are related to the distribution of distances. Thinner peaks reveal more ordered structures (peaks becomes very thin in the case of crystals, of course). These studies allowed detecting a change in the RDF of titanium-doped tantala as the level of doping is changed. The width of the first peak becomes thinner indicating that titanium replaces the tantalum in the metal–oxygen bond thus producing a more ordered structure. At the same time the mechanical losses of the tantala coating decrease. Studies and simulations are ongoing to understand this correlation.

The study of mechanical losses in amorphous materials is very important to reduce the coating thermal noise. The measurements available in the literature show that most amorphous materials have mechanical losses in the range of 10^{-4}–10^{-3} [42]. A striking exception is the case of bulk silica which, at room temperature, has very low losses, in the range of 10^{-6}–10^{-9} (depending on the type of silica). Silica coatings show mechanical losses lower than those measured in tantala coatings. Still the losses in silica coatings are considerably higher than those seen in bulk silica. The reasons for this difference are not understood but it has been shown that losses in silica coatings can be reduced somehow by annealing at temperatures approaching 1000 K [43]. Unfortunately these high annealing temperatures are not possible for multilayer coatings since high index materials such as Ta_2O_5 (or others) would crystalize. Crystallization will give rise to a higher level of scattering and probably absorption. The problem of having high index materials with losses as low as those observed in silica remains open. Some measurements made on amorphous silicon have shown mechanical losses as low as a few parts in 10^{-5} at cryogenic temperatures [44]. In principle, these coatings could be candidate for cryogenic

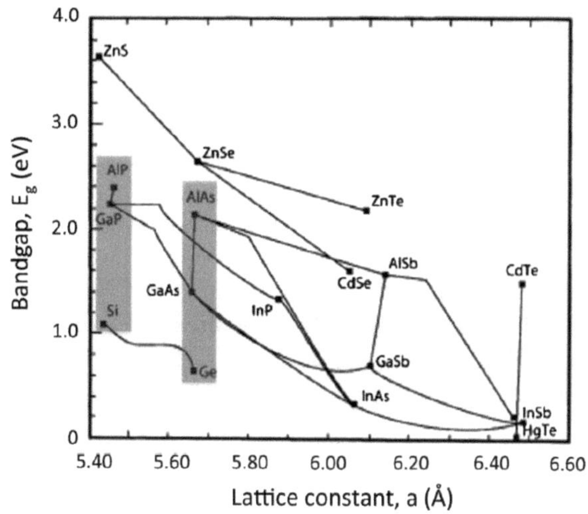

Fig. 8.13 Bandgap versus lattice constant of different kinds of crystals. The *light blue* areas show the materials being evaluated as coatings for gravitational wave laser interferometers

mirrors deposited on silicon substrates and operating at 1550 nm. Unfortunately so far amorphous silicon coatings appear to have a rather large optical absorption [45].

An alternative path to the reduction of coating thermal noise is in the development of coatings made of monocrystals. Crystals are known to have better mechanical properties and lower mechanical losses. Crystal coatings are made by Molecular Beam Epitaxy (MBE) or Metal Organic Chemical Vapor Deposition (MOCVD) and are widely used for applications in microelectronics. Their applications in optics are much more limited. One of the reasons is that their growth is very slow and thus not convenient compared to other technologies when coatings several microns thick are needed. Moreover, the growth of crystal coatings requires the matching of the lattice constants of the substrate and of the coating materials (low index and high index materials). As a consequence the choice of materials one could use is limited. Figure 8.13 shows the lattice constant of different kinds of crystals. Two kinds of crystal coatings have been investigated so far.

The first is a multilayer using GaAs as high index material and AlAs as low index material. Actually, since the GaAs layer includes some Al and the AlAs includes some Ga [46] one refers to this multilayers as $Al_xGa_{1-x}As$. The layers are grown on a GaAs substrate which is then removed leaving the free standing multilayer coating. By measuring the quality factor of the resulting oscillators it has been possible to show coating mechanical losses as low as 2.5×10^{-5} at room temperature and 4.5×10^{-6} at 10 K [47]. Reflectivities as high as 99.982 % at 1064 nm have been measured [48]. The absorption of these coatings have been measured at a level of about 10 ppm.

The second type of crystalline mirror that has been investigated is a multilayer of GaP (as high index material) and AlGaP (as low index material). In this case the stack is grown on a silicon substrate. A ten doublets stack with a reflectivity of 83 % (well in agreement with predictions) has been grown on a silicon substrate and tested. Preliminary measurements show the coating mechanical losses to be around 1.4×10^{-5} at 12 K [44]. The optical performances of the coating in terms of scattering and absorption remain to be measured.

One of the main limitations to crystalline coatings is that one cannot grow the coating on any substrate since the substrate and coating lattices need to match. One technique that has been developed consists of transferring the coating from the initial substrate on which it was grown to another substrate. This is done in two steps: first the coating is separated from the original substrate via chemo-mechanical substrate removal or epitaxial lift-off processes. Then the multilayer is bonded on the final mirror substrate. This technology has already been used to transfer GaAs epitaxial layers to sapphire, silicon, and quartz on surfaces as large as 150 mm for microelectronics applications. In all these application the transfer was done on a flat substrate. Attempts are ongoing to transfer AlGaAs multilayers stack on fused silica substrates to be used as mirrors of a high finesse Fabry–Perot cavity [47]. Preliminary measurements show finesses approaching 10^5.

8.8 Conclusions

Thermal noise will be one of the main limitations to the sensitivity of advanced detectors such as Advanced LIGO, Advanced Virgo and KAGRA. To overcome as much as possible this limitation these detectors will use monolithic suspensions for the mirror last stage suspension and last generation dielectric coatings in their optimized design. In addition the KAGRA detector will open the field of km-scale laser interferometers operating at cryogenic temperature. To further reduce thermal noise and allow reaching the sensitivities envisaged for the Einstein Telescope a strong R&D program on new coating and substrate materials and on cryogenic suspensions is ongoing. From the success of this program will depend the future improvements in the field of laser interferometers devoted to gravitational wave detection.

References

1. T. Accadia et al., J. Instrum. doi:10.1088/1748-0221/7/03/P03012
2. B.P. Abbott et al., Rep. Prog. Phys. doi:10.1088/0034-4885/72/7/076901
3. T. Accadia et al., Advanced Virgo Technical Design Report VIR-0128-12 (2012) https://tds.ego-gw.it/itf/tds/file.php?callFile=VIR-0128A-12.pdf
4. G.M. Harry, Class. Quantum Gravity. doi:10.1088/0264-9381/27/8/084006
5. K. Somiya, Class. Quantum Gravity **29**, 124007 (2012)
6. H.B. Callen, R.F. Greene, Phys. Rev. **86**, 702 (1952)
7. H.B. Callen, T.A. Welton, Phys. Rev. **83**, 34 (1951)
8. A. Einstein, Ann. Phys. **322**, 549 (1905)
9. J.B. Johnson, Phys. Rev. **32**, 97 (1928)
10. H. Nyquist, Phys. Rev. **32**, 110 (1928)
11. R. Dolesi et al., Phys. Rev. D. doi:10.1103/PhysRevD.84.063007

12. P.R. Saulson, Phys. Rev. D **42**(8), 2437 (1990)
13. A. Gillespie, F. Raab, Phys. Rev. D **52**(2), 577 (1995)
14. F. Bondu, J.Y. Vinet, Phys. Lett. A **198**(2), 74 (1995)
15. Y. Levin, Phys. Rev. D **57**(2), 659 (1998)
16. S.D. Penn et al., Phys. Lett. A **352**(1–2), 3 (2006)
17. M. Abernathy et al., Einstein gravitational wave Telescope conceptual design study ET-106C-10. https://tds.ego-gw.it/itf/tds/file.php?callFile=ET-0106C-10.pdf
18. V.B. Braginsky, M.L. Gorodetsky, S.P. Vyatchanin, Phys. Lett. A **264**, 1 (1999)
19. S. Rowan et al., Phys. Lett. A **265**(1–2), 5 (2000)
20. R. Nawrodt et al., J. Phys. Conf. Ser. **122**(1), 012008 (2008)
21. E.D. Black et al., Phys. Rev. Lett. **93**(24), 241101 (2004)
22. S. Rowan et al., in *Gravitational Wave Detection*, ed. by M. Adrian Cruise, P. Saulson. SPIE proceedings, vol. 4856, p. 292 (2003)
23. Y. Levin, Phys. Lett. A **372**(12), 1941 (2008)
24. V.B. Braginsky, S.V. Vyatchanin, Phys. Lett. A **324**, 345 (2004)
25. B. Benthem, Y. Levin, Phys. Rev. D **80**, 062004 (2009)
26. A.M. Gretarsson et al., in *3rd Edoardo Amaldi Conference on Gravitational Waves*. AIP Conference proceedings, vol. 523, p. 306 (2000)
27. G.M. Harry et al., Class. Quantum Gravity **19**(5), 897 (2002)
28. S.D. Penn et al., Class. Quantum Gravity **20**(13), 2917 (2003)
29. D.R.M. Crooks et al., Class. Quantum Gravity **21**, S1059 (2004)
30. G.M. Harry et al., Appl. Opt. **45**(7), 1569 (2006)
31. E.D. Black et al., Phys. Lett. A **328**, 1 (2004)
32. A.D. Ludlow et al., Opt. Lett. **32**(6), 641 (2007)
33. R. Flaminio et al., Class. Quantum Gravity **27**(8), 084030 (2010)
34. J. Agresti et al., in *Advances in Thin-Film Coatings for Optical Applications*, 3rd ed. by M.J. Allison. Proceedings of SPIE, vol. 6286, p. 28608 (2006)
35. A.E. Villar et al., Phys. Rev. D **81**(12), 122001 (2010)
36. R. Vacher, E. Courtens, M. Foret, Phys. Rev. B **72**, 214205 (2005)
37. K.S. Gilroy, W.A. Phillips, Phil. Mag. B **43**(5), 735 (1981)
38. I.W. Martin et al., Class. Quantum Gravity **26**, 155012 (2009)
39. I.W. Martin et al., Class. Quantum Gravity **27**, 225020 (2010)
40. K. Evans et al., J. Phys. Conf. Ser. **371**, 012058 (2012)
41. R. Bassiri et al., Appl. Phys. Lett. **98**(3), 03194 (2011)
42. K.A. Topp, D.G. Cahill, Z. Angew. Phys. B **101**, 235 (1996)
43. S.D. Penn, Amorphous Coatings LIGO Document G1200622–v1 (2012)
44. S. Rowan, Studies of materials to reduce coating thermal noise ET-0038A-12. https://tds.ego-gw.it/itf/tds/file.php?callFile=ET-0038A-12.ppt
45. R. Bassiri, Investigating coating material properties for future generations of gravitational wave detectors LIGO Document G1200620–v1. https://dcc.ligo.org/cgi-bin/DocDB/ShowDocument?docid=92236
46. G.D. Cole et al., Appl. Phys. Lett. **92**(26), 261108 (2008)
47. G.D. Cole, in *Optical Trapping and Optical Micromanipulation IX*, ed. by K. Dholakia, G.C. Spalding, Proceeding of SPIE, vol. 8458, p. 845807 (2012)
48. G.D. Cole et al., in *Proceedings of 23rd IEEE International Conference on Microelectromechanical Systems (MEMS 2010)*, p. 847 (2010)

Chapter 9
Thermal Effects and Other Wavefront Aberrations in Recycling Cavities

Alessio Rocchi

Abstract Thermal effects in the test masses of gravitational waves interferometric detectors may result in a strong limitation to their operation and sensitivity. Already in initial LIGO and Virgo, these effects have been observed and required the installation of dedicated compensation systems. Based on CO_2 laser projectors, the thermal compensators heat the peripheral of the input test masses to reduce the lensing effect. In advanced detectors, the power circulating in the interferometer will increase, thus making thermal effects more relevant. Moreover, other sources of optical defects, due to the fabrication process of the optics, will play a crucial role. High precision beam shaping of the CO_2 laser beam and high sensitivity wavefront sensors will be implemented to face these challenges, moving the thermal compensation system toward adaptive optical systems.

9.1 Introduction to Thermal Effects

In interferometric GW detectors, the sensitivity at frequencies above 200 Hz is limited by the photon shot noise as [1]

$$h_{\text{shot}}^{\text{FP}}(f) = \frac{\lambda}{4LF}\sqrt{\frac{h\nu}{2P}}\sqrt{1+(4\pi f \tau_s)^2} \tag{9.1}$$

where τ_s is the storage time of the light inside the Fabry–Perot cavity. This feature is shown in Fig. 9.1, where the Virgo design sensitivity is plotted with the fundamental noise sources.

The obvious method to decrease this fundamental limit is to increase the power circulating in the interferometer. However, this is limited by the nonzero optical

A. Rocchi (✉)
INFN Sezione di Roma Tor Vergata, Via della Ricerca Scientifica 1, 00133 Rome, Italy
e-mail: alessio.rocchi@roma2.infn.it

M. Bassan (ed.), *Advanced Interferometers and the Search for Gravitational Waves*, Astrophysics and Space Science Library 404, DOI: 10.1007/978-3-319-03792-9_9,

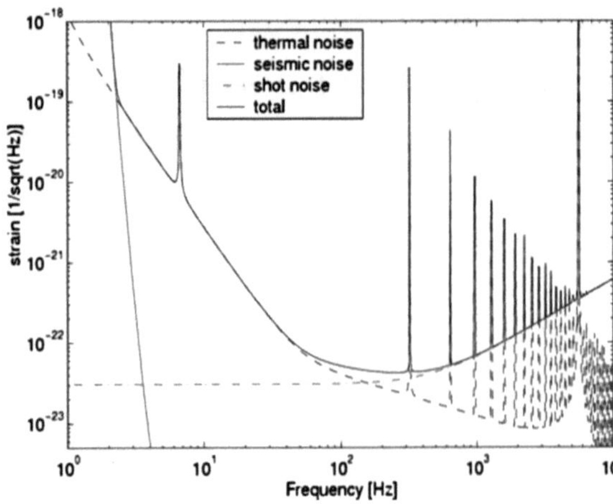

Fig. 9.1 Virgo design sensitivity plotted with its fundamental noise sources

absorption in the substrate and coatings of the test masses and the beam splitter. The temperature-dependent index of refraction and a nonzero thermal expansion coefficient of optical materials suitable for laser interferometric gravitational wave detectors ensures that nonuniform temperature increases induced by the absorption of the Gaussian-profiled probe light will result in nonuniform optical path length distortions which will affect both the controllability and sensitivity of the instrument.

9.1.1 Thermal Lensing

Let us consider a body at uniform temperature T_0, thickness l, refraction index n, and nonzero thermo-optic coefficient ($\mathrm{d}n/\mathrm{d}T \neq 0$). The optical path length through the body is $\mathrm{OPL} = n \cdot l$. If we increase uniformly the temperature of the body by a quantity ΔT, since the refraction index depends on the temperature (neglecting at first approximation the thermal expansion), the new OPL through the body becomes:

$$\mathrm{OPL}_1 = n_1 \cdot l \tag{9.2}$$

where

$$n_1 = n + \frac{\mathrm{d}n}{\mathrm{d}T}\Delta T \tag{9.3}$$

thus

$$\mathrm{OPL}_1 = \left(n + \frac{\mathrm{d}n}{\mathrm{d}T}\Delta T\right) \cdot l = n \cdot l + \frac{\mathrm{d}n}{\mathrm{d}T}\Delta T \cdot l = \mathrm{OPL} + \Delta\mathrm{OPL} \tag{9.4}$$

and we can define the quantity *optical path length increase* as

Fig. 9.2 Some of the many applications of finite element models

$$\Delta \text{OPL} = \frac{\mathrm{d}n}{\mathrm{d}T} \Delta T \cdot l \tag{9.5}$$

If we include the thermal expansion, Eq. 9.5 becomes [2]

$$\Delta \text{OPL} = \frac{\mathrm{d}n}{\mathrm{d}T} \Delta T \cdot l + \alpha(1+\sigma)(n-1)\Delta T \cdot l \tag{9.6}$$

where α is the thermal expansion coefficient and σ the Poisson's ratio. In the most general case of a nonuniform heating, the expression for the optical path length increase reads as

$$\Delta \text{OPL}(r,\theta) = \frac{\mathrm{d}n}{\mathrm{d}T} \int_0^l \Delta T(r,\theta,z)\mathrm{d}z + \alpha(1+\sigma)(n-1) \int_0^l \Delta T(r,\theta,z)\mathrm{d}z \tag{9.7}$$

Another contribution to the optical path length increase comes from the elasto-optic effect: the local refractive index changes with mechanical strain resulting from thermal expansion. Generally, this contribution is negligibly small compared to the thermo-optic and the thermoelastic terms and it is usually not accounted for.

In practice, most of our efforts are devoted to determining the temperature field inside the optics of the interferometer. This can be accomplished by following two methods: looking for an analytical solution or through Finite Element Method (FEM).

The FEM (also known as finite element analysis (FEA)) is a numerical technique for finding solutions to partial differential equations systems, when the analytical approach would be impossible or extremely difficult to implement. In simple terms, FEM divides a very complicated problem into small elements that can be solved in relation (usually linear or quadratic) to each other. Thus, FEM replaces partial differential equations with systems of algebraic equations.

Developed for solving problems in mechanical engineering, nowadays FEM is used in a variety of scientific disciplines, ranging from physics to biology (see Fig. 9.2).

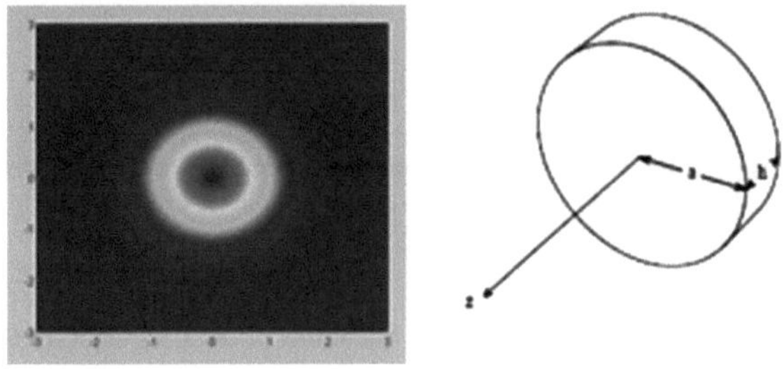

Fig. 9.3 *Left* plot of the intensity of a Gaussian laser beam. *Right* sketch of the cylindrical coordinate system used for the calculations

9.1.1.1 Analytical Approach: Hello-Vinet Theory

This section summarizes in brief the Hello-Vinet theory and a full explanation of the method can be found in the Virgo Physics Book, Chap. 2[1] and in [2].

The laser beam in the interferometer crossing the mirrors has a Gaussian profile (fundamental TEM_{00} mode), and the intensity as a function of the radial coordinate can be expressed as

$$I(r) = \frac{2P}{\pi w^2} e^{-\frac{2r^2}{w^2}} \tag{9.8}$$

It does not depend on any angular coordinate nor on z. Thus, we can exploit the symmetry of the problem and use cylindrical coordinates (r, z, θ), with $0 < r < a$ and $-h/2 < z < h/2$, with a the radius and h the thickness of the mirror, as shown in Fig. 9.3.

In the most general case, the time evolution of the temperature field is found by solving the Fourier equation:

$$\left[\rho C \partial_t - K_c \nabla^2\right] T(r, z, t) = p(r, z, t) \tag{9.9}$$

where T is the temperature increase (it was ΔT), ρ the density, C the specific heat, K_c the thermal conductivity, p the possible heat source inside the mirror, and $\partial_t = \partial/\partial t$.

We want to determine the steady-state ($\partial_t \to 0$) temperature field, when the heat lost by radiation balances the power absorbed by the mirror. If we limit ourselves only to the heat absorbed by the mirror surface, then there are no internal heat sources and the Fourier equation becomes the Laplace one

$$\nabla^2 T(r, z) = 0 \tag{9.10}$$

which in cylindrical coordinates can be rewritten as

$$\left(\partial_r^2 + \frac{1}{r}\partial_r + \partial_z^2\right) T(r, z) = 0 \tag{9.11}$$

[1] Available at https://wwwcascina.virgo.infn.it/vpb/vpb2.ps.gz.

The general solution to Eq. 9.11 is a harmonic function of the kind

$$T(r, z) = J_0(kr)\left(Ae^{kz} + Be^{-kz}\right) \tag{9.12}$$

where $J_0(x)$ are the Bessel functions of the first kind, A, B e k the constants determined by the boundary conditions.

On the mirror surface hit by the laser beam ($z = -h/2$), the boundary condition will be

$$-K_c\left[\frac{\partial T(r,z)}{\partial z}\right]_{z=-h/2} = -\sigma_e\left([T_0 + T(r, -h/2)]^4 - T_0^4\right) + \varepsilon I(r) \tag{9.13}$$

ε is the absorption coefficient of the mirror high reflectivity coating and σ_e is the emissivity of fused silica. Since $T \ll T_0$, because T represents the temperature increase and T_0 the room temperature, it is possible to linearize the term relative to irradiance and rewrite Eq. 9.13 as

$$-K_c\left[\frac{\partial T(r,z)}{\partial z}\right]_{z=-h/2} = -4\sigma_e T_0^3 T(r, -h/2) + \varepsilon I(r) \tag{9.14}$$

On the opposite surface ($z = h/2$), it will be

$$-K_c\left[\frac{\partial T(r,z)}{\partial z}\right]_{z=h/2} = 4\sigma_e T_0^3 T(r, h/2) \tag{9.15}$$

And finally, the irradiance boundary condition on the barrel of the mirror ($r = a$) reads as

$$-K_c\left[\frac{\partial T(r,z)}{\partial r}\right]_{r=a} = 4\sigma_e T_0^3 T(a, z) \tag{9.16}$$

Let us consider Eq. 9.16: we know that the general solution is described by Eq. 9.12 and that a Bessel function has the following property:

$$\partial_x J_0(x) = -J_1(x) \tag{9.17}$$

By inserting Eqs. 9.12 into 9.16 and using the property (9.17), it is straightforward to get

$$K_c k J_1(\text{ka}) = 4\sigma_e T_0^3 J_0(ka) \tag{9.18}$$

If we define $\chi \equiv \frac{4\sigma_e T_0^3 a}{K_c}$ and $\zeta = \text{ka}$, the previous equation becomes

$$\zeta J_1(\zeta) - \chi J_0(\zeta) = 0 \tag{9.19}$$

which has a discrete and infinite set of solutions $\{\zeta_n,\ n = 1, 2 \ldots\}$. Thus, the constant k gets values $k_n = \zeta_n/a$ and the general solution to Eq. 9.11 can be written as

$$T(r, z) = \sum_n \left(A_n e^{k_n z} + B e^{-k_n z}\right) J_0(\zeta_n r/a) \tag{9.20}$$

It can be shown from the Sturm–Liouville theorem that the functions $\{J_0(\zeta_n r/a),\ n = 1, 2 \ldots\}$ form a complete orthogonal basis for functions defined in the interval $[0,\ a]$. Thus, any function $I(r)$ can be expanded in series of J_0 in $[0,\ a]$ (Fourier–Bessel series).

$$I(r) = \sum_{n=0}^{\infty} I_n J_0(\zeta_n r/a) \tag{9.21}$$

and

$$I_n = \frac{2\zeta_n^2}{a^2\left(\chi^2 + \zeta_n^2\right) J_0(\zeta_n)^2} \int_0^a I(r) J_0(\zeta_n r/a) r \mathrm{d}r \tag{9.22}$$

In the case of interferometric detectors, $I(r)$ is given by Eq. 9.8, so that

$$I_n = \frac{2\zeta_n^2}{a^2\left(\chi^2 + \zeta_n^2\right) J_0(\zeta_n)^2} \int_0^a \frac{2P}{\pi w^2} e^{-\frac{2r^2}{w^2}} J_0(\zeta_n r/a) r \mathrm{d}r \tag{9.23}$$

Moreover, to minimize clipping losses the radius of the mirrors is chosen to be much larger than the size of the Gaussian laser beam, so that in Eq. 9.23 it is possible to replace the upper integration limit with ∞, getting

$$I_n = \frac{P}{\pi w^2} \frac{2\zeta_n^2}{a^2\left(\chi^2 + \zeta_n^2\right) J_0(\zeta_n)^2} e^{-\frac{w^2 \zeta_n^2}{8a^2}} \tag{9.24}$$

and finally, the intensity profile of the laser Gaussian beam becomes

$$I(r) = \frac{P}{\pi w^2} \sum_{n=0}^{\infty} \frac{2\zeta_n^2}{a^2\left(\chi^2 + \zeta_n^2\right) J_0(\zeta_n)^2} e^{-\frac{w^2 \zeta_n^2}{8a^2}} J_0(\zeta_n r/a) \tag{9.25}$$

Using Eqs. 9.20 and 9.25, the boundary conditions (9.14) and (9.15) reduce to a linear system of equations, where A_n and B_n are the unknowns

$$(\zeta_n - \chi)\, \Gamma_n^2 A_n - (\zeta_n + \chi)\, B_n = \frac{-\varepsilon I_n a \Gamma_n}{K_c} \tag{9.26}$$

$$(\zeta_n + \chi)\, A_n - (\zeta_n - \chi)\, \Gamma_n^2 B_n = 0 \tag{9.27}$$

having defined, for the sake of simplicity, $\Gamma_n = e^{-\frac{\zeta_n h}{2a}}$. Solving for A_n and B_n is straightforward and we are finally able to write the equation describing the

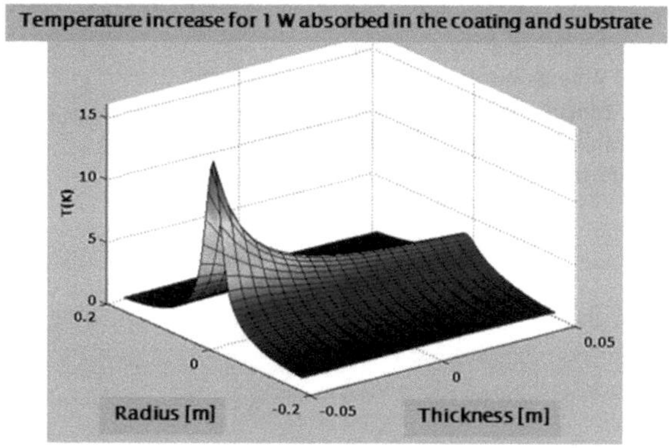

Fig. 9.4 Temperature field inside a Virgo mirror for 2 W total absorbed power

temperature field into a cylindrical mirror that absorbs a fraction ε of the power stored in a Gaussian-profiled laser beam

$$T(r,z) = \sum_n \frac{\varepsilon I_n a}{K_c} e^{-\frac{\zeta_n h}{2a}} \frac{(\zeta_n - \chi)\, e^{-\frac{\zeta_n (h-z)}{a}} + (\zeta_n + \chi)\, e^{-\frac{\zeta_n z}{a}}}{(\zeta_n + \chi)^2 - (\zeta_n - \chi)^2\, e^{-\frac{2\zeta_n h}{a}}} J_0(\zeta_n r/a) \quad (9.28)$$

In a similar manner, it is possible to determine the temperature field when the heating occurs inside the bulk of the mirror [2]. Figure 9.4 shows an example of temperature field inside a Virgo mirror for 1 W of power absorbed on the high reflectivity surface and in the substrate.

We are now able to evaluate the optical path length increase. Inserting Eqs. 9.28 into 9.7 and performing the integration along the z direction, that represents the thickness of the mirror, one obtains

$$\Delta\mathrm{OPL}(r) = \left(\frac{\mathrm{d}n}{\mathrm{d}T} + \alpha(1+\sigma)(n-1)\right) \sum_n \frac{\varepsilon I_n a^2}{K_c \zeta_n} \frac{1 - e^{-\frac{\zeta_n h}{a}}}{\zeta_n + \chi - (\zeta_n - \chi)\, e^{-\frac{\zeta_n h}{a}}} J_0(\zeta_n r/a) \quad (9.29)$$

An analog expression could be found for the optical path length increase due to substrate absorption [2]. The total ΔOPL is the sum of these two contributions. Figure 9.5 shows the optical path length increase due to coating absorption in a Virgo mirror for 1 W of absorbed power. One important thing to notice is that thermal lensing does not create a spherical lens.

9.1.2 Consequences of Thermal Effects

An important point to clarify is where thermal lensing is occurring in a GW interferometric detector. Let us consider the scheme shown in Fig. 9.6, where the

Fig. 9.5 Optical path length increase inside a Virgo mirror for 2 W total absorbed power as a function of the radial coordinate (0 is the center of the mirror)

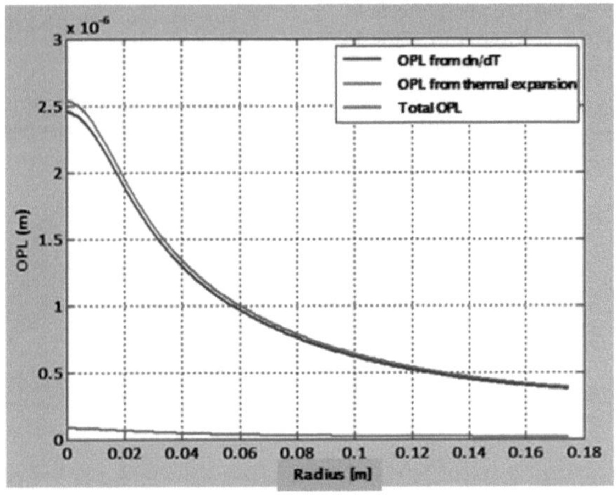

Fig. 9.6 Linear scheme of a power-recycled Michelson interferometer with Fabry–Perot arm cavities

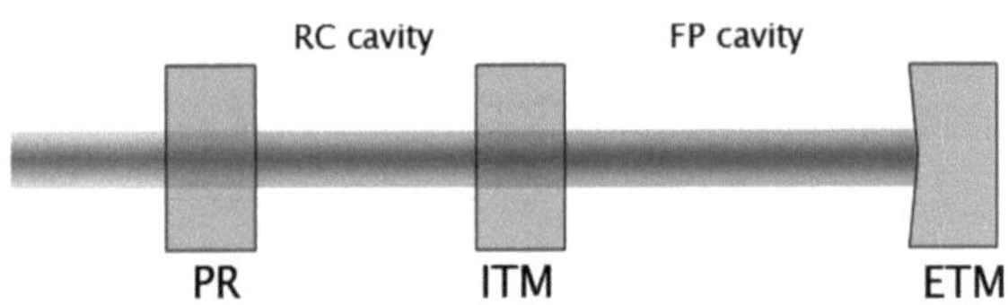

interferometer has been "linearized," i.e., only two coupled cavities, the power recycling and one Fabry–Perot, are drawn.

The Power Recycling (PR) mirror will absorb for sure some power on its high reflectivity surface and in its bulk. So, the refraction index inside the substrate will change due to the temperature increase, thus there will be some thermal lensing. However, the PR bulk lies outside the interferometer. In fact, the Power Recycling cavity is defined by the High Reflectivity (HR) surfaces of the PR mirror and input test mass. We can thus conclude that the thermal lensing in the PR mirror does not affect the interferometer.

Similar considerations apply to the end test masses (ETM): their substrate is out of the Fabry–Perot cavity, thus out of the interferometer and is not crossed by the beam.

Input Test Masses (ITM) absorb some of the power stored in the recycling cavity in their substrates, plus they absorb some of the Fabry–Perot cavity power on the high reflectivity coating. So, the ITM substrate is where the thermal lensing occurs, as it is inside the power recycling cavity.

We know that three fields circulate inside the interferometer: the carrier and the lower and upper radio frequency sidebands. Thermal lensing will affect only the control sidebands. To help understand this sentence, we can take a look at Fig. 9.7. The carrier resonates both in the recycling and in the Fabry–Perot cavities. It will sense the thermal lensing in the ITM substrate, but when it enters the FP cavity it will be cleaned by the Finesse of the cavity itself that acts as a filter.

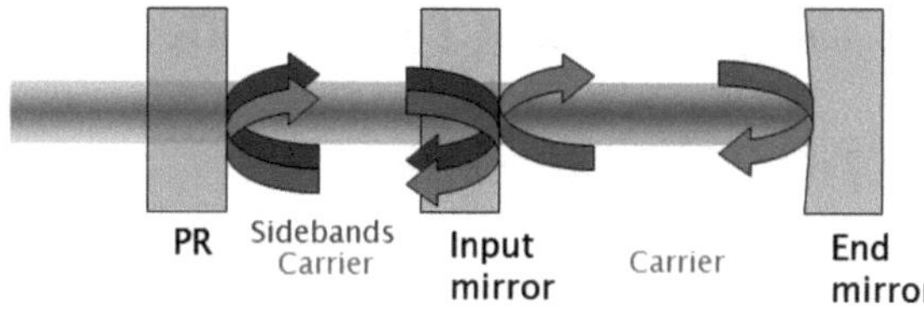

Fig. 9.7 Resonant conditions of the fields circulating in the interferometer

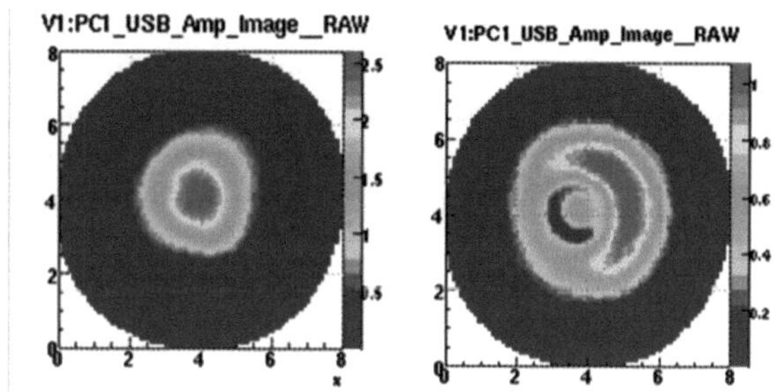

Fig. 9.8 Amplitude images of a sideband field. *Left* with no thermal lensing. *Right* in presence of strong thermal lensing. Lack of cylindrical symmetry may be due to some tilted optic

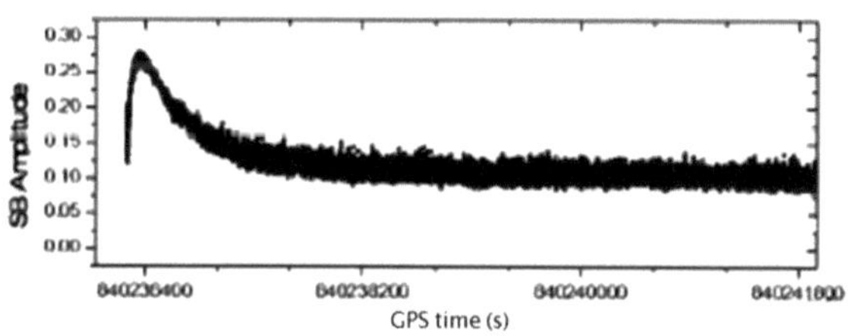

Fig. 9.9 Typical transient decrease of the sidebands amplitude as observed during a Virgo lock acquisition

On the contrary, the sidebands only resonate in the recycling cavity and, so, they do not take advantage of the filtering effect of the FP cavity. Radio frequency sidebands fully sense the thermal lens in the ITM substrate. Since thermal effects give rise to a nonspherical lens, the sidebands fields in the recycling cavity are strongly aberrated (see Fig. 9.8).

The cavity, seen by the sidebands, is "less" resonant and so the sidebands power decreases (as shown in Fig. 9.9). At a certain level, the sidebands amplitude becomes too low to keep the interferometer locked. In this situation, the only solution is to decrease the power at the input of the interferometer, to reduce the strength of the thermal lens. However, in this way, the sensitivity of the detector is degraded because of the increase in the shot noise.

9.2 Compensation of Thermal Effects in Virgo

9.2.1 General Guidelines

To correct the thermal lens, one could think, at first glance, to introduce in the recycling cavity two spherical lenses, one for each ITM. However, we have seen that thermal lensing generates a lens that is not spherical. Moreover, the "strength" of the lens, that is the focal length, changes according to the interferometer input power.

Somehow, we must induce in the input mirrors a lens equal but opposite to the thermal one. We must use an adaptive (flexible) system in order to be able to change at will the strength and shape of the corrective lens, to follow the different interferometer (ITF) operating conditions.

Another important point to establish is to what extent the OPL must be flattened. The input mirrors are cylinders with a radius of 175 mm, so is it necessary to have a flat OPL up to the edge? The answer to this question comes from optical simulations. In fact, we must recall that the size of the ITF beam is much smaller than that of the mirror. So, the beam will sense optical path length distortions only where it exists. In practice, it turns out that the extent over which the OPL must be corrected is approximately 1.5 times the size of the beam on the ITMs. For instance, in Virgo the YAG laser radius on the ITM was about 22 mm, thus the area to be compensated had a dimension of about 33 mm.

Finally, from Eq. 9.7, we can see that thermal lensing is only given by the radial component of the temperature gradient. Thus, to flatten the OPL, we need to null this component. This can be done by heating the peripheral of the mirrors. Not all the techniques are suitable for this goal as the mirrors are the free-fall test masses of our GW detector. A "touchless" way to heat the mirror is by shining it with a radiation that is completely absorbed ($>5\,\mu$m for fused silica). For this reason, LIGO and Virgo thermal compensation systems (TCS) used CO_2 ($\lambda = 10.6\,\mu$m) lasers to heat the peripheral of the input test masses.

9.2.2 The Virgo Scheme

The wavefront distortions in the recycling cavities have been corrected in Virgo by shining an annular heating pattern generated by a CO_2 laser, directly on the input test masses [3]. Figure 9.10 shows the layout of the Virgo TCS optical bench. Two identical benches were installed, one for each input test mass.

An axicon lens (an optic with a conical surface) is used to convert the laser Gaussian beam into an annular beam. The left image of Fig. 9.11 shows the working principle of an axicon lens, while on the right a measured intensity profile of a CO_2

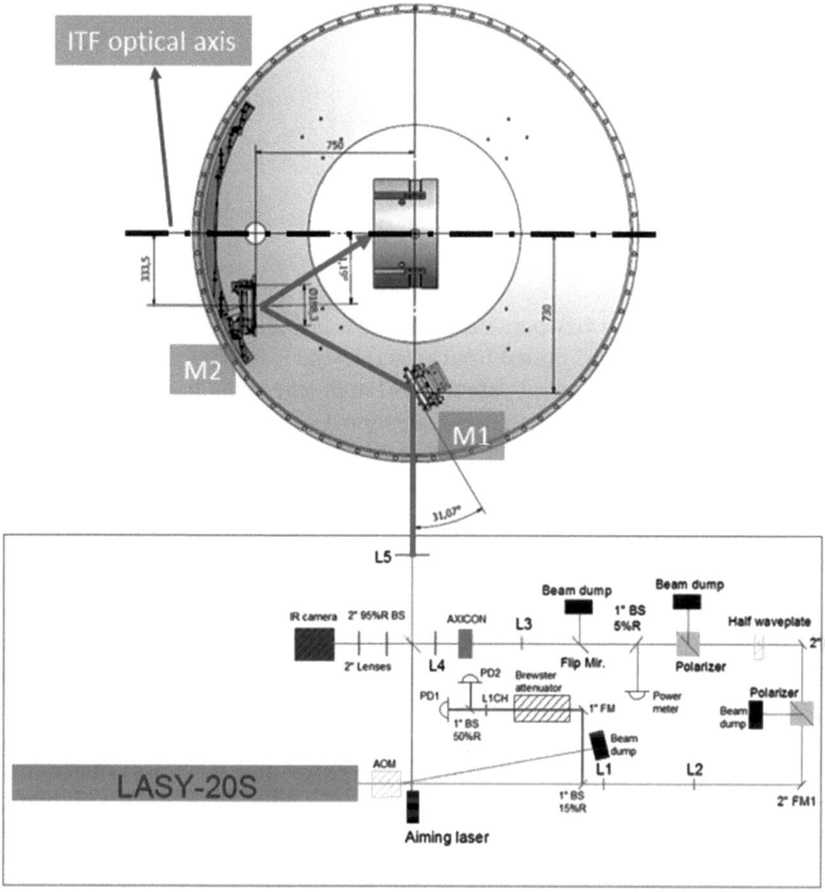

Fig. 9.10 Layout of the Virgo TCS bench. The beam emitted by the laser crosses an acousto-optic modulator for intensity stabilization, polarization optics (half wave plate and polarizer) for DC power control, the axicon, and an imaging lens

laser after going through an axicon is reported. An acousto optic modulator[2] is used to reduce the intensity noise of the laser, while DC control of the power content of the beam is accomplished using a half-wave plate and a fixed polarizer. Such a system allows to significantly reduce the deviation of the beam impinging on the axicon, which would otherwise result in a nonhomogeneous intensity distribution inside the

[2] An acousto-optic modulator (AOM), also called a Bragg cell, uses the acousto-optic effect to diffract and shift the frequency of light using sound waves (usually at radio-frequency). A piezoelectric transducer is attached to a material such as glass. An oscillating electric signal drives the transducer to vibrate, which creates sound waves in the glass. These can be thought of as moving periodic planes of expansion and compression that change the index of refraction. Incoming light scatters off the resulting periodic index modulation and interference occurs similar to Bragg diffraction. The interaction can be thought of as four-wave mixing between phonons and photons.

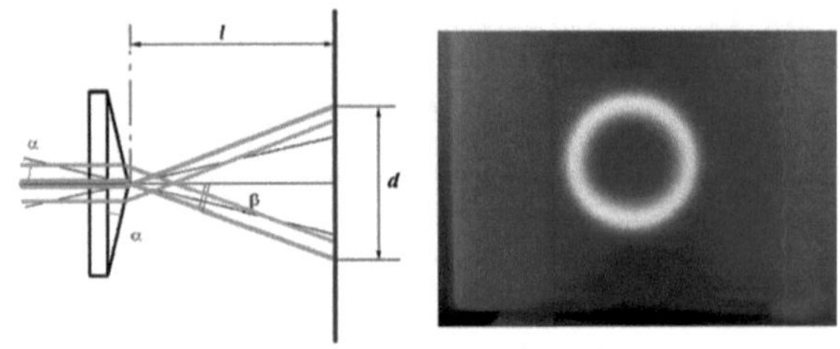

Fig. 9.11 *Left* working principle of an axicon lens. *Right* intensity profile of a laser beam after passing through an axicon

annular pattern. Since no viewport facing the ITMs is available, the TCS beam enters the tower where the ITM is hosted from a lateral window and two in-vacuum optics (M1 and M2 in Fig. 9.10) steer it onto the ITM high reflectivity face. Mirror M2 can be remotely actuated to allow a precise alignment of the heating pattern onto the test mass. A high sensitivity infrared camera is installed on each bench to constantly monitor the quality of the heating beam sent to the test masses. Moreover, an active control loop is used to reduce the CO_2 laser intensity fluctuations. The signal from a high sensitivity infrared photodiode is band-passed and fed back through an acousto-optic modulator.

The efficiency of the Virgo TCS has been tested by progressively increasing the CO_2 power, while looking at the behavior of the interferometer and at the optical phase and amplitude of the carrier and modulation sidebands recorded by a phase camera [4]. The TCS power was increased until the thermal effects in the input mirrors were totally compensated [3, 5], as if the interferometer was operating at low power. In the upper left plot of Fig. 9.12 it is possible to see that when no correction is applied the sideband field is strongly aberrated. The other plots in the same figure show that, as the TCS power is increased, the sidebands amplitude recovers the expected Gaussian shape.

Another indicator that the TCS is working correctly is represented by the sidebands power. As it can be seen from Fig. 9.13, when the TCS is on, the sidebands power increases by about a factor of 5.

The improvement of the interferometer operating point could be confirmed also by the increase of the optical gain (about 50 %) and by the shape of the transfer function of the second stage frequency stabilization loop, which looks very close to what should be in absence of thermal effects. The performances of the TCS allowed the interferometer to run VSR2 [6] with 17 W input power compared to 8 W during VSR1 [7].

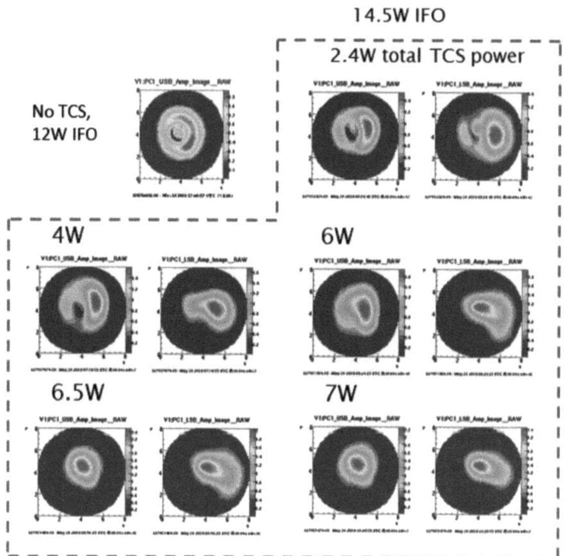

Fig. 9.12 Amplitude images of the sidebands recorded with a phase camera. *Upper left* no TCS (one only is shown for simplicity). The other plots show the change in shape as a function of the TCS power

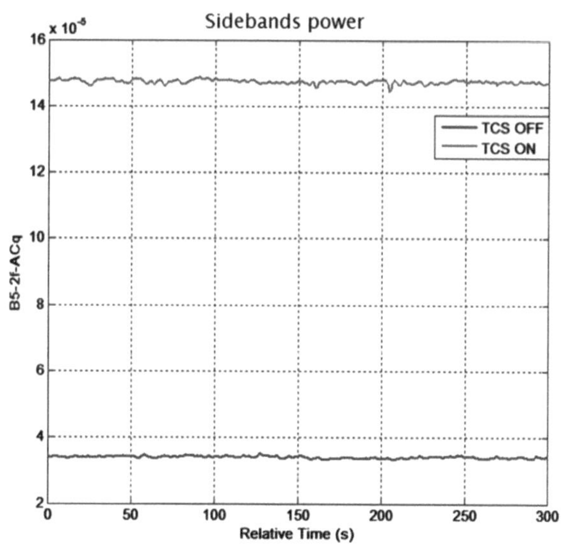

Fig. 9.13 Sidebands power as a function of time when the TCS is off (*bottom, blue line*) and when it is on (*top, green*). The increase is about a factor of 5

9.3 Adaptive Optical System for Advanced Virgo

9.3.1 Thermal Effects in Advanced Detectors

Advanced detectors aim to improve the sensitivity by a factor of 10 at all frequencies. Thus, also where the shot noise is limiting. Figure 9.14 shows the expected sensitivity

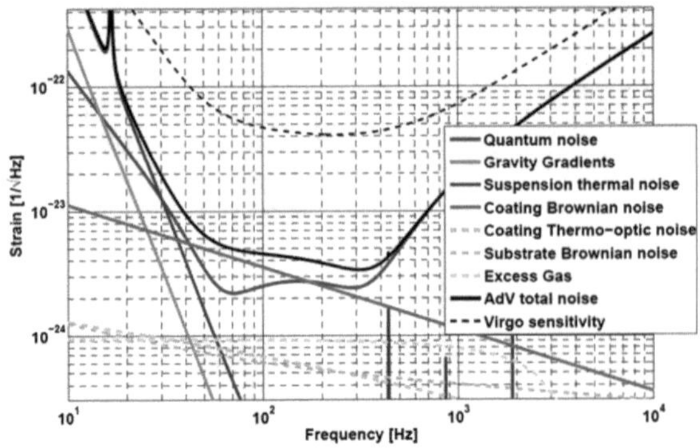

Fig. 9.14 Advanced Virgo design sensitivity plotted against the Virgo sensitivity (*dashed curve*). The major noise contributions are also evidenced

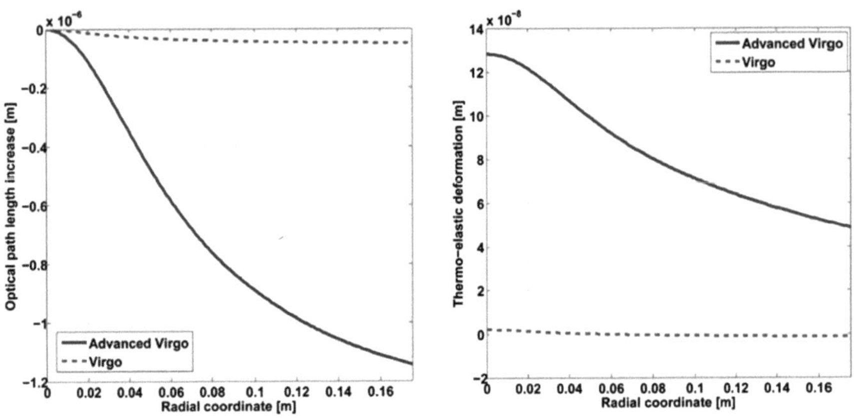

Fig. 9.15 *Left* optical path length increase in Advanced Virgo (*blue*) and Virgo (*green dashed*). *Right* thermoelastic deformation of the HR surface in Advanced Virgo (*blue*) and Virgo (*green dashed*)

curve for Advanced Virgo [8] compared to the Virgo one. Several technical upgrades will be implemented to achieve this result. Among others, what is relevant for the issue of thermal effects is represented by the increase in the ITF input power (from 25 to 125 W) and in the arm cavity finesse (from 150 to 450). These improvements lead to a considerable increase in the optical power stored in the detector, from about 20 to about 700 kW. Considering an absorption of 1 ppm on the HR coating, it means that the absorbed power in advanced detectors test masses is about 0.7 W with respect to 20 mW in initial detectors.

The consequence is an increase in the amplitude of thermal lensing by more than one order of magnitude, as shown in the left plot of Fig. 9.15.

Moreover, another effect, that so far has been neglected, becomes relevant. The right plot of Fig. 9.15 shows the thermoelastic deformation of the high reflectivity surface of the Advanced Virgo test masses, again compared to Virgo. The deformation will increase the Radius of Curvature (RoC) of all mirrors (by about 40 m in Advanced Virgo) and, thus, also the Fabry–Perot cavities will be affected by thermal aberrations. As a consequence, the Fabry–Perot cavity will become less concentric, and the spot sizes at the mirrors will shrink, leading to an increase in thermal noise. Moreover, higher order modes can become resonant in the arm cavities, considerably increasing round trip losses. To maintain the arm cavity mode structure, it will be then necessary to control the radii of curvature of all test masses (within ± 2 m from the initial RoC in Advanced Virgo).

The need to control the radius of curvature of the test mass in GW interferometric detectors has already been faced in the past: the GEO detector [9] used a ring heater (RH) to change the RoC of one of the two test masses [10]. The GEO RH is placed on the back of the mirror, radiatively coupled with the optic.

Compensation and control of the test mass high reflectivity (HR) surfaces will be accomplished in Advanced Virgo with the same technique. The TCS baseline design considers four ring heaters, one around each test mass. The input mirror RH also provides limited compensation of thermo-optic effect in the recycling cavities. Unlike the GEO heater, the Advanced Virgo RH will be equipped with a reflecting shield to maximize the amount of power reaching the test mass.

9.3.2 TCS Noise Coupling

There is an issue that so far has been neglected: TCS can inject displacement noise into the detector. In fact, the intensity fluctuations of the CO_2 laser "shake" the test mass through several coupling mechanisms [11].

There are four distinct mechanisms that do contribute (see also Chap. 8):

- **Thermoelastic (TE)**: fluctuations in locally deposited heat cause fluctuations in local thermal expansion. This effect mostly changes the FP cavity length. The change in the optical path length outside the cavity is small for this effect.
- **Thermo-refractive (TR)**: fluctuations in locally deposited heat cause fluctuations in local refractive index. This effect changes the optical path length of the light outside the FP cavity.
- **Flexure (F)**: fluctuations in locally deposited heat cause fluctuations in global shape of the optic. The deposited heat effectively curls the optic like a bimetallic strip. The area read out by the arm cavity beam will move even if there is no overlap with the heating beam.
- **Radiation pressure**: negligible.

When the heating pattern has an annular shape, the dominant coupling mechanism is the Flexure.

Thus, the total displacement noise due to thermo-optical intensity noise coupling is [11]:

$$\langle \Delta z_{\mathrm{TCS}} \rangle = \frac{P_{\mathrm{CO_2}}}{2\pi f C \rho} \left(\Theta \left[(1+\sigma)\alpha \left(1 - \frac{\pi}{2F}(n-1)\right) - \frac{\pi}{2F}\frac{\mathrm{d}n}{\mathrm{d}T} \right] + \frac{6\alpha}{h^2} C_{\mathrm{num}} \right) \times \mathrm{RIN} \left[\frac{\mathrm{m}}{\sqrt{\mathrm{Hz}}} \right] \tag{9.30}$$

where $P_{\mathrm{CO_2}}$ is compensating power, C the specific heat, ρ the density, σ the Poisson's ratio, α the linear thermal expansion coefficient, $\mathrm{d}n/\mathrm{d}T$ the thermo-optic coefficient, and n the refraction index of the substrate's material. Θ is a constant proportional to the overlap integral between the ITF and TCS beams, h the thickness of the mirror, and F the finesse of the arm cavity. C_{num} is a parameter that depends on the geometric dimensions of the TCS beam and that is numerically evaluated. Finally, the RIN is the relative intensity noise of the CO_2 laser, $\delta P/P$ measured in $1/\sqrt{\mathrm{Hz}}$.

Intensity stabilization servos have already been implemented in initial detectors, capable of reducing the CO_2 laser fluctuations by one order of magnitude over the whole frequency band of interest [3]. However, the limit of $10^{-7}/\sqrt{\mathrm{Hz}}$, due to the intrinsic noise of the photodetectors, is not enough for advanced interferometers. The stabilization servo should gain another factor of 10 to bring the TCS noise below the sensitivity requirements. This is not possible with the present technology.

Thus, it is no more possible to shine the input test masses directly with the CO_2 laser, as it is done in initial detectors. This implies the need of an additional transmissive optic, named Compensation Plate (CP) to act on with the compensating beam. The CPs are placed in the recycling cavity, the thermoelastic and flexure terms become zero and the total displacement noise is reduced by a factor equal to the inverse of the Fabry–Perot cavity gain ($\pi/2F$).

9.3.3 TCS in Advanced Virgo

Following the above considerations, the conceptual actuation scheme of the compensation system foreseen for Advanced Virgo [12, 13] is shown in Fig. 9.16. A pictorial view of the input payload, with the ITM, RH, and CP, is shown in Fig. 9.17.

This scheme also allows to reduce the coupling between the two degrees of freedom (lensing and RoC), so as to have a control matrix as diagonal as possible.

The thickness of the CP has been optimized by minimizing the heat escaping from its barrel and taking into account the need to accumulate enough optical path length. The distance between CP and ITM is 20 cm, this allows to minimize the radiative coupling between the two optics. In fact, the heated CP radiates heat toward the test mass. The heating of the TM is uniform, but since the barrel of the input mirror radiates away a part of the heat, a radial temperature gradient is established. This gives rise to an increase in optical path length that adds to the thermal lensing.

The position of the RH along the barrel of the TM is such as to maximize its efficiency.

Fig. 9.16 Actuation scheme of the Advanced Virgo TCS: *blue rectangles* represent the CPs (heated by the CO_2 lasers) while the *green dots* around the test masses are the ring heaters

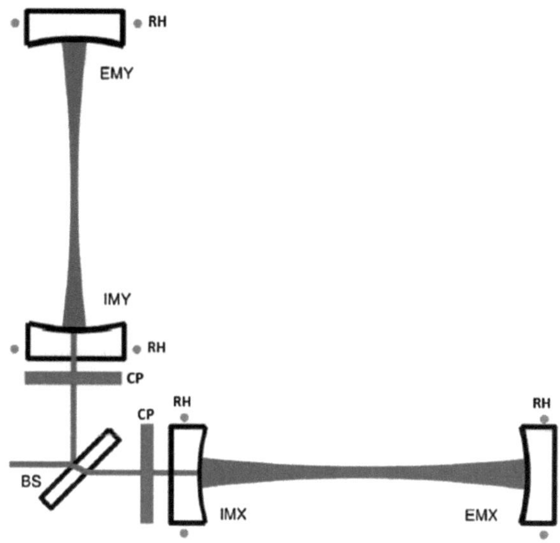

Fig. 9.17 Picture of the input payload, comprising the input test mass with the ring heater and the compensation plate

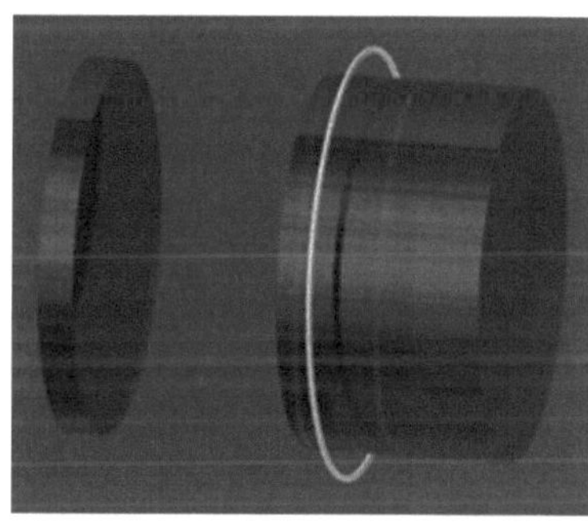

A useful way to picture the effect of thermal distortions is the fractional power scattered out from the TEM_{00} mode [14], termed "coupling losses" L:

$$L = 1 - A^* A, \tag{9.31}$$

where

$$A = \frac{\langle E_0 | E \rangle}{\langle E_0 | E_0 \rangle} = \frac{\langle E_0 | e^{i2\pi/\lambda h(x,y)} | E_0 \rangle}{\langle E_0 | E_0 \rangle} = 2\pi \int_0^{r_{\mathrm{ITM}}} e^{i2\pi/\lambda h(r)} |E_0(r)|^2 r dr. \tag{9.32}$$

E_0 represents the normalized undisturbed cavity field before being subjected to the phase distortion $2\pi/\lambda \cdot h(x, y)$, E is the distorted field, r_{ITM} is the radius of the input test mass, and $h(x, y)$ is the optical path length increase in the recycling cavity. The last equality in Eq. 9.32 holds if cylindrical symmetry is assumed. A phase distortion in a cavity acts to scatter power out of the fundamental mode into higher order modes, and thus out of the cavity, and can be viewed as a simple coupling loss term.

Another figure of merit often used to describe the amplitude of aberrations is the Gaussian-weighted Root-Mean-Squared (RMS) of the optical path length increase. The weighting function represents the intensity of the beam circulating in the interferometer, which, indeed, is "sensing" the aberrations. This quantity is defined as [15]

$$\sigma_g^2 = \frac{\sum_{m,n} G(x_m, y_n)[h(x_m, y_n) - h_g]^2}{\sum_{m,n} G(x_m, y_n)} \tag{9.33}$$

where $G(x, y)$ is the interferometer beam intensity at (x, y), and h_g is given by

$$h_g = \frac{\sum_{m,n} G(x_m, y_n) h(x_m, y_n)}{\sum_{m,n} G(x_m, y_n)} \tag{9.34}$$

With no thermal compensation, in Advanced Virgo, the sideband fields coupling losses would amount to approximately $5 \cdot 10^5$ ppm and the corresponding RMS is about 125 nm. For a comparison, in Virgo, the losses due to thermal lensing are of the order of 10^4 ppm. The Advanced Virgo TCS needs to reduce to coupling losses at least by a factor of 10^3 (corresponding, roughly speaking, to a maximum RMS of about 2 nm [16, 17]) to allow the correct operation of the detector at design sensitivity.

Enhanced interferometers used axicon-based optical projectors to convert a CO_2 laser Gaussian beam into an annular beam [3]. This heating pattern is not adequate for the advanced detectors, since it can reduce the residual coupling losses only by a factor of 10^2, while at least a factor of 10^3 is required. Studies on the optimization of the heating pattern, made with Finite Element Model (FEM), have shown that it is possible to reduce the residual coupling losses to about 6 ppm, thus leading to a reduction factor of about 10^5, with about 18 W of CO_2 power shined on the compensation plate.

9.3.4 *Other Sources of Wavefront Aberrations in Advanced Virgo Recycling Cavities*

So far, we have focused our attention only on thermal effects, both for the optical path length increase in the recycling cavity and for the increase of the test masses RoCs in the Fabry–Perot cavity. Moreover, up to now we have considered only effects arising from uniform absorption, i.e., with cylindrical symmetry.

However, in an ITF there are several sources of optical defects, often without any symmetry, arising from imperfections in the production and polishing of the glass used for the various substrates in the recycling cavity. Surface figure errors on reflective and transmissive surfaces do contribute to the aberrations as well as spatial variations in the index of refraction of the substrates (see Fig. 9.18).

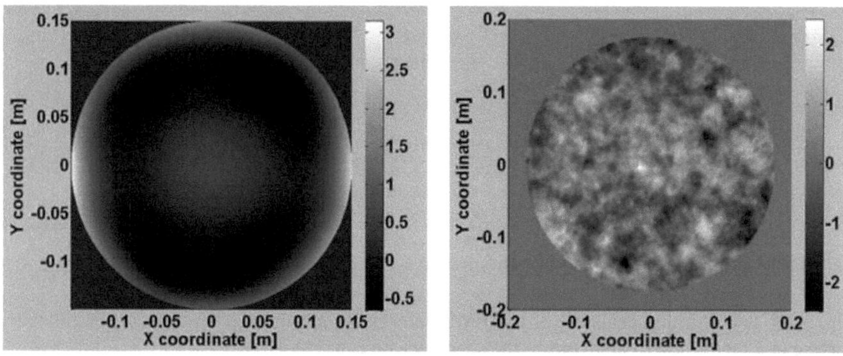

Fig. 9.18 *Left* substrate transmission map measured at LMA on an advanced LIGO test mass, the color scale is 10^{-7} m. *Right* simulated surface map. *Color scale* in nanometers

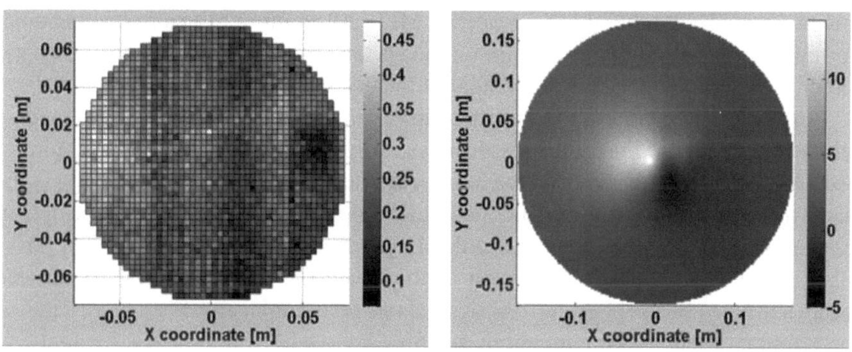

Fig. 9.19 *Left* coating absorption map measured at LMA on an advanced LIGO test mass. *Right* resulting nonuniformity of the OPL for 125 W of ITF input power. *Color scale* in nanometers

Finally, we also have to consider the nonhomogeneity of the high reflectivity coating absorptions which generates a nonsymmetric optical path length increase, as shown in Fig. 9.19.

The sum of these aberrations on all the surfaces in the recycling cavity exceeds the 2 nm RMS requirement. Thus, the Advanced Virgo recycling cavity gain for the sidebands is strongly reduced [18]. As a consequence also these sources of aberrations must be corrected. As already done for the symmetric contribution, the Finite Element Method can be used to evaluate the corrective heating pattern to be applied to the compensation plates. Due to the lack of symmetry of the optical defects, full 3D modeling is required, making such simulations rather computationally expensive (few days of CPU time).

The results of the optimization procedure show that, by depositing heat in the right CP positions, the residual optical path length RMS can be reduced by a factor of 20 for spatial frequencies below 40 m^{-1} [19] and amounts to 0.35 nm, well within the requirements.

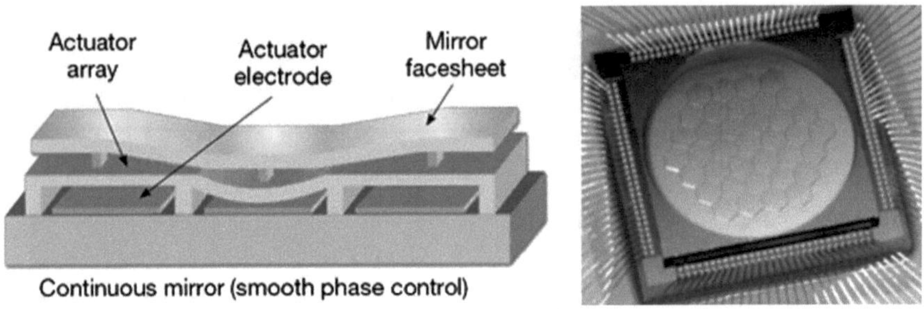

Fig. 9.20 *Left* working principle of a MEMS deformable mirror. *Right* picture of the reflecting surface of a mems deformable mirror, the segments are clearly visible

The method selected to generate the heating patterns is based on a CO_2 laser scanning system. This technique, developed at MIT [20], comprises a pair of galvanometer mirrors, to move the laser beam on the surface of the CP, and an acousto-optic modulator to modify the power content of the beam.

An alternative method under investigation exploits the new possibilities offered by the development of Micro-Electro-Mechanical-System (MEMS) deformable mirrors (see Fig. 9.20), often used in adaptive optical systems. The working principle is the same as for Liquid-crystal-on-silicone spatial light modulators (or diffractive optical elements) used to generate higher order Laguerre-Gauss modes [21, 22]: a phase plate imprints the phase modulation to the wavefront of the incoming Gaussian beam and converts the beam into the desired one holographically. For a more detailed description of the working principle, see for example [23]. In practice, by applying a known phase modulation, it is possible to change the intensity distribution.

9.3.5 Sensing Wavefront Aberrations in Recycling Cavities

The aberrations in the recycling cavity optics will be sensed by several complementary techniques. The amplitude of the optical path length increase will appear in some ITF channels, such as the power stored in the radio frequency sidebands. These are scalar quantities that can only give a measurement of the amount of power scattered into higher order modes. Moreover, phase cameras [4] will sense the intensity distribution and phase of the fields in the recycling cavity (carrier and sidebands). However, active control of the aberrations using error signals coming from such detectors seems a difficult task.

Therefore, each optic with a significant thermal load will be independently monitored. The HR face of each test mass will be monitored in reflection for deformation. The input test mass/compensation plate phase profile will be monitored on reflection on-axis from the recycling cavity side. The TCS control loop will then use a blend of all the signals from the different channels.

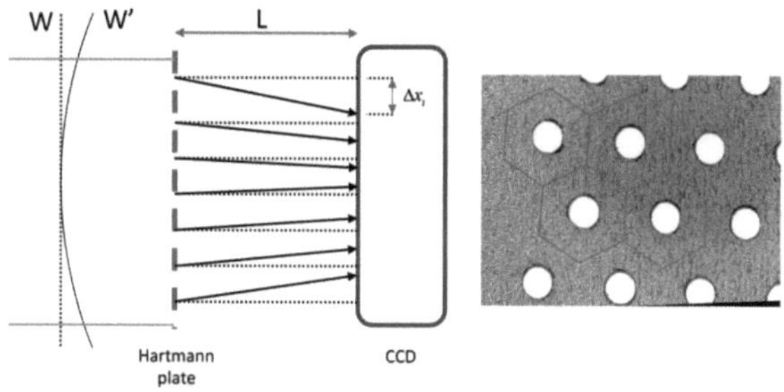

Fig. 9.21 *Left* working principle of the Hartmann Wavefront Sensor. *Right* close view of a Hartmann plate

The TCS sensors, dedicated to the measurement of thermally induced distortions, consist of a Hartmann Wavefront Sensor (HWS) and a probe beam (at a different wavelength than the ITF beam) whose wavefront contains the thermal aberration information to be sensed. The working principle of the device is shown in the left image of Fig. 9.21. An aberrated wavefront W' is incident on a Hartmann plate (essentially a plate containing a series of apertures, see right image of Fig. 9.21). The resulting rays propagate a distance L, normal to the wavefront, and are incident on a CCD. The new spot position, x_i', is measured and compared to a reference spot positions, x_i, determined using a non-aberrated wavefront W. The wavefront gradient in the ith position is given as

$$\frac{\partial \Delta W}{\partial x} = \frac{\Delta x_i}{L} \tag{9.35}$$

The Hartmann sensor selected for Advanced Virgo is that already developed and characterized on test bench experiments and in the Gingin High Optical Power Test Facility for the measurement of wavefront distortion [24].

This sensor has been demonstrated to have a shot-to-shot reproducibility of λ/1450 at 820 nm, which can be improved to λ/15500 with averaging, and with an overall accuracy of λ/6800 [25].

Due to the requirement on the maximum residual optical path length rms of 2 nm, the precision of the wavefront sensing must be of the order of 0.2 nm, corresponding to λ/4000 @ 800 nm (the wavelength of the probe beam) and must be compared with the Hartmann sensor measured noise level of λ/15500.

The key to the wavefront measurement is to relay the image of the measured optical surface onto the Hartmann plate. In other words, if M is the magnification of the optical system between the measured optical surface and the wavefront sensor surface, then the wavefront gradient is

$$\left(\frac{\partial \Delta W}{\partial x}, \frac{\partial \Delta W}{\partial y}\right) \tag{9.36}$$

at coordinate (x, y) on the optic will result in a measurement of the gradient

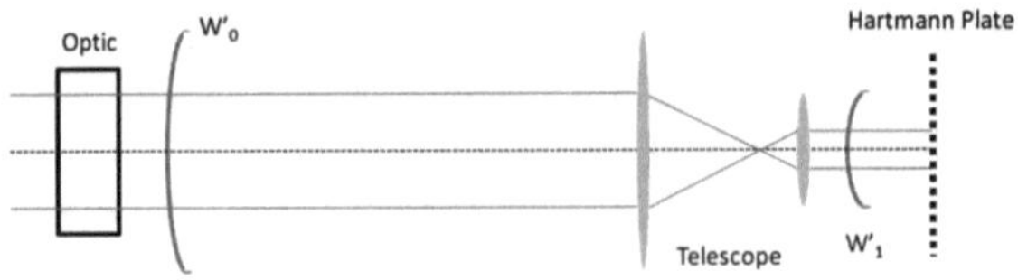

Fig. 9.22 Working principle of the optical system for wavefront measurement

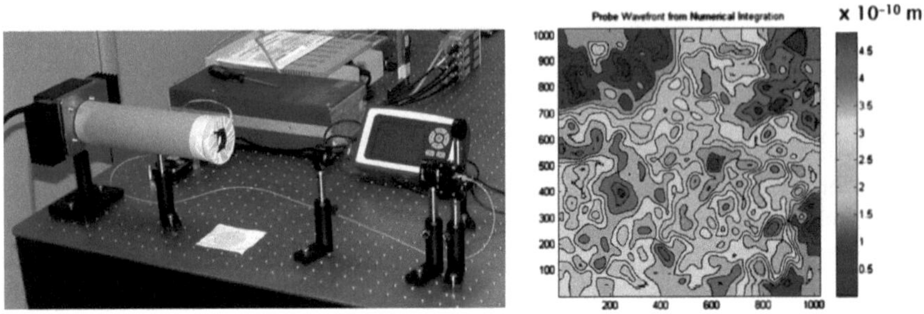

Fig. 9.23 *Left* experimental setup for the characterization of the Hartmann sensor. *Right* measured wavefront, 100 frames were averaged for this measurement

$$\left(M\frac{\partial \Delta W}{\partial x}, M\frac{\partial \Delta W}{\partial y}\right) \tag{9.37}$$

at sensor coordinates $(x/M, y/M)$. The general optical scheme for such measurements is shown in Fig. 9.22.

A prototype of a Hartmann sensor is currently being characterized in the Tor Vergata laboratories. Figure 9.23 shows, on the left, the experimental setup, while on the right a measured wavefront averaged over 100 frames, at a speed of 60 frames/s. The peak-to-valley optical path length increase is about 0.5 nm, corresponding to a RMS of 0.1 nm or $\lambda/6500$ at 650 nm.

The conceptual layout of the HWS beams in the recycling cavity is shown in Fig. 9.24. In the figure, the beams have different colors for the sake of clarity; indeed the wavelength is the same for both beams. The beams will be injected/extracted from the injection and detection suspended benches, superposed to the main ITF beam with a dichroic mirror[3] at the level of the mode matching telescopes. This scheme allows for on-axis double pass wavefront measurement (that increases the signal-to-noise ratio by a factor of 2) and for probing of all the optics in the recycling cavity.

Additional optics are necessary to fulfill the main optical requirements for HWS beams: to image a plane around the ITM HR surface on the Hartmann plate, to illuminate the ITM with a Gaussian beam 10 cm in size and to match the optimal beam size on the sensor. The two sensing beams are kept separated by the BS HR

[3] Dichroic mirrors are color filters often used to reflect visible light and transmit infrared radiation.

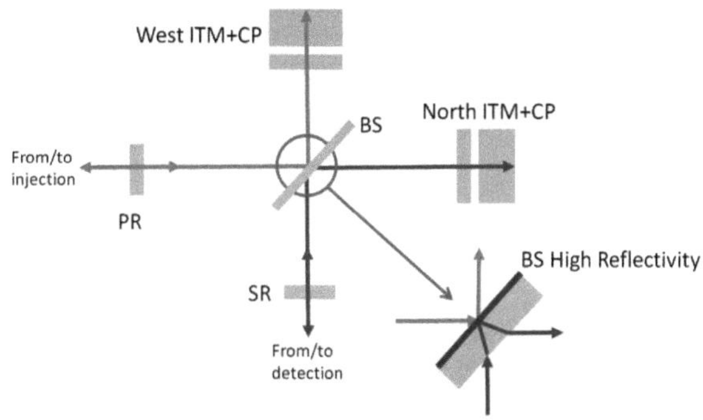

Fig. 9.24 Conceptual layout of the HWS probe beams in the recycling cavity. For the sake of clarity the beams have different colors, but the wavelength is the same

coating: with a reflectivity @ 800 nm of about 98 %, the cross coupling[4] becomes less than 2 % and the cross-sampling[5] less than 0.04 %. Beam from the detection bench will also sense the BS thermal lensing, thus allowing its correction on the NI CP.

9.4 Conclusions

Thermal effects are an unavoidable annoying presence in our interferometric GW detectors: tiny fractions of the power stored in the ITF are absorbed by the mirrors and the low thermal conductivity of SiO_2 generates temperature gradients that, in turn, result in optical path length increase. The wavefronts of the sideband fields in the recycling cavity get strongly aberrated and finally the ITF cannot acquire the lock at high input powers.

Thermal effects can be compensated using thermal actuators (Far-Infra-Red lasers, ring heaters) as shown by the Virgo and LIGO experiences.

However, in Advanced Virgo, the strength of thermal effects will increase by more than one order of magnitude with respect to Virgo and another effect will be relevant: change of the RoC of all the TMs due to thermal expansion, requiring one more actuator (ring heater). More sources of wavefront aberrations, i.e., intrinsic optics defects (nonthermally induced), introducing the necessity to generate also nonsymmetric heating patterns. Finite element modeling becomes an essential tool to derive the optimized heating patterns to be shined on the compensation plates.

[4] Cross-coupling is defined as the amount of light from one source that reaches the other (wrong) sensor. For example, how much radiation emitted from the source on the injection bench reaching the Hartmann sensor on the detection bench.

[5] Cross-sampling defines the quantity of light from one source coming back to its (right) sensor after probing the wrong optic. For instance, the quantity of light emitted by the source on the injection bench that returns to the Hartmann sensor on the injection bench after passing through the North ITM.

Increase in sensitivity to aberrations and in complexity of the compensating system will require a rather complex control loop and extremely sensitive sensors.

References

1. P.R. Saulson, *Fundamentals of Interferometric Gravitational Wave Detectors* (World Scientific, Singapore, 1994)
2. J.Y. Vinet, Living Rev. Relat. **12**, 5 (2009)
3. T. Accadia et al., in *A thermal compensation system for the gravitational wave detector Virgo*, ed. by T. Damour, R.T. Jantzen, R. Ruffini. Proceedings of the 12th Marcell Grossmann Meeting (World Scientific, Singapore, 2011)
4. K. Goda, D. Ottaway, B. Connelly, R. Adhikari, N. Mavalvala, A. Gretarsson, Opt. Lett. **29**, 1452 (2004)
5. T. Accadia et al., Class. Quantum Grav. **27**, 084002 (2010)
6. T. Accadia et al., Class. Quantum Grav. **28**, 114002 (2011)
7. F. Acernese et al., Class. Quantum Grav. **25**, 184001 (2008)
8. T. Accadia et al., in *Advanced Virgo*, ed. by E. Auge, J. Dumarchez, J.T.T. Van. Proceedings of the 46th Rencontres de Moriond (The Gioi Publishers, Vietnam, 2011)
9. H. Grote, LIGO scientific collaboration. Class. Quantum Grav. **27**, 084003 (2010)
10. H. Luek et al., Class. Quantum Grav. **21**, S985–S989 (2004)
11. V. Fafone, A. Rocchi, TCS noise: general concepts and application to the Virgo/Virgo+ case, VIR-0615B-09 (2009)
12. A. Rocchi et al., in *Thermal effects and their compensation in the interferometric gravitational wave detector Advanced Virgo*, ed. by E. Auge, J. Dumarchez, J.T.T. Van. Proceedings of the 46th Rencontres de Moriond (The Gioi Publishers, Vietnam, 2011)
13. A. Rocchi et al., J. P. Conf. Series **363**, 012016 (2012)
14. P. Hello, Eur. Phys. J. D **15**, 373–383 (2001)
15. R. Day, E. Genin, J. Marque, Influence of mirror roughness on clipping losses in optical cavities, VIR-0398A-10 (2010)
16. J. Degallaix, Idea for the size and residual aberration amplitude on the ITMs for Advanced Virgo MSRC, VIR-0365A-11 (2011)
17. R. Day, Update on simulations of ITF with roughness maps, VIR-0575A-11 (2011)
18. R. Ward, A study of cold ITF defects in MSRC with SIS, VIR-0326A-11 (2011)
19. V. Fafone, A. Rocchi, AdV TCS: considerations on actuation and sensing for the correction of high order aberrations in MSRC, VIR-0373A-11 (2011)
20. R.C. Lawrence, Ph.D. thesis, MIT, 2003, LIGO-P030001-00-R
21. N. Matsumoto et al., J. Opt. Soc. Am. A **25**, 7 (2008)
22. M. Granata et al., Phys. Rev. Lett. **105**, 231102 (2010)
23. S.A. Kennedy et al., Phys. Rev. A **66**, 043801 (2002)
24. Y. Fan et al., Rev. Sci. Instrum. **79**, 104501 (2008)
25. A.F. Brooks et al., Opt. Express **15**(16), 10370–10375 (2007)

Chapter 10
Stray Light Issues

Julien Marque and Gabriele Vajente

Abstract Stray light in gravitational wave interferometric detectors is a possible source of limitation for the low frequency performance of these instruments. It has to be "controlled," i.e., the instrument has to be carefully designed and baffled in order to overcome this potential issue. We will see in this chapter that this problem has already been identified and taken into account in the design phase of the detectors in the 1980s. However, all the first- generation instruments had to deal with unexpected issues regarding stray light. Commissioners have set up systematic investigations to trace the various noise sources and fix optical setups. The new generation of advanced detectors will actually be even more sensitive to displacement noise due to stray light because of the increase of power stored in the cavities and the enhanced sensitivity to amplitude field fluctuation. We will review all kinds of mechanisms that can lead to "stray light noise," we will see how to quantify it, and mitigate it. Despite all the precautions that have been taken, we will see that the detectors will actually be affected by this noise at least for a limited period of their data taking time.

10.1 The Stray Light Displacement Noise

Stray light is light in an optical system, which was not intended in the design. In the case of gravitational wave detectors, the light is from the laser source, but follows different paths with respect to the main beam. This light will be seen by the detector

J. Marque (✉)
European Gravitational Observatory, Via Edoardo Amaldi, S. Stefano a Macerata, 56021 Cascina (Pisa), Italy
e-mail: julien.marque@ego-gw.it

G. Vajente
I.N.F.N Sezione di Pisa Largo B. Pontecorvo, 3 - Polo Fibonacci Edificio, 56127 Pisa, Italy
e-mail: gabriele.vajente@pi.infn.it

M. Bassan (ed.), *Advanced Interferometers and the Search for Gravitational Waves*, Astrophysics and Space Science Library 404, DOI: 10.1007/978-3-319-03792-9_10,

as a spurious displacement noise which can limit its sensitivity. There are many possible sources of stray light:

- Secondary beams/spurious reflection (generated by imperfect anti-reflective coatings of optics used in transmission),
- Scattering from rough surface (diffuse reflection from imperfect mirror surface),
- Scattering from micro-defects (like dust, ditch, or scratch),
- Diffraction on the edge (all optics have a limited aperture).

Stray Light Control (SLC) is the name of the advanced detector subsystem that takes all the precautions in order to mitigate the displacement noise induced by the stray light sources. Actually, the stray light issue is not something new and peculiar to the advanced detector generation. It was already a concern in the design phase of the first generation, back in the 1980s. Indeed, the issue was put in evidence with the first prototypes. In [1], a 3 m arm length prototype was used to identify and study various noise sources. Evidences of parasitic interferences were observed, leading to large amplitude noise at the frequency of the suspension pendulum and harmonics. Therefore, an active damping of the pendulum resonance was introduced and the use of Faraday Isolators was suggested.

In the 30 following years, the calculation of the stray light coupling mechanism was further investigated, using different approaches. This will be discussed in Sect. 10.2, and the most common approach will be developed. During the commissioning phase of the first-generation instruments, many unexpected issues came up. We will review this experience and report about the measurements made to verify the models in Sect. 10.3.

There exists many, often complementary, ways to deal with stray light noise issue. Here is a not exhaustive list (these items will be further detailed in Sect. 10.4):

- Optimization of low scattering optics/beam dumps/baffles,
- Mitigation of the motion of the stray light sources (suspension and motion control),
- Optimization of Anti-Reflective coatings,
- Optical isolation (by using Faraday Isolators).

Section 10.5 will conclude this chapter by reviewing the most recent works in the field. We will see that this question is far from being closed. Indeed, the duty cycle of advanced detectors at optimal performance will likely be limited in a non-negligible manner due to unavoidable periods of high microseismic ground activity.

10.2 Stray Light Coupling Mechanisms

Stray light can affect the sensitivity and the operations of a gravitational wave interferometer when it couples back with the main laser beam which is used to extract information on the motion of the test masses. It is therefore important to understand with which mechanism stray light can affect the interferometric signals.

The first step is to compute the fraction of light that couples back with the fundamental mode. We consider one output port of the interferometer and the corresponding laser beam, which is, typically, mainly a fundamental Gaussian mode. An optical element in the laser path is responsible for the creation of stray light which partially propagates back into the interferometer. The key parameter to compute is the amount of this power that is superimposed with the mode of the exiting beam:

$$f_{sc} = \left|\langle \Psi_{stray} | \Psi_0^* \rangle\right|^2, \tag{10.1}$$

where Ψ_{stray} is the stray light field and Ψ_0^* is a normalized field in the interferometer mode. The star indicates that the mode should be reversed, in order to correspond to the beam entering from the port. We can consider two different processes that create the stray light: specular reflection and incoherent scattering. In the first case, it is possible to exactly compute the Gaussian beam parameter of the reflected beam, using for example the ABCD matrix formalism [2]. Then the superposition integral in Eq. 10.1 can be computed analytically [3, 4]. The case of scattering due to surface roughness is more complex. The method that can be used to tackle this phenomenon mainly depends on the spatial frequency of the surface defects that create the scattering. We first consider the case of low and medium frequencies, i.e., when the typical spatial length of the surface defects is of the same order of magnitude or larger than the laser beam size. In this case it is possible to carry out optical simulation with realistic surface maps, using techniques based on FFT propagation of the laser field [5]. These simulations directly provide the field propagating back into the interferometer and this allows to compute the superposition integral of Eq. 10.1. In the case of large spatial frequencies, the light is scattered mainly at large angles and therefore an FFT approach is not able to properly simulate it. In this case a statistical approach can be followed, based on the expected angular distribution of the scattered light. See [4] for an example of this approach.

Finally, more complex configurations might arise, when the light undergoes multiple scattering before recoupling into the main beam. A typical example is the scattering of light at large angles by one of the cavity mirrors, the subsequent scattering off the vacuum tube wall, and finally the recoupling with the main beam. Such complex paths are important and estimates of f_{sc} can be obtained with analytical or statistical approaches [6–8].

The fact that part of the light recouples with the main beam might not seem a problem in itself. However, this light will carry a phase information which depends on the length of the scattering path, and that is modulated by the motion of the scattering element $z_s(t)$:

$$\phi_s(t) = \phi_0 + \frac{4\pi}{\lambda} z_s(t) \tag{10.2}$$

The stray light field E_s sums with the unperturbed field E_{in} at a reference position inside the interferometer:

$$E_{in} + E_s = E_{in} + \sqrt{f_{sc}}\, E_{out} e^{i\phi_s(t)}, \tag{10.3}$$

where E_{out} is the field exiting from the interferometer port being considered and impinging on the scattering element. Equation 10.3 can be rewritten in terms of the ratio of the laser powers exiting and into the interferometer:

$$E_{\text{in}} + E_s = E_{\text{in}} \left[1 + \sqrt{f_{\text{sc}}} \frac{E_{\text{out}}}{E_{\text{in}}} e^{i\phi_s(t)} \right] = E_{\text{in}} \left[1 + \sqrt{f_{\text{sc}} \frac{P_{\text{out}}}{P_{\text{in}}}} e^{i\phi_s(t)} \right] \tag{10.4}$$

and in the last equality we factored out the phase difference of the two fields and included it in the static contribution to the scattering phase ϕ_0 of Eq. 10.2. At this point we make the assumption that the amount of scattering is small, which should be always true in our interferometric detectors. To be more precise, we assume that

$$\sqrt{f_{\text{sc}} \frac{P_{\text{out}}}{P_{\text{in}}}} \ll 1 \tag{10.5}$$

This allows us to recognize that the leftmost term in Eq. 10.4 is the first-order expansion of an exponential:

$$E_{\text{in}} + E_s = E_{\text{in}} \exp \left[\sqrt{f_{\text{sc}} \frac{P_{\text{out}}}{P_{\text{in}}}} \left(\cos \phi_s(t) + i \sin \phi_s(t) \right) \right] \tag{10.6}$$

where we have expanded the phase exponential of Eq. 10.4 in terms of sine and cosine.

The interpretation of Eq. 10.6 is simple: the effect of the recoupling stray light is embedded in the exponential term and therefore it consists in an additional modulation of the laser field. Since the exponent contains both a real and an imaginary part, both the phase and the amplitude of the field are modulated. The description of such modulations on the field is detailed in Chap. 5. A comparison with the results of that chapter allows us to write the phase and amplitude modulation of the laser field inside the interferometer, varying with time depending on the motion of the scattering element:

$$n_\phi(t) = \sqrt{f_{\text{sc}} \frac{P_{\text{out}}}{P_{\text{in}}}} \sin \phi_s(t) \tag{10.7}$$

$$\frac{\delta P(t)}{P} = \sqrt{f_{\text{sc}} \frac{P_{\text{out}}}{P_{\text{in}}}} \cos \phi_s(t) \tag{10.8}$$

It is important to note that so far we have not made any assumption on the amplitude of the scattering element motion $z_s(t)$. Indeed the above equation is valid for every magnitude of this motion: the only assumption was that the amount of stray power which recouples into the interferometer is small, in order to be able to treat it as a perturbation. The dependency of phase and amplitude noise on the scattering element

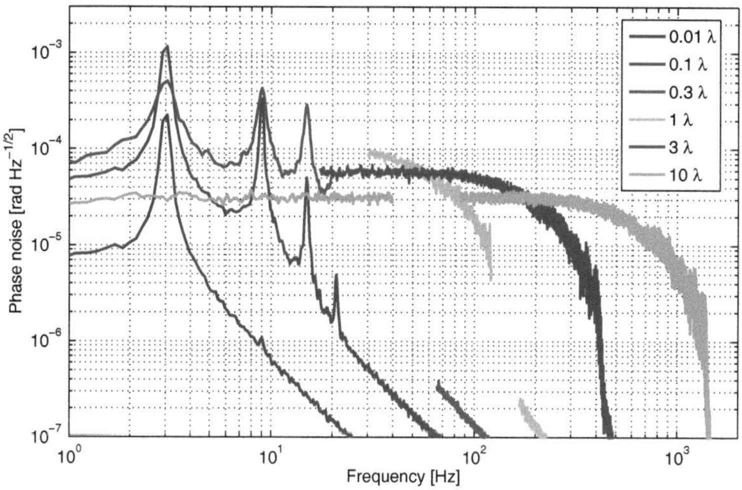

Fig. 10.1 Simulation of coupling of stray light into phase noise. We assumed a scattering element with a total coupling $f_{sc}\frac{P_{out}}{P_{in}}$ of 10^{-6}, moving as a pendulum with resonant frequency of 3 Hz and quality factor of 100, excited by white noise. The phase noise induced by stray light is shown as a function of the RMS amplitude of the motion

motion depends a lot on the amplitude of the latter. If the motion is small, then we can approximate the sine and cosine with their argument and recover a linear dependence of the added noises on the scattering element motion. However, as the amplitude of the motion increases, the first-order approximation is no longer valid, and therefore we expect to start seeing the effect of quadratic and higher order couplings. Figure 10.1 shows a simulation of the phase noise induced by a scattering element with $f_{sc}\frac{P_{out}}{P_{in}} \simeq 10^{-6}$. The element is moving like a pendulum with a resonant frequency of 3 Hz and a quality factor of 100, driven by a white noise. The motion is therefore almost monochromatic at 3 Hz. When the amplitude of the motion is much smaller than one wavelength, the coupling is mostly linear and the phase noise is also a simple monochromatic oscillation at the pendulum frequency. When the amplitude increases, we see that the coupling at 3 Hz is reduced and higher order harmonics appears. When the motion is comparable with one wavelength, it makes no more sense to describe it in terms of harmonics and indeed the noise assumed a typical flat shape, ending abruptly at a corner frequency f_{max} which depends on the motion amplitude. Indeed, if we assume that the scattering element has a sinusoidal motion with amplitude A_s and at a frequency f_s, the maximum frequency can be calculated by counting how many times the phase crosses zero over one second [9]:

$$f_{max} = 2\pi \frac{2A_s}{\lambda} f_s \tag{10.9}$$

A different picture of the effect of stray light is obtained when we look at the time-frequency spectrogram of the phase noise of Eq. 10.7. As shown in Fig. 10.2,

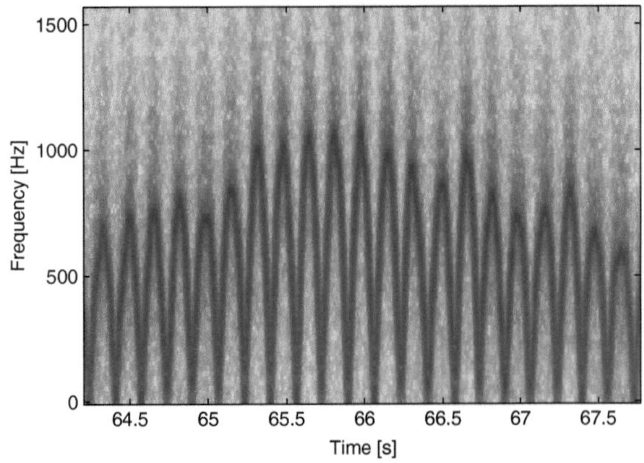

Fig. 10.2 Simulation of coupling of stray light into phase noise, with the same parameters described in Fig. 10.1. Here, a fast spectrogram of the field phase noise is shown. The scattering element is moving with an amplitude of 10λ

it is possible to see a clearly fast nonstationary behavior of the noise. The smooth bump visible in the spectrum of Fig. 10.1 appears to be composed of a narrow feature moving up and down in frequency. Indeed, a scattering element moving with a constant velocity will create a linear ramp in the phase $\phi_s(t)$ with a slope proportional to its speed. When insert the sine or cosine of Eq. 10.7, this will create a pure oscillation at a frequency which depends again on the velocity of the scattering element. Therefore, what we see in Fig. 10.2 is the oscillatory motion of the scattering element which has a velocity that goes from zero to maximum and back to zero in half a period. Indeed the arches in the spectrogram are spaced exactly by one-sixth of a second, since the object in this simulation is moving at 3 Hz.

One important point to stress here is that all the nonlinear behavior of stray light is encoded in the above equations which converts possibly large motions to additional noises. These noises are however always very small, in the hypothesis of small coupling of the stray light. This means that once the noises are computed from the estimated motion of the scattering element, they can be propagated into the interferometer and to the output signal using simply the known linear response of the detector (see Chap. 3):

$$\tilde{h}(f) = K_\phi(f)\sqrt{f_{\mathrm{sc}}\frac{P_{\mathrm{out}}}{P_{\mathrm{in}}}}\mathscr{F}\left[\sin\phi_s(t)\right] + K_{\delta P/P}(f)\sqrt{f_{\mathrm{sc}}\frac{P_{\mathrm{out}}}{P_{\mathrm{in}}}}\mathscr{F}\left[\cos\phi_s(t)\right] \tag{10.10}$$

where we indicated with $\mathscr{F}$ the Fourier transform and with $K_\phi(f)$ and $K_{\delta P/P}(f)$ the transfer functions from, respectively, phase and amplitude noise to the detector output h. We see that in general the effect of stray light can couple through both phase and amplitude noise, the relative importance depending on the detector response. For example, in first-generation detectors, the coupling of amplitude noise was negligible with respect to the phase noise, and therefore the latter was the dominant process.

The situation is different in the case of advanced detectors. The first reason is that they will use DC readout, which means that the gravitational wave signal is encoded in power variations at the antisymmetric port. Therefore, amplitude noise coming from stray light coupling at this port will be important. The second reason is that amplitude noise can create fluctuation of the stored power inside the main arm cavities, which will couple to real motion of the test masses through radiation pressure effects. For this reason it is normally necessary to use detailed optical simulation of the interferometer to properly compute the coupling of amplitude and phase noise to the sensitivity. In addition, it is important to keep in mind that the response to these noises might be changed by the action of the feedback control loops. Let us consider, for example, stray light entering from the symmetric port, together with the main input beam. Phase noise at this port is for all purposes equivalent to input laser frequency noise: this is however measured inside the interferometer and actively suppressed by the frequency stabilization loop. Since this loop can have very large gain, up to some kHz, the coupling of stray light through phase noise is largely suppressed. Therefore, we can expect the coupling of stray light at the symmetric port to be dominated through the amplitude noise mechanism only.

10.3 Experimental Evidences and Measurements

Evidence of stray light effect was present in all first-generation detectors. Figure 10.3 shows a collection of measured gravitational wave detector sensitivities, all limited by phase noise introduced by stray light. From these four plots it is apparent that all coupling were quite nonlinear, since the noise was shaped like a big bump in the sensitivity. The noise was coming from large motion of some element at low frequency, up-converted into the detector sensitive bandwidth. This predominant nonlinear effect made it very difficult to understand the origin of the noise and to track it down to the correct culprit. Fortunately, the stray light noise amplitude and maximum frequency depends on how much the offending element is moving, which is typically not constant in time but changing with environmental conditions. It proved to be very important to track the evolution over time of the noise, to identify any nonstationary behavior and correlate it with environmental and seismic activity. In many cases, it is possible to look at the fast variation of the noise level and to recognize the characteristic behavior already shown in Fig. 10.2.

An important step forward in the understanding of stray light coupling came with the first campaigns of systematic measurements, that tried to artificially increase in a controlled way the stray light noise coupling. The simplest way to do this is to shake one optical element at a known frequency and see if the increase of its motion has a visible effect on the detector sensitivity. This kind of measurement, first reported in TAMA [10], was extended to almost all possible points of the interferometers that could generate stray light: vacuum tubes, optical elements placed on benches sitting on ground, walls, etc.

In GEO600, a scattering element was added on purpose on the antisymmetric port of the detector and its motion was controlled with a shaker. Figure 10.4 shows the result of this campaign of measurement. The effect of the scattering element is

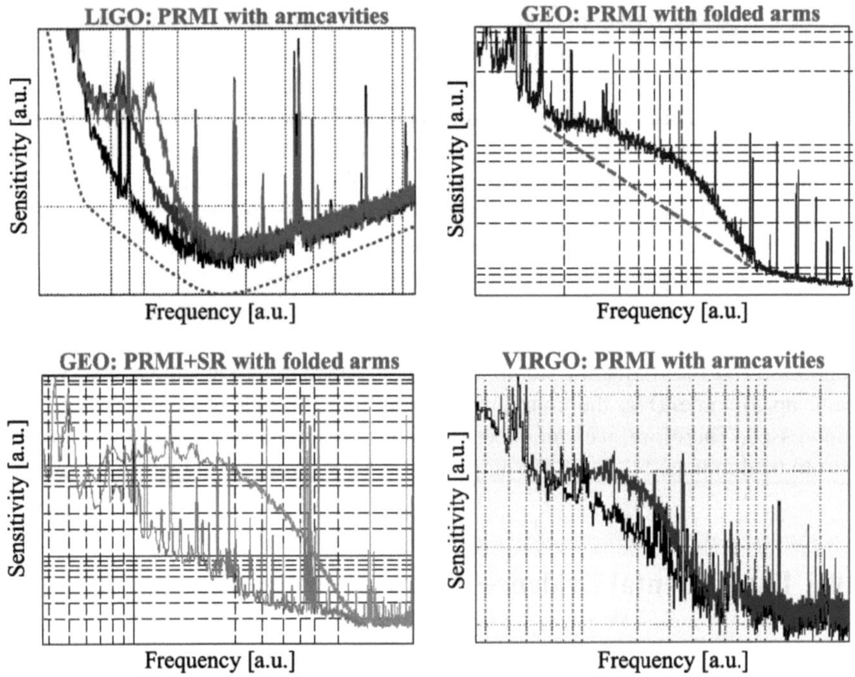

Fig. 10.3 All figures represent the reference sensitivity of various detectors as well as the sensitivity when affected by phase noise originating from a scattering element. *Upper left plot*: stray light shoulder in the LIGO Livingston detector, a Power Recycled Michelson interferometer (PRMI) with 4 km Fabry Perot arm cavity length. *Upper right plot*: stray light shoulder in the GEO600 detector in the configuration PRMI with a single-folded arms of 600 m length. *Lower left plot*: stray light shoulder in the GEO600 detector in the configuration Power and Signal-Recycled Michelson interferometer (dual recycled DRMI) with single-folded arms of 600 m length. *Lower right plot*: stray light shoulder in the Virgo detector, a PRMI with 3 km Fabry Perot arm cavity length. (Courtesy S. Hild [11])

very clear, even when it is not actively driven. In this case the element was moving because of the local microseismic activity. The large nonlinear effect, often called up-conversion, and the increase of the noise bandwidth proportionally to the motion amplitude is very clear.

In a similar way, the linear coupling regime at higher frequency was also tested, as shown in Fig. 10.5, where a sinusoidal excitation at about 800 Hz was added to one of the optical element in transmission of the arm mirrors. The coupling shows a predominant linear term. However, the presence of sidebands around the main peak is an indication of a nonlinear mechanism, which mixes low and high frequency components. In all first-generation detectors stray light was, at some point, a dominant noise source, which was searched for and solved with many ad hoc interventions [12].

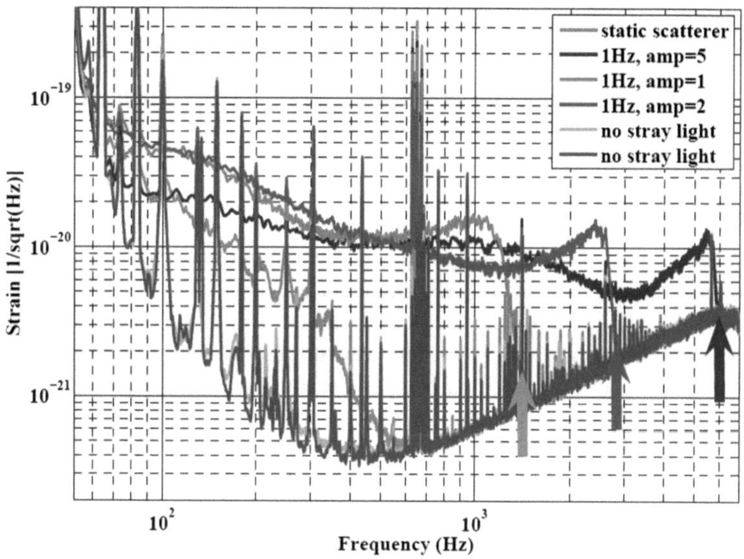

Fig. 10.4 Injection of stray light from the signal recycling port. The stray light source motion is driven with a triangular voltage of 1 Hz and different amplitude in the millimeter range. The stray light shoulders have different cut-off frequencies depending on the scatterer relative displacement amplitude. (Courtesy S. Hild [11])

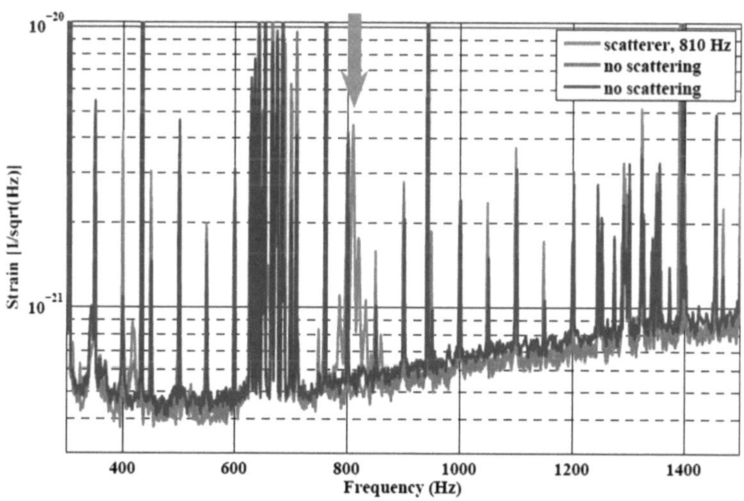

Fig. 10.5 Injection of stray light from the transmission of one arm mirror. The stray light source motion is driven with a triangular voltage of 810 Hz and an amplitude of less than one laser wavelength. A noise peak at 810 Hz surrounded by some sideband structure is observed. (Courtesy S. Hild [11])

10.4 Stray Light Noise Mitigation

First of all, it is important to underline how much the issue is delicate. Indeed, if a scatterer object is ground connected, the maximum amount of light back-scattered into the arm of the interferometer that is allowed before this noise becomes dominant in advanced detectors is very tiny: only about 10^{-24} W per Watt in the arm [13]. At full interferometer power, it means less than 5 photons per second!

Besides, most of the light of Advanced Detectors will be lost in the Fabry–Perot arm cavities. Indeed, the recycling technique maximizes the power in the arm cavities (600–800 kW): the recycling cavity is optimally coupled, meaning that it does not reflect the input power which is entirely transmitted to the arms, and then actually lost into the arms. The light is mainly lost by scattering off the surface of the Fabry–Perot mirrors [14]. Given the current state of the art of the coating and polishing techniques, 30 cm diameter mirrors absorb a bit less than 1 ppm, scatter 10 ppm of light at wide angle, and 10–20 ppm at small angles. Small angle scattering comes from the surface figure error of the mirror, currently limited by the nonuniformity of the coating deposition. Wide angle scattering is limited by point defects located at the surface and/or within the layers of the coating (the surface micro-roughness contribution to wide angle scattering is negligible). The limit between small angle and wide angle scattering, where both scattering mechanisms have equal contribution, is about 0.5°.

Because the noise coming from the arms themselves must involve at least three scattering events, there are other locations where scattering processes can be even more critical. Indeed, some of the light is also transmitted by the arm cavities (a few ppm), reflected by the recycling cavity (a few per cent), and will leak to the dark fringe since the contrast defect is expected to be of the order of one per thousand (see Chap. 3). The external optics at these ports are critical since a single scattering event is enough to recouple part of the light into the interferometer.

10.4.1 Core Optics Small Angle Scattering

Very small angle scattering due to imperfect mirror surface can be modeled as high-order Hermite-Gauss modes [15]. This interpretation is quite important for advanced detectors. Indeed, the resonance properties of these higher order modes in the various cavities can play a significant role. These modes being larger than the fundamental modes, their clipping losses will also be larger, and this will enhance very small angle scattering in the cavities [16].

But, since the small angle scattering process would require a very large number of higher order modes to be modeled in a realistic way, it is more common to interpret mirrors as diffractive elements and use Fast Fourier Transform propagation computations [17]. This method has actually been used to predict the distribution of power scattered around the core optics of advanced detectors. And the results have been

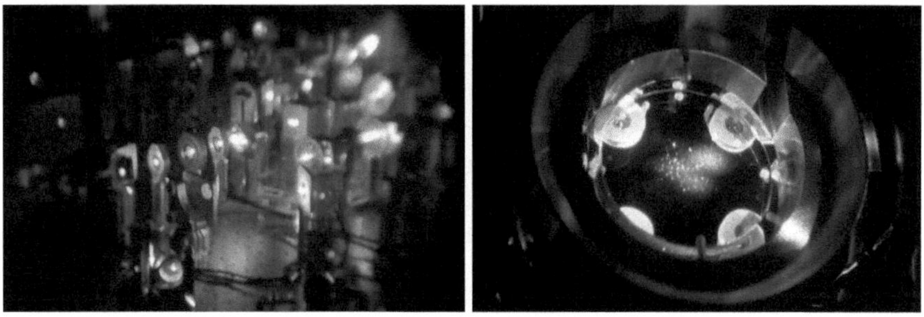

Fig. 10.6 *Left* picture of the Virgo laser bench, taken through an Infrared viewer (*green light* is exclusively light scattered by the optics on the bench). *Right* picture of the Virgo Beam Splitter mirror (*white light* is exclusively light scattered by optics in the interferometer). Credit: B. Canuel

used to specify the requirements of the baffles placed around the core optics of the arm cavity mirrors. These baffles have in general a diaphragm shape and are made of some absorbing material in order to prevent the stray light to recombine with the main beam.

For Advanced Virgo, the arm cavity baffle will sustain a maximum power density of $0.25\,\text{W/cm}^2$. An 80 cm diameter baffle would catch 95 % of the small angle scattering [14]. The advanced detector baffles that will catch the smallest angle scattering will be suspended. They will be made of absorbing material (as Diamond-like Carbon or oxidized stainless steel [18]) in order to absorb as much light as possible.

The rest of the small angle scattering needs to be caught directly inside the kilometers long tubes. For this purpose, in Virgo for instance, 160 baffles with a clear aperture of 920–990 mm have been pre-installed in each arm. They are made of stainless steel and have serrated edges. Serrated edges mitigate the back-scattering from the edge facets normal to the impinging beam, and the diffraction toward the center of the mirrors. These baffles have been laser machined in order to get very sharp edges.

10.4.2 Core Optics Wide Angle Scattering

The wide angle scattering of optics, as seen on the pictures of Fig. 10.6, is commonly characterized by the Total Integrated Scattering (TIS), which the total amount of light that is scattered off the surface of the optics. A more detailed characterization can be done by measuring the scattering at each angle: this is called the Bidirectional Reflectance Distribution Function (BRDF). It is usually admitted that the scattering at large angles for common optics (micro-roughness larger than 3Å) depends only on the micro-roughness of the surface. Moreover, a Lambertian law is usually admitted for large angles: the BRDF does not depend on the angle and is equal to the TIS divided by π.

The best mirrors available today are superpolished: they have a sub-Angstrom micro-roughness. However, their typical TIS, as seen by a millimeter beam width, is 4 ppm, at best, which cannot be explained just by the micro-roughness of the surface [19]. BRDF measurements in LIGO arm cavities have confirmed the mismatch between measured wide angle scattering and expected scattering due to micro-roughness [20]. The analysis of the data measurement has clearly identified point defects in the coating, or at its surface, to be the dominant source of scattering. While often thought to be dust or particle encapsulated in the coating, the origin of the point defects has never been clearly demonstrated.

In order to deal with this issue, large baffles are being positioned in advanced detectors either in the vacuum chambers of the core optics or in the large vacuum links of the central interferometer. These baffles are made of absorbing glass Anti-Reflective coated, or mirror polished oxidized stainless steel. The oxidation process allows to decrease the reflectivity of stainless steel to about 10 % at 1064 nm.

10.4.3 External Optics Back-Scattering

At every port of the advanced detector interferometers (symmetric, antisymmetric, and in transmission to the arms), there will be complicated optical setups that have as main function to collect some light in order to build the error signals required for the main interferometer longitudinal and alignment controls. These external optics systems essentially consist of telescopes to adapt the beam size (typically from 50 to 1 mm beam width). All these elements are critical as they can back-reflect or back-scatter light into the interferometer. Some detailed analysis was done in [4] for the input optics of Advanced Virgo. It is shown that some of the optics of the input telescope are required to have the best possible Anti-Reflective coating (<100 ppm), while the roughness of many other components should be of very high quality.

In many cases (at the symmetric and antisymmetric ports in particular), once the beam width has been reduced to few millimeters, a Faraday isolator is used to further reduce the amount of light coming from the following optics (see Chap. 5). These Faraday isolators have been developed along the past 10 years in order to provide a very high isolation factor (about 40–50 dB) even at high power (up to 200 W) [21].

A more original technique has been proposed in [22], then further developed for GEO600 and demonstrated in [23] to deal with scattered light noise from external optics. This technique relies on an optomechanical phase shifter placed between the interferometer and the external optics. By modulating the optical path length at high frequency, the noise due to the light back-scattered from the external optics can be shifted out of the frequency band of interest. This technique, implemented in GEO600, allowed for a noise reduction of one order of magnitude.

Besides, there is the need to block some beams which can have a high power, up to 100 W, when the lock of the interferometer is lost. Various materials have been investigated for the construction of these beam dumps. High absorbing material at 1064 nm with high thermal conductivity, that implies high damage threshold, as well as low out-gasing properties are required. Three materials have been proposed and are currently used (the damage threshold in air is indicated between parenthesis): Diamond-like Carbon [24] (0.5 kW/cm^2), Silicon (6 kW/cm^2) and Silicon Carbide (30 kW/cm^2).

Finally, the best way to mitigate back-scattering noise from the external optics is actually to seismically isolate the optics. This will be the case for most of the optics for the advanced detector generation. In advanced LIGO, more optics are integrated in the large vacuum chambers with respect to the enhanced LIGO optical scheme, like the optics bench in transmission of the arms. In advanced Virgo, new vacuum chambers are being constructed to host optical benches suspended to a small seismic isolator at every port. A relative motion control at low frequencies, between these benches and the interferometer, is envisaged in order to prevent any nonlinear process to occur.

10.5 Future Prospects

Despite all the efforts to prevent stray light noise from limiting the performance of advanced detectors, the most recent calculations combined with seismic data collected over more than one year [13] show that the low frequency sensitivity of the instruments will likely be compromised during the highest microseismic periods. Figure 10.7 represents the ground motion amplitude spectrum measured at the LIGO Livingston site. More precisely, each of the 3 solid curves are the amplitude spectrum contour for which 50, 90, and 95 % of the collected data have an equal or lesser amplitude. On the same figure, dashed curves represent the corresponding amplitude spectrum of the scattered field (assuming the scatterer is moving as the ground at microseismic frequencies). It is clear that during 5 % of the time, the shoulder shape of the scattered light field will have a cut-off frequency higher than 10 Hz, so that the noise will possibly affect the detector performance at low frequency. Figure 10.8 represents a tentative noise projection for an hypothetic scatterer for advanced LIGO. Various assumptions were made for the recoupling amplitude depending on the surface figure error of the core optics (see [13] for more details). In conclusion, it seems very likely that the displacement noise due to stray light will very likely affect the sensitivity of advanced detectors during at least a few percent of their total data taking time.

As we have seen in Sect. 10.4, the stray light issue in advanced detectors would also greatly benefit from lower scattering optics, both in terms of surface figure error and wide angle scattering. The Round-Trip losses of an optical cavity is limited by the scattering performance of the optics for a wide range of cavity lengths (see [25] for more details). This in turn limits the performance of many instruments that use a Fabry–Perot cavity for very accurate measurements in physics.

In this chapter, we intentionally neglected to discuss the possibility to install systems that can monitor the performance of the baffling system of advanced detectors. Indeed, these systems have not been implemented yet. However, there is an

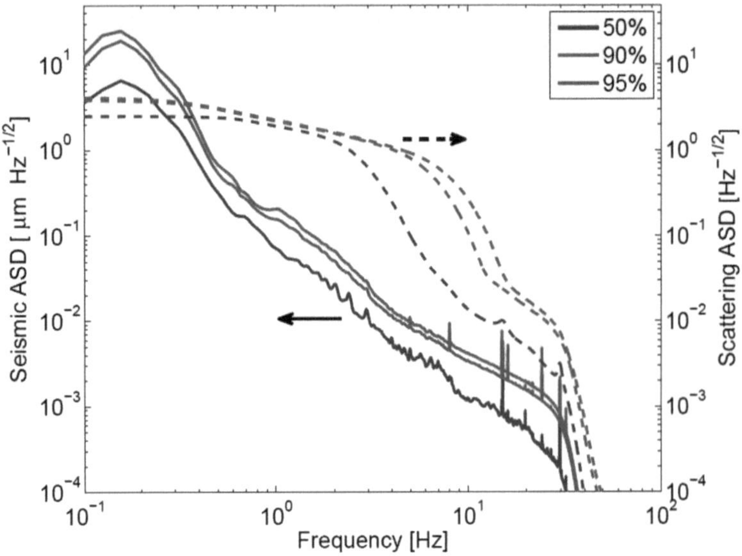

Fig. 10.7 Each contour (*solid lines*) represents the fraction of time for which the ground motion was smaller than the contour. The respective *dashed lines* represents the associated amplitude spectral density of the unconverted scattered field. The data have been collected on the Livingston observatory for about 3000 days. (Courtesy D. Ottaway [13])

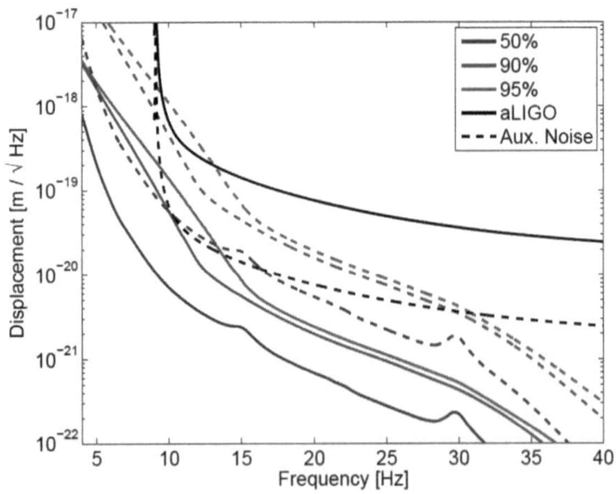

Fig. 10.8 Advanced LIGO design sensitivity and predicted displacement noise due to unconverted scattered light for the Livingston interferometer. *Solid* and *dashed curves* represent a pessimistic and an optimistic projection depending on the expected surface figure error performance of the core optics. (Courtesy D. Ottaway [13])

obvious need to setup diagnosis systems to quickly identify potential issues given the high number of baffling hardware. During the commissioning of the first-generation interferometers, this task was difficult and time-consuming.

Finally, it is also important to veto from the scientific data short events related to stray light noise (due, for instance, to an object moving suddenly fast). Simple techniques have already been used during the science runs of the first-generation detectors [26]. But more ambitious techniques [27] that involve the injection of squeezed light (see Chap. 11) have been proposed for the future. This technique would not only allow to veto noisy periods but also to subtract the noise from the gravitational wave channel.

Acknowledgments The authors would like to acknowledge S. Hild, D. Ottaway, and OSA for the authorization to reproduce Figs. 10.3, 10.4, 10.5, 10.7 and 10.8 in this chapter.

References

1. H. Billing, K. Maischberger, A. Rudiger, R. Schilling, L. Schnupp, L. Winkler, An argon laser interferometer for the detection of gravitational radiation. J. Phys. E: Sci. Instrum. **12**, 1043 (1979)
2. A. E. Siegman, *Lasers* (University Science Books, California, 1986)
3. J.-Y. Vinet, *The Virgo Physics Book, Optics and Related Topics*, https://wwwcascina.virgo.infn.it/vpb/
4. B. Canuel, E. Genin, G. Vajente, J. Marque, Displacement noise from back scattering and specular reflection of input optics in advanced gravitational wave detectors. Opt. Express **21**, 10546 (2013)
5. J.-Y. Vinet, P. Hello, C.N. Man, A. Brillet, A high accuracy method for the simulation of non-ideal optical cavities. J. Phys. 1 France **2**, 1287 (1992)
6. E. Flanagran, K. Thorne, Noise due to backscatter of baffles, the nearby wall and objects at the far end of the beam tube; and recommended actions. LIGO, Tech. Rep., LIGO-T940063-00 (1994)
7. J.-Y. Vinet, V. Brisson, S. Braccini, Scattered light noise in gravitational wave interferometric detectors: Coherent effect. Phys. Rev. D **54**, 1276 (1996)
8. J.-Y. Vinet, V. Brisson, S. Braccini, I. Ferrante, L. Pinard, F. Bondu, E. Tournie, Scattered light noise in gravitational wave interferometric detectors: a statistical approach. Phys. Rev. D **56**, 6085 (1997)
9. B. Canuel, I. Fiori, J. Marque, E. Tournefier, Diffused light mitigation in Virgo and constraints for Virgo+ and AdV. Virgo Internal Note, VIR-0792A-09 (2009)
10. R. Takahashi, K. Arai, S. Kawamura, M.R. Smith, Direct measurement of the scattered light effect on the sensitivity in TAMA300. Phys. Rev. D **70**, 62003 (2004)
11. S. Hild, Beyond the First Generation: Extending the Science Range of the Gravitational Wave Detector GEO600. PhD Thesis (2006)
12. I. Fiori et al., Noise from scattered light in Virgo's second science run data. Class. Quantum Gravity **27**, 194011 (2010)
13. D.J. Ottaway, P. Fritschel, S.J. Waldman, Impact of upconverted scattered light on advanced interferometric gravitational wave detectors. Opt. Express **20**, 8329–8336 (2012)
14. T. Accadia et al., Advanced Virgo Technical Design Report. Virgo Internal Note, VIR-0128A-12 (2012)
15. W. Winkler, R. Schilling, K. Danzmann, J. Mizuno, A. Rudiger, K.A. Strain, Light scattering described in the mode picture. Appl. Opt. **33**(31), 7547–7550 (1994)
16. R.A. Day, G. Vajente, M. Kasprzack, J. Marque, Reduction of higher order mode generation in large scale gravitational wave interferometers by central heating residual aberration correction. Phys. Rev. D **87**, 082003 (2013)
17. R. Day, E. Genin, J. Marque, Influence of mirror roughness on clipping losses in optical cavities. Virgo Internal Note, VIR-0398A-10 (2010)
18. V. Sannibale, L. Austin, H. Kelman, T. Nguyen, M. Ruiz, M. Smith, H. Yamamoto, L. Zhang, The Mitigation of the Scattering Light for the Advanced LIGO Interferometers. Poster at Amaldi 9th, Cardiff (2011)

19. F. Magana-Sandoval, R.X. Adhikari, V. Frolov, J. Harms, J. Lee, S. Sankar, P.R. Saulson, J.R. Smith, Large-angle scattered light measurements for quantum-noise filter cavity design studies. JOSA A **29**, 1722 (2012)
20. W. Kells, C. Vorvick, Imaged Scattered light from LIGO Resonant Cavities: Micro-roughness versus Point Scatter Loss. Ligo Internal Note, G080078–01-D (2008)
21. O.V. Palashov, D.S. Zheleznov, A.V. Voitovich, V.V. Zelenogorsky, E.E. Kamenetsky, E.A. Khazanov, R.M. Martin, K.L. Dooley, L. Williams, A. Lucianetti, V. Quetschke, G. Mueller, D.H. Reitze, D.H. Tanner, E. Genin, B. Canuel, J. Marque, High-vacuum-compatible high-power Faraday isolators for gravitational-wave interferometers. J. Opt. Soc. Am. B **29**, 1784 (2012)
22. R. Schilling, L. Schnupp, W. Winkler, H. Billing, K. Maischberger, A. Rudiger, A method to blot out scattered light effects and its application to a gravitational wave detector. J. Phys. E: Sci. Instrum. **14**, 65 (1981)
23. H. Luck, J. Degallaix, H. Grote, M. Hewitson, S. Hild, B. Willke, K. Danzmann, Opto-mechanical frequency shifting of scattered light. J. Opt. A Pure Appl. Opt. **10**, 085004 (2008)
24. T. Tomoru et al., Study of optical dumpers used in high vacuum system of interferometric gravitational wave. J. Phys. Conf. Ser. **32**, 476 (2006)
25. M. Evans, L. Barsotti, J. Harms, P. Kwee, H. Miao, Realistic Filter Cavities for Advanced Gravitational Wave Detectors. arXiv:1305.1599 (2013)
26. J. Aasi et al., The characterization of Virgo data and its impact on gravitational-wave searches. Class. Quantum Gravity **29**, 155002 (2012)
27. S. Steinlechner, J. Bauchrowitz, M. Meinders, H. Muller-Ebhardt, K. Danzmann, R. Schnabel, Quantum-Dense Metrology. arXiv:1211.3570 (2012)

Chapter 11
A Basic Introduction to Quantum Noise and Quantum-Non-Demolition Techniques

Stefan Hild

Abstract For the past 120 years the Michelson interferometer has served as the standard tool for high precision length measurements. Starting from the very first Michelson interferometer to today's large-scale gravitational wave observatories, the peak sensitivity of the Michelson interferometer has improved by about 10 orders of magnitude. Advanced gravitational wave detectors, such as Advanced LIGO or Advanced Virgo, will achieve a measurement precision limited by the Heisenberg Uncertainty Principle, giving rise to the interferometric Standard Quantum Limit for 40 kg test masses. This chapter will give a basic outline of the concepts currently under consideration for surpassing the Standard Quantum Limit.

11.1 How to Approach This Chapter

In this chapter we will discuss the notion of *quantum noise* and some methods by which it can be reduced in laser-interferometric Gravitational Wave (GW) detectors[1]. It needs to be pointed out that a comprehensive discussion of this topic is far beyond what is possible within the limited space here. So, the reader might wonder what can be gained from reading this chapter? In this chapter I will give my best, trying to approach quantum noise and quantum-non-demolition (QND) techniques from the standpoint of an experimental physicist who aims at thinking in terms of clear and lightweight models with easy applicability, rather avoiding excessive theoretical formalism; hence, my main goal is to convey an intuitive understanding of the underlying concepts of quantum noise and interferometry techniques aiming to

[1] Please note, that if not stated explicitly otherwise, we neglect in this chapter all other fundamental noises, such as thermal noise, as well as the myriad of technical noises that commissioners of GW detectors have to battle with.

S. Hild (✉)
SUPA, Institute for Gravitational Research, University of Glasgow, Glasgow G12 8QQ, UK
e-mail: stefan.hild@ligo.org

M. Bassan (ed.), *Advanced Interferometers and the Search for Gravitational Waves*, Astrophysics and Space Science Library 404, DOI: 10.1007/978-3-319-03792-9_11,

reduce this noise. While this approach requires certain simplifications to be made, the presented formalism will help to foster a qualitative understanding of the underlying phenomena and may serve as a springboard for an in-depth study of the topic.

If you are looking for a mathematically exact description of quantum noise and QND techniques, please stop reading here and please be deferred to one of the excellent books and articles already existing [1–4]. Otherwise, please lean back and enjoy the beauty of interferometry experiments at and beyond the quantum limit!

11.2 The Basics of Quantum Noise

The notation *quantum noise* originates from the fact that this noise is a direct consequence of the quantum nature of light. For two reasons quantum noise plays a very special role in the noise-cocktail of advanced laser-interferometric GW detectors:

First of all, quantum noise is of a different nature compared to most other limiting noise sources, because quantum noise originates directly from measurement and readout processes (in contrast to, for example, seismic excitation or thermally driven fluctuations which directly change the position of test mass surfaces). Secondly, as can be seen in Fig. 11.1, quantum noise is the noise source directly limiting the sensitivity of advanced GW detectors over most frequencies of the detection band.

Hence, reducing the quantum noise contribution by means of clever interferometry concepts is of the highest priority if we want to further improve the sensitivity of future GW detectors. So, let us start our quest!

11.2.1 Principles for Building a GW Detector

Before we have a closer look at the nature of quantum noise and what processes lead to its presence in the GW channel of the interferometer, it is helpful to recall what is required for conceiving a GW detector with good performance. Although the current laser-interferometric GW detectors are extremely complex machines, a successful design can be boiled down to two basic design principles:

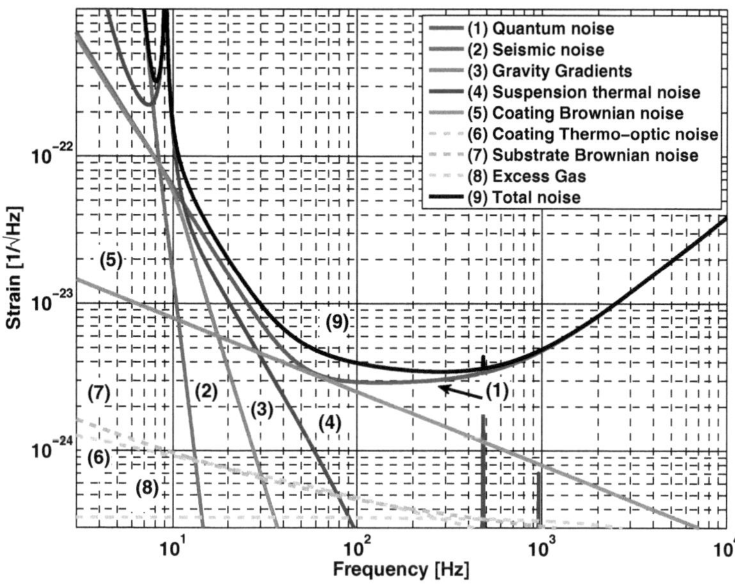

Fig. 11.1 Noise budget of the advanced LIGO broadband configuration as described in Abbott et al. [5]. The *coloured lines* (*1–8*) represent the amplitude spectral density of various noise components, whereas the trace labelled "Total Noise" shows the overall instrument sensitivity. For all frequencies above about 12 Hz quantum noise is the dominating noise source (This figure was produced using the GWINC software [6])

1. You need to make sure your test masses are quieter than the signal you want to observe.
2. You need to make sure that the test mass position can be read out to the required accuracy, without introducing any significant level of back action noise.

In order to satisfy the first of the requirements stated above we employ a myriad of sophisticated techniques and we exercise the greatest care when it comes to providing seismic isolation of the test masses, put them into ultra-high vacuum to reduce their acoustic coupling, and use ultra-low loss materials for the test masses as well as for their suspensions to reduce the influence of thermal noise. The second design principle relates fundamentally the optical readout and the interferometric measurement of the differential arm length degree of freedom of the GW detector.

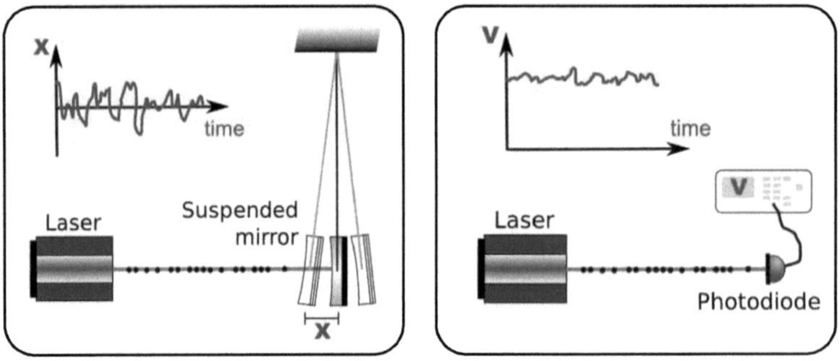

Fig. 11.2 Illustrations of radiation pressure noise (*left*) and shot noise (*right*)

11.2.2 Photon Shot Noise and Quantum Radiation Pressure Noise

So, finally let us answer the question of what quantum noise is actually composed of. It is ultimately made up from two components: photon shot noise at high frequencies and photon radiation pressure noise at the low frequency end of the detection band. The most straightforward way to understand the origin of the two quantum noise components is illustrated in Fig. 11.2. The photons in a laser beam are not equally spaced in time, but follow a Poissonian distribution. Consequently, when a laser is shone onto a photodetector, the resulting time-series of the photo current is not constant over time but is subject to fluctuations. These fluctuations, literally also called shot noise, increase proportional to the *square root* of the optical power in the laser beam. In contrast, the signal strength in the GW detector increases *linearly* with the optical power. Hence, by increasing the optical power in a GW detector the signal-to-shot-noise ratio can be effectively increased, proportional to the square root of the optical power inside the GW interferometer. The amplitude spectral density of strain equivalent of photon shot noise in a simple Michelson interferometer (without arm cavities or recycling) can be expressed as

$$h_{\mathrm{sn}}(f) = \frac{1}{L}\sqrt{\frac{\hbar c \lambda}{2\pi P}}\,, \tag{11.1}$$

where f is the frequency, L is the arm length of the GW detector, c is the speed of light, λ the laser wavelength, and P the optical power in the interferometer arms.

The second effect of the temporally inhomogeneous distribution of the photons in a laser beam is illustrated on the left panel of Fig. 11.2. Photons carry momentum and when they are reflected from a free-falling test mass they transfer momentum, or better, a radiation pressure force onto the mirror. Since the photons do not arrive at the test mass with exactly similar separation in time, the force onto the mirror and therefore also its position fluctuate over time, giving rise to photon radiation pressure noise. The amplitude spectral density of the quantum radiation pressure noise in a simple Michelson is given by

$$h_{\text{rp}}(f) = \frac{1}{mf^2L}\sqrt{\frac{\hbar P}{2\pi^3 c\lambda}}, \tag{11.2}$$

where m is the mass of the mirror. In contrast to shot noise we see that the radiation pressure noise spectrum is not white, but its contribution falls off as $1/f^2$. From this it is obvious that radiation pressure noise will dominate at low frequencies, while shot noise is the dominant contribution at high frequencies.

We define the quantum noise of an interferometer to be the (uncorrelated) sum of photon shot noise and photon radiation pressure noise.

11.2.3 The Standard Quantum Limit

It is important to note in Eqs. 11.1 and 11.2 that shot noise and radiation pressure noise scale inversely with the optical power inside the Michelson interferometer; so, while it is possible to improve the strain sensitivity of an interferometer at high frequencies by increasing the circulating light power, inevitably at the same time the sensitivity will decrease at low frequencies as the radiation pressure increases with the light power. This relation is illustrated in Fig. 11.3. Ultimately, for every observation frequency there exists an optimal power, which results in identical magnitude contributions from shot noise and quantum radiation pressure noise at this frequency. For example, the best sensitivity at a frequency of 8 Hz is achieved for 1 MW (see trace 2c in Fig. 11.3). The Standard Quantum Limit (SQL) is defined as the lower bound envelope of the quantum noise spectra for all optical powers circulating in interferometer with a specific set of design parameters [7]. As illustrated in Fig. 11.3, the SQL for a simple Michelson interferometer follows a slope inverse to the observation frequency.

The SQL has originally been suggested as an ultimate limit for the sensitivity of laser interferometers that cannot be surpassed. However, it was quickly realized that the SQL only imposes a limit on the achievable sensitivity of classical interferometers but that there are several techniques, such as detuned signal recycling and quantum non-demolition configurations, which can enable interferometric measurements with displacement sensitivities below the SQL. The basic principles of the most prominent techniques that allow us to beat the SQL will be discussed in the following sections.

11.3 A Simple Graphical Tool to Understand Quantum Noise

11.3.1 The Quadrature Picture

In order to gain an intuitive understanding of quantum noise in interferometric measurements, in the following we will introduce a rather simple, yet powerful graphical

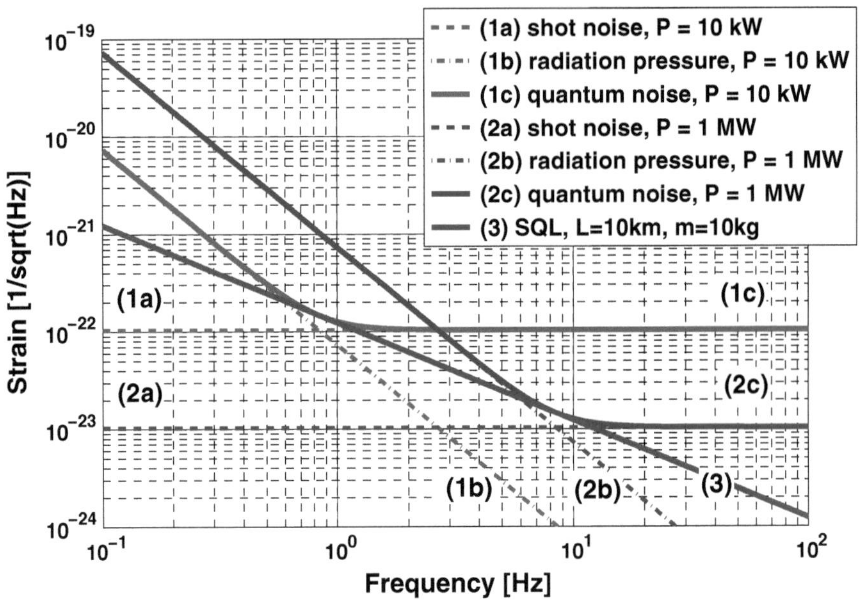

Fig. 11.3 The SQL (trace 3) and the quantum noise contribution for a simple Michelson interferometer, featuring 10 km arm length and test masses of 10 kg weight, for low (traces 1a–1c) and high (traces 2a–2c) circulating power

tool, often formally referred to as the *quadrature picture* or informally sometimes referred to as the *ball on a stick picture*.

In general we can describe an electric field **E** at a position **r** and at the time t by

$$\mathbf{E}(\mathbf{r}, t) = E_0 \left[a(\mathbf{r}) e^{-i\omega t} - a^*(\mathbf{r}) e^{+i\omega t} \right] \mathbf{p}(\mathbf{r}, t), \tag{11.3}$$

where a is the complex amplitude of the electro-magnetic field, ω its angular frequency and **p** the polarization. We can introduce the following two new properties:

$$X_1(\mathbf{r}) = a^*(\mathbf{r}) + a(\mathbf{r}) \tag{11.4}$$

$$X_2(\mathbf{r}) = i \left[a^*(\mathbf{r}) - a(\mathbf{r}) \right]. \tag{11.5}$$

X_1 and X_2 are usually referred to as the amplitude and phase quadrature, respectively. Using the quadrature representation we can now rewrite Eq. 11.3 to express the electromagnetic field in terms of the amplitude and phase quadratures:

$$\mathbf{E}(\mathbf{r}, t) = E_0 \left[X_1 \cos(\omega t) - X_2 \sin(\omega t) \right] \mathbf{p}(\mathbf{r}, t). \tag{11.6}$$

In close analogy to Eqs. 11.4 and 11.5 we can formulate the so-called quadrature operators which are the foundation of a description of light fields in the realm of quantum mechanics as

$$\hat{X}_1(\mathbf{r}) = \hat{a}^\dagger(\mathbf{r}) + \hat{a}(\mathbf{r}) \tag{11.7}$$

$$\hat{X}_2(\mathbf{r}) = i\left[\hat{a}^\dagger(\mathbf{r}) - \hat{a}(\mathbf{r})\right], \tag{11.8}$$

where $\hat{X}_1(\mathbf{r})$ is the **amplitude quadrature operator** and $\hat{X}_2(\mathbf{r})$ is the **phase quadrature operator**.

11.3.2 The Ball on a Stick Picture

A common way to visualize the quadrature operators introduced above is the "ball on the stick" picture. In the following we use a concept similar to that which has been introduced for example by Chen [8, 9]. The left-hand side of Fig. 11.4 shows an example of a coherent light field. Let us assume the light field consists of a huge number of photons and we continuously perform measurements over a finite duration to determine the *light states*. We can represent each measurement by a single dot in the $\hat{X}_1$, $\hat{X}_2$ plane. If we perform a large number of measurements, then we can measure the probability distribution of the light state, which is indicated by the "ball" or "cloud" shown in Fig. 11.4. The solid arrow points at the center of that ball which is the point in the $\hat{X}_1$, $\hat{X}_2$ plane featuring the highest probability to encounter the field in this state if a measurement is carried out. So we can graphically represent the coherent part of a light field (with a certain amplitude and phase) by the arrow shown, while the uncertainty (or noise) of the field is represented by the ball; hence the phrase *ball on a stick*. The quantum nature of light forbids us to reduce the area of the ball below a certain limit, in the following referred to as the uncertainty limit. This limit can be considered to be a direct consequence of the *Heisenberg Uncertainty Principle* which also applies in a quantum-mechanical description of light.

While the uncertainty principle dictates the minimal area of the ball, we have the freedom to manipulate its shape. One way to change the shape of the ball is a technique with the figurative name *squeezing* [10]. The right-hand panel of Fig. 11.4 shows a light field that is squeezed in phase quadrature, i.e., the uncertainty of the light state is reduced in one quadrature (in this case the phase quadrature), while at the same time one has to pay the price of at least a proportionally increased uncertainty in the orthogonal quadrature (in this case the amplitude quadrature). As we will see later in Sects. 11.4 and 11.6, squeezed light can be used to improve the quantum noise limited sensitivity of laser-interferometric gravitational wave detectors.

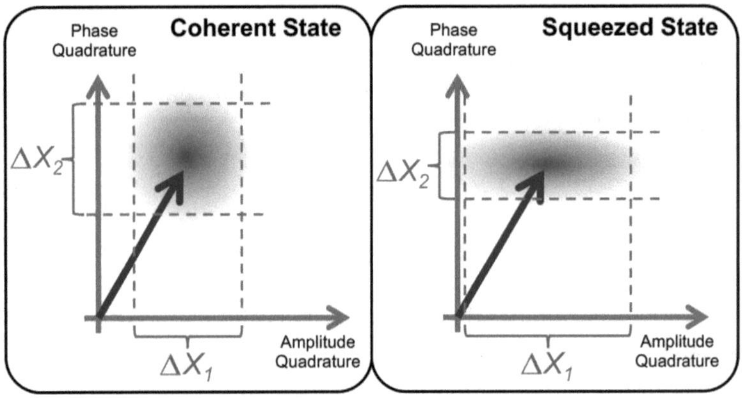

Fig. 11.4 *Left* Graphical representation of a coherent state in the quadrature picture. Due to the quantum nature of light the photons in a laser beam do not all have the same amplitude and phase, but follow a probability distribution indicated by the ball or cloud. The minimal area of this ball is limited by the uncertainty principle. *Right* Graphical representation of a squeezed state. While the uncertainty in the phase quadrature is reduced, the uncertainty in the amplitude quadrature is increased, so that the ball now takes the shape of an ellipse, i.e., the so-called *squeezing ellipse*

However, for the moment let us go back to the quadrature picture of a conventional, unsqueezed light field and explore how our graphical tool relates to the strain spectrum of quantum noise we previously discussed in Sect. 11.2. The strain spectral density of quantum noise can simply be understood as an inverse signal-to-noise ratio (SNR), or better as a noise-to-signal ratio, where "noise" refers to the amplitude of the quantum fluctuations in the instrument and "signal" refers to the gravitational wave-induced phase change of the light which manifests as the change in the differential arm length scaled with the frequency-dependent signal gain in the instrument. The lower the quantum noise at constant signal amplitude or the higher the signal amplitude at a constant noise level, the lower is the quantum noise limited strain spectral density of the interferometer, i.e., the better is the sensitivity of the instrument.

So, where do we find the quantum noise-limited strain spectral density in our graphical picture? Let us have a look at Fig. 11.5: The second panel shows the uncertainty ball of the light state entering our interferometric system of interest, but in contrast to the previous section we have omitted the filling of the ball and just represent its outline by a circle.[2] In addition, we have drawn two arrows, E_1 and E_2, representing the noise in the two orthogonal quadratures. Please note that for a coherent state the noise in the two quadratures is completely uncorrelated which is the reason why the two arrows here are displayed in different colors.

When a gravitational wave causes the length of an interferometer arm to change or if a local disturbance moves the position of an interferometer mirror, then this will cause a signal to show up in the phase quadrature, indicated in the third panel

[2] Please note, that the circle does not indicate 100 % of the distribution function, but the radius usually corresponds to 1 σ.

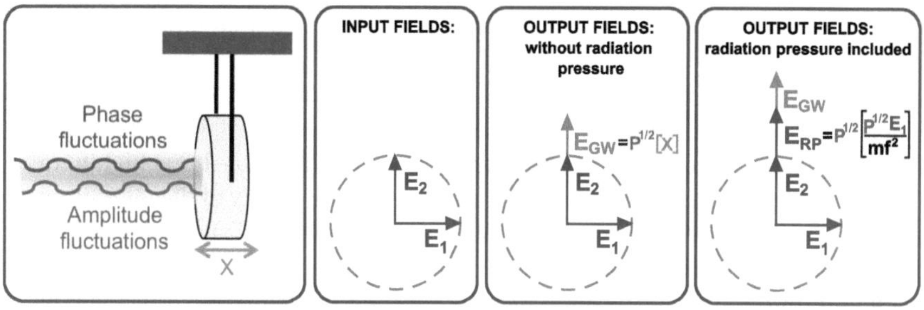

Fig. 11.5 Illustration of optomechanical coupling of quantum noise on suspended mirrors. The input field is described by the uncorrelated noise contributions E_1 and E_2 in the amplitude and phase quadrature, respectively. Without optomechanical coupling the noise contributions in the phase and amplitude quadrature are uncorrelated. In case the optomechanical coupling becomes dominant, fluctuations in the amplitude quadrature (E_1) are converted into fluctuations in the phase quadrature (E_{RP}), thereby introducing correlations of the noise components in the two quadratures. P is the optical power inside the system, m is the reduced mass of the ensemble of test masses, f is the frequency, and X symbolises a test mass displacement or the equivalent strain

of our picture by the short arrow E_{GW}. To measure our gravitational wave signal we have to decide on a readout angle or quadrature angle. Obviously, in the simple case displayed in the third panel of Fig. 11.5, we obtain the best SNR exactly reading out the phase quadrature; while the magnitude of the noise is similar for all possible readout angles the signal is maximal in the phase quadrature (vertical direction). So let us assume for the moment that, if not stated explicitly otherwise, we always choose a readout angle that perfectly coincides with the phase quadrature.

So far we have not included any optomechanical coupling in our picture, i.e., the situation displayed in the third panel of Fig. 11.5 is only valid for scenarios with negligible radiation pressure (i.e., for high frequencies or very low ratio of optical power and mirror mass). Let us now include the effects of quantum radiation pressure acting on our suspended test masses. As illustrated on the left-hand panel of Fig. 11.5, both phase and amplitude fluctuations are "impinging" onto the test mass. While the phase fluctuations have no mechanical effect on the mirror, the amplitude fluctuations couple via radiation pressure into position fluctuations of the test mass. These position fluctuations then show up, similar to the GW signal, as an additional component in the phase quadrature (see E_{RP} on the right-hand panel of Fig. 11.5). The key difference between the original fluctuations in the phase quadrature (E_2) and the fluctuations optomechanically coupled from the amplitude quadrature into the phase quadrature (E_{RP}) is the fact that now we have introduced correlated fluctuations in the amplitude and phase quadrature. As we will see in later sections, these introduced correlations between E_1 and E_{RP} can be used for beating the SQL. However, for the moment let us get back to the right-hand panel of Fig. 11.5: The magnitude of the radiation pressure induced fluctuations in the phase quadrature, i.e., the length of E_{RP}, scales directly with the light power, is inverse to the mass of the mirror, and due to the response function of a suspended mirror to an external force inverse to the frequency squared.

So, now we have completely assembled our graphical tool and it is time to check whether it delivers useful results and predictions. Let us see if we can explain the individual traces in Fig. 11.3. For the low power case (traces 1a–1c) at high frequencies radiation pressure does not play any role and the sensitivity of the interferometer is limited by the ratio of the shot noise in phase quadrature (E_2) and the GW signal (E_{GW}). At the low frequency end the coupling of fluctuations from the amplitude quadrature into the phase quadrature increases and the length of E_{RP} showing up in the phase quadrature becomes more and more dominant with lower frequency. While the GW signal stays constant in frequency, the change of E_{RP} significantly reduces the SNR of our measurement and therefore decreases the sensitivity towards the low frequency end of the measurement band. So, our picture allows us to explain the shape of the overall quantum noise spectrum (trace 1c). What will happen if we increase the optical power by a factor of 100? First of all the length of the GW arrow will increase by a factor of 10. At the same time the length of E_1 and E_2 stays constant, so that the overall signal-to-noise ratio at high frequencies increases by a factor of 10, i.e., the quantum noise contribution is reduced by a factor 10. However, since E_{RP} scales directly with the optical power, its length is now 100 times longer. So, at low frequencies the noise is increased by a factor of 100, whereas the length of the signal arrow E_{GW} is only increased by a factor of 10, so that in total we lose SNR at the low frequency end. Hence, our simple model predicts that for a factor 100 power increase we will obtain a factor 10 improvement with respect to the trace 1c at high frequencies and a factor 10 sensitivity reduction at low frequencies. These predictions match exactly the shape of the trace 2c in Fig. 11.3. So, our simple graphical tool has passed its first crucial test.

11.4 Squeezed Light Injection

In the previous section we already briefly discussed squeezed light and described it in the quadrature picture (see right-hand panel of Fig. 11.4). The squeezing ellipse can be characterized by three properties: (i) the squeezing level, i.e., ratio of diameter of the "unsqueezed" ball and the length of the minor axis of the "squeezed" ellipse, as well as the level of *anti-squeezing* which is given by the ratio of the ball diameter and the length of the major axis of the squeezing ellipse; (ii) the orientation of the squeezing ellipse in the quadrature plane, also referred to as *squeezing angle*; and (iii) the frequency of the squeezed light field. Squeezed states of light can in general be created by means of nonlinear optical effects, such as for instance an optic parametric oscillator (OPO). During the past decade there have been major advances in the generation of squeezed light for gravitational wave detectors such as the demonstration of squeezing levels of more than 12 dB [11] as well as squeezing down to frequencies of a few Hz [12]. Due to limited space in this chapter we will omit a detailed description of the generation of squeezed light states here. The interested

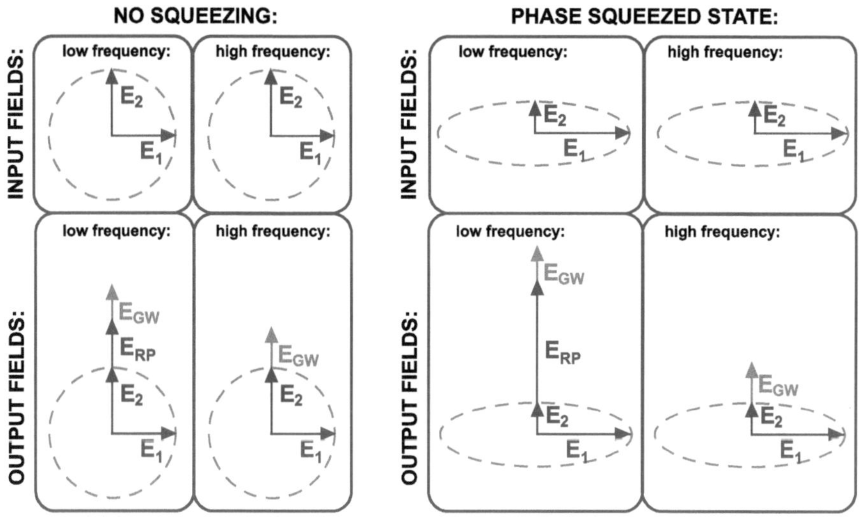

Fig. 11.6 Quadrature picture for a simple Michelson interferometer with phase squeezing (*right*) and without squeezing (*left*) as reference

reader can find good overviews of the technical principles involved in the generation of squeezed light in [13, 14].

Let us now investigate how the application of squeezing can improve the sensitivity of a GW detector. Assuming we have produced sufficiently strong squeezing at the desired frequency, then the only other parameter we need to decide on is the orientation of the squeezing ellipse. If we want to increase the sensitivity of our GW detector at high frequencies then we should inject phase squeezing, i.e., the minor axis of the squeezing ellipse is oriented in parallel to the phase quadrature. Figure 11.6 illustrates the application of phase squeezing. The ball is squeezed so that E_2 is shortened and E_1 becomes longer. Since the magnitude of the GW signals stays constant, at high frequencies we can improve the signal-to-noise ratio and therefore improve the sensitivity of our interferometer. However, as indicated in the third panel of Fig. 11.6, at low frequencies the noise in the phase quadrature is strongly increased when applying phase squeezing, because the increased amplitude fluctuations originating from the anti-squeezing couple via radiation pressure into the phase quadrature (E_{RP}). So, whereas the GW signal is independent of the squeezing level, the quantum noise at low frequencies increases with the application of phase squeezing and therefore overall we lose sensitivity at the low frequency end of the spectrum.

However, if we are keen to improve the interferometer sensitivity at low frequencies, this can also be accomplished by the injection of squeezed light. In this case we have to apply amplitude squeezing, so that the length of E_1 is reduced and therefore also the radiation pressure noise coupling into the phase quadrature decreases. The consequence of this is improved low frequency sensitivity at the expense of reduced

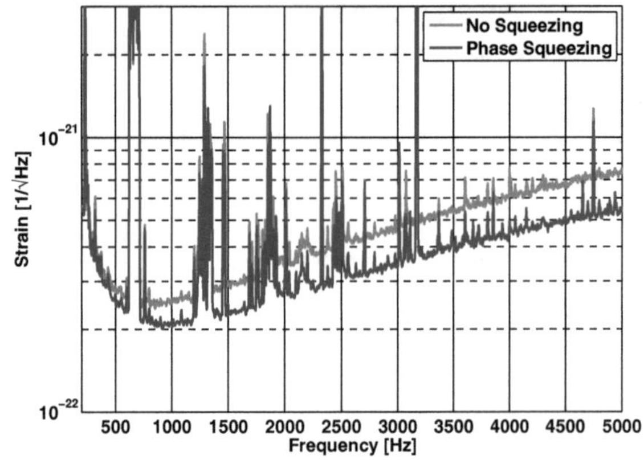

Fig. 11.7 Improvement of the GEO 600 sensitivity achieved by the injection of phase squeezing at the output port of the interferometer

sensitivity at high frequencies, because due to the anti-squeezing in that case E_2 will be longer than in the unsqueezed case. Hence, we see that the application of purely phase-squeezed or purely amplitude-squeezed light only allows us to improve the sensitivity over a certain part of the detection band. From a qualitative point of view such squeezing in the ideal case (i.e., if the squeezing level is exactly the same as the anti-squeezing level) allows exactly the same sensitivity variations as is possible by increasing or decreasing the light power inside the interferometer.

It is not possible to surpass the SQL by application of purely phase-squeezed or purely amplitude-squeezed light.

GEO 600 was the first large-scale GW detector to routinely improve its sensitivity by the application of squeezed light [15]. The implemented squeezing source is able to produce phase squeezing of almost 10 dB, which would be equivalent to a factor 3 improvement in strain sensitivity. However, due to high optical losses (in particular in the output mode cleaner) the obtained sensitivity improvement is only about a factor $\sqrt{2}$. Please note that at low frequencies GEO 600 is limited by seismic and control noise, masking the expected increase of quantum noise at the low-frequency end which is a consequence of the injection of phase squeezed light.

The experience with long-term injection of squeezed light states at GEO 600 [16] together with the recent tests of squeezed light in the Hanford 2 km interferometer [17], transform squeezing from "science fiction" to an established tool for gravitational wave detection.

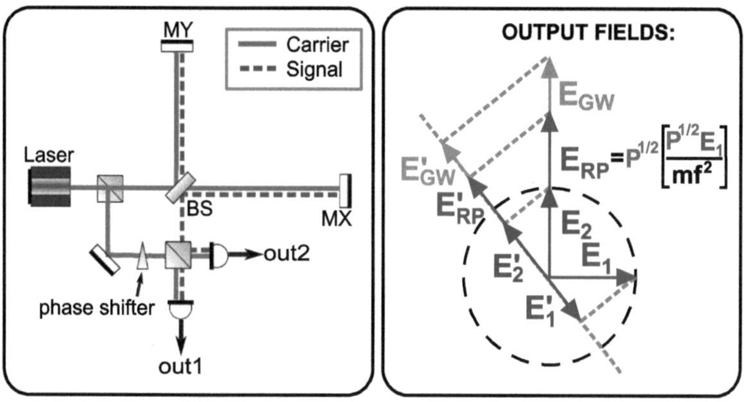

Fig. 11.8 *Left panel* Schematic layout of a simple Michelson interferometer with homodyne readout. The output port is kept on destructive interference so that only the signal can leave towards the photodiodes, where it beats with light picked off before the interferometer. In order to avoid open ports a homodyne detector consisting of a pair of photodiodes is used. *Right panel* Illustration of radiation pressure noise cancellation via homodyne readout

11.5 Homodyne Readout and Variational Readout

So far we have always assumed that we would read out our interferometer exactly in the phase quadrature. Naively thinking, this choice seems to make most sense because this gives us the strongest GW signal. However, as we will see in this section it can be beneficial to use a different readout angle, i.e., to readout a superposition of phase and amplitude quadrature.

From a technical point of view it is possible to vary this readout angle by utilizing a homodyne readout scheme as illustrated on the left-hand part of Fig. 11.8. Instead of using DC-readout [18, 19] in which a deliberate offset is introduced to make a fraction of the carrier field leak out through the output port forming the local oscillator for the gravitational wave signal. We can also pick off some light before the main interferometer and guide it to the output port where it can serve as local oscillator. The latter scheme is referred to as "homodyne readout." By shifting the phase of the local oscillator with respect to the light leaving the main interferometer, for instance by microscopically changing the local oscillator path length, one can choose whether to read out the amplitude quadrature, the phase quadrature, or an arbitrary superposition of these.

So, how does the homodyne readout help us to reduce quantum noise? In order to understand this our graphical tool comes in handy again, as displayed on the right-hand panel of Fig. 11.8. Recall that the length of E_{RP} depends on the observation frequency. However, the length of E_1, which is correlated with E_{RP} and represents the noise in the amplitude quadrature is not frequency dependent. This means that for every frequency there is a specific readout angle for which the projections, E'_{RP} and E'_1, of the two correlated arrows (E_{RP}, E_1) onto the readout axis have exactly

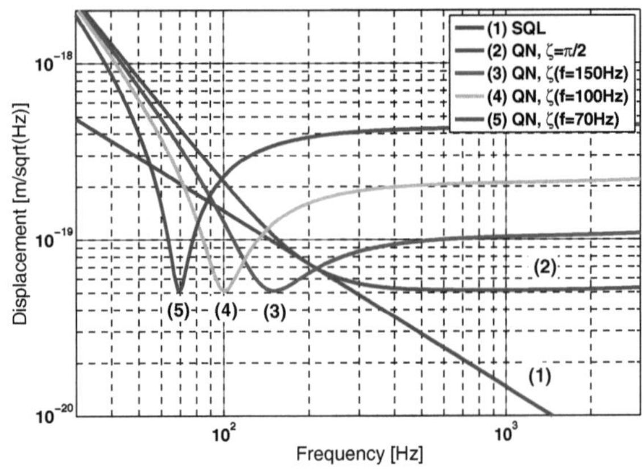

Fig. 11.9 SQL and simulated quantum noise of the AEI 10 m sub-SQL interferometer for different readout angles. Using homodyne readout it is possible to exactly cancel radiation pressure noise at one frequency and to surpass the SQL in a narrow frequency range. The figure is taken from [20]

the same length. Since E'_{RP} and E'_1 are correlated but point in opposite directions they exactly cancel out; thus our sensitivity is determined by the ratio of the lengths of the E'_{GW} and E'_2. In this case we have managed to cancel completely the radiation pressure noise at our measurement frequency. Please note that the ratio of the E'_{GW} and E'_2 stays constant for any readout angle different from pure amplitude quadrature readout.

The experimental demonstration of radiation pressure noise cancellation via homodyne readout is one of the goals of the AEI 10 m sub-SQL interferometer [21, 22]. Figure 11.9 shows the simulated quantum noise for the AEI-10m interferometer for 4 different readout angles [20]. Trace 2 represents the standard case of pure phase quadrature readout. We see a flat shot noise spectrum at frequencies above a few 100 Hz (plus the additional roll-off at 10 kHz originating from the limited bandwidth of the arm cavity), and at low frequencies we see the characteristic increase in quantum noise due to radiation pressure noise, following a $1/f^2$ characteristic. The other three traces show the quantum noise for a certain readout angle. As one can see it is possible to cancel the radiation pressure noise at one frequency and to dip down below the SQL achieving exactly the same peak displacement sensitivity as in the shot noise limited frequency regime of trace 2, i.e., the potential sensitivity improvement is limited by shot noise which was buried under quantum radiation pressure noise before.

A simple homodyne readout is one of the possibilities to achieve narrowband sub-SQL sensitivities.

Let us have a closer look at trace 4 in Fig. 11.9 and try to better understand its shape. We have already explained that we can completely cancel radiation pressure noise at one frequency, in this case 100 Hz. However, at frequencies below and above this "sweet spot" quantum noise is only partly reduced or even increased. Below 100 Hz

E'_1 is shorter than E'_{RP} and therefore the cancellation of the correlated noise terms works only partly. Finally we should explain why the noise represented by trace 4 is higher than the trace 2 for all frequencies above a few 100 Hz where the trace 2 is made up of shot noise. At these high frequencies E'_{RP} is of negligible length. However, due to the rotated readout angle amplitude noise from E_1 is mixed into our readout signal (E'_1), so that the sensitivity is determined by the ratio of E'_{GW} and the sum of the uncorrelated noise contributions E'_2 and E'_1.

So, we have seen that changing the readout quadrature (homodyne readout) allows us to cancel radiation pressure noise at a single frequency at the expense of reduced sensitivity at high frequencies. If it would be possible to change the readout angle in a frequency-dependent way, then it should be possible to cancel radiation pressure noise at low frequencies, i.e., where it dominates, without introducing additional noise at high frequencies. This technique, called *variational readout* can actually be realized by making use of the dispersion of the light reflected from a cavity [23], in analogy to the generation of frequency-dependent squeezing discussed in the following section.

11.6 Frequency-Dependent Squeezing

We have seen above that the injection of simple amplitude or phase squeezing cannot improve the sensitivity over the whole frequency range and therefore with such squeezed light states it is not possible to surpass the SQL. However, if we were to manage to change the squeezing angle of the injected light in a frequency-dependent way, then it should be possible to obtain a reduction of quantum noise at all frequencies, which in turn would allow broadband measurements with a sensitivity below the SQL.

Using our graphical tool, in Sect. 11.3.2 we learned that our sensitivity is limited at high frequencies by shot noise in the phase quadrature, while at low frequencies the dominant noise term originates from amplitude fluctuations coupled to the phase quadrature via radiation pressure-induced mirror motion. Hence, if we could rotate the squeezing ellipse as a function of the detection frequency to always provide squeezing in the optimal quadrature we could reduce quantum noise in the entire detection band.

Injection of squeezed light for which the squeezing angle is a function of frequency is in the following referred to as *frequency-dependent squeezing*.

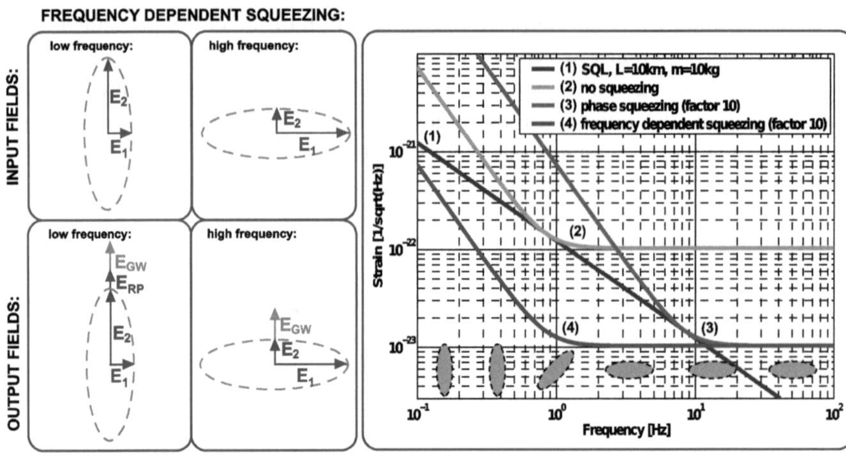

Fig. 11.10 *Left* Illustration of frequency-dependent squeezing using the quadrature picture. *Right* Quantum noise of a simple Michelson interferometer with pure phase squeezing as well as frequency-dependent squeezing. Please note that even though theoretically a squeezing factor of 10 is possible, in real interferometers optical losses and control noise of the squeezing will degrade the sensitivity improvement. It is assumed that in future GW detectors quantum noise reduction by squeezing injection is limited to a factor of 3 to 4 [24]

The two left panels of Fig. 11.10 illustrate frequency-dependent squeezing in the quadrature picture. For all frequencies we readout the phase quadrature. At high frequencies we apply phase squeezing and obtain a quantum noise reduction exactly the same as for injection of frequency-independent phase squeezing. In contrast to this, towards the low frequency end of the detection band we arrange for the squeezing ellipse to continuously rotate from phase quadrature squeezing to squeezing in the amplitude quadrature. The key here is that the length of the original E_1 arrow in the amplitude quadrature reduces, which results in less noise coupled to the phase quadrature via radiation pressure-induced mirror motion. Therefore, also the length of the E_{RP} arrow in the phase quadrature is reduced. As a consequence the SNR at low frequency is larger than in the unsqueezed case (compare to left panel of Fig. 11.6). The right-hand plot of Fig. 11.10 compares the quantum noise spectrum of a simple Michelson interferometer without squeezing (trace 2), pure phase squeezing (trace 3), and frequency-dependent squeezing (trace 4), where the gray ellipses at the bottom of the plot illustrate the optimized orientation of the squeezing ellipse of the injected squeezing at different frequencies in the detection band.

Frequency-dependent squeezing can provide a broadband reduction of quantum noise significantly below the SQL.

Frequency-dependent squeezing has been identified as one of the key technologies for the Einstein Telescope [25, 26] as well as for upgrades to the Advanced LIGO

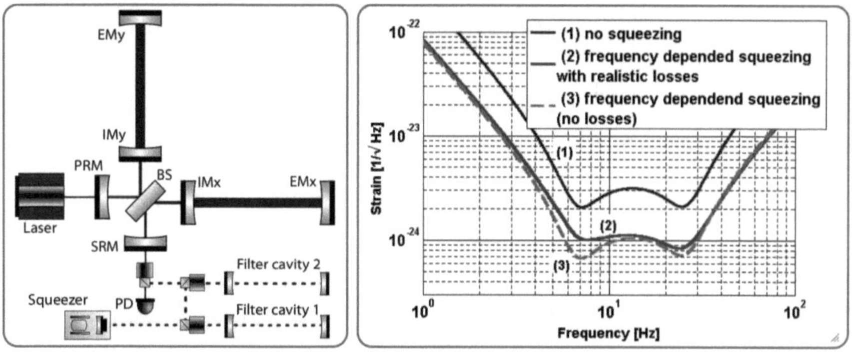

Fig. 11.11 *Left* Cartoon of the optical layout of the ET low-frequency interferometer, which uses detuned signal recycling and exploits the dispersion of the squeezed light being reflected from two filter cavities to provide squeezing at the optimal frequency-dependent angle for all frequencies. *Right* Quantum noise for an ET low-frequency interferometer without squeezing (trace 1), with 6 dB of frequency-dependent squeezing injected into an ideal interferometer (trace 3) and frequency-dependent squeezing considering realistic losses (trace 2)

instruments [27, 28]. Figure 11.11 illustrates the implementation of frequency dependent squeezing into an ET low frequency interferometer. The squeezed light leaving the squeezing source is reflected from two slightly differently detuned cavities. The dispersion of the detuned cavities introduces the frequency dependent rotation of the squeezing ellipse. Afterwards the squeezed light is injected into the output port of the interferometer. For the squeezing injected into the detection port the interferometer acts as a reflector. The injected squeezing leaves the interferometer and co-propagates with the GW signal towards the detection stage where the squeezed output field is converted into an electronic signal.

From the technical point of view the key difficulties in realizing frequency-dependent squeezing are the design and low-noise operation of the filter cavities with the required bandwidth and low enough losses in order not to destroy the squeezing. In the case of the ET low-frequency interferometer the required bandwidth of the filter cavity is similar to the bandwidth of the signal recycling. As one can see from the right-hand plot of Fig. 11.11, for an ET low frequency interferometer the signal recycling bandwidth is about 10 Hz, which in turn requires to employ filter cavities of the same finesse and length as the main interferometer. If shorter filter cavities are preferred it is necessary to increase the finesse which, depending on the chosen length, may take on extremely high values.

In high-finesse filter cavities the presence of even very low optical losses may strongly degrade the squeezing strength of the generated frequency-dependent squeezing, as is illustrated in Fig 11.11. Whereas trace 3 shows the quantum noise level for filter cavities without any losses, trace 2 represents the quantum noise for 10 km long filter cavities with 75 ppm round-trip loss. Why are losses so important, even though we only use the filter cavities in reflection? The problem is that some of the squeezing (mainly at and around the resonances of the detuned filter cavities)

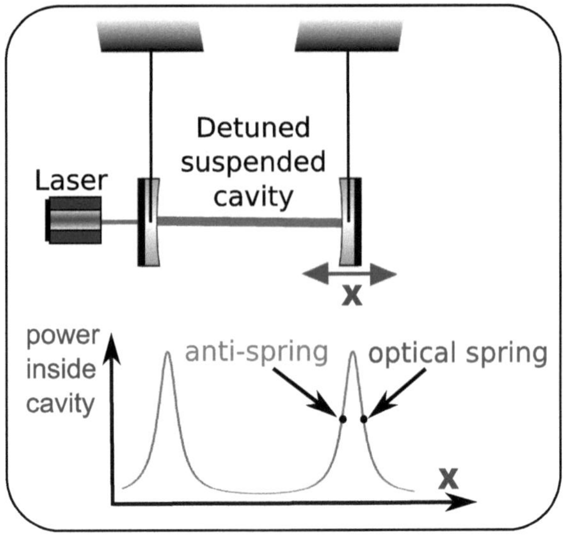

Fig. 11.12 Illustration of the optical spring principle. In a cavity tuned slightly away from its resonance, the optical power strongly depends on the microscopic position of the mirrors or more accurately the cavity length. If one or both of the cavity mirrors are suspended, then they are susceptible to radiation pressure forces, which depend on the position of the mirror. Radiation pressure, together with gravity, acts as a linear restoring force, resembling the dynamics of a mechanical spring

actually enters the filter cavities and hence is subject to internal optical loss. Loss in this context means that a fraction of the squeezed field is lost and replaced by a coherent unsqueezed state. For constant optical loss per filter cavity mirror, the higher we choose the finesse of the filter cavity, the stronger is the squeezing degradation.

11.7 Optical Springs and Optomechanical Rigidity

An optical spring is a fascinating physical phenomenon, which allows to connect two or more suspended mirrors with springs completely made out of photons (also compare Chap. 12 of this book). These springs can be made kilometers long and stiffer than diamond [29, 30], while at the same time, in contrast to any mechanical spring, they are somewhat free from classical noise.

So how can we create an optical spring? The basic principle is shown in Fig. 11.12: in a detuned Fabry-Perot cavity featuring high optical power and at least one suspended mirror, the equilibrium position of the suspended mirror is given by radiation pressure trying to push the mirrors further apart and gravity pulling the suspended mirror to shorten the cavity. Now, let us consider the case labelled "optical spring" in Fig. 11.12 and let us try to push the right mirror away from equilibrium. In case we push the mirror to the right two things will change: first, gravity will try to pull the mirror back towards its equilibrium and secondly as the cavity length has increased the cavity power will decrease, as the cavity is pushed further away from resonance. Therefore, the radiation pressure force will decrease. Both these effects will cause the mirror to move towards the left back into its equilibrium position. Now let us try to push the mirror towards the left. As this means that we shorten the cavity and move closer to resonance, the cavity power increases and the increased radiation pressure force pushes the mirror back towards the equilibrium. So, we have a system exhibiting a linear restoring force, which can in simple terms be described by Hooke's law $F = -kx$, where F is the superposition of forces acting on the mirror, x is the distance of the mirror from its equilibrium point, and k is the effective spring constant.

In a cavity with two suspended mirrors connected by an optical spring, when we push one mirror (and keep it fixed so that it cannot return to its original position), then the other mirror will move exactly by the same amount, so that the pathlength for the light stays constant. Using this phenomenon one can control the position of a mirror over several kilometers of distance, just by acting on the second cavity mirror.

There are two more interesting aspects about optical springs worth to be mentioned here: When setting the cavity operation point onto the opposite slope of the Airy peak (labeled as "anti-spring" in Fig. 11.12), then it is easy to see that there will be an optical anti-spring, i.e., the radiation pressure force acts in the same direction as a mirror displacement. The other interesting aspect is related to the damping of optical springs: When changing the cavity length, the optical power does not settle immediately at its new value, but it takes some time for the light in the cavity to build up or relax. This kind of "delay" causes the optical spring to feature anti-damping. The anti-damping can render optical systems unstable if no counter measures are applied. Two ways have been proposed to avoid this problem: First of all one can use active feedback to introduce damping or one can use two different springs acting on the same system and combining them in a way such that the result is an effective spring with positive damping [31].

So, how can optical springs actually help to surpass the SQL? In this respect two concepts are of interest. First of all optical springs play an important role in the context of advanced GW detectors employing detuned signal recycling [32], which has already been discussed in detail in Chap. 3 of this book. The other class of concepts employs optical spring systems to act as transducer of the GW. Figure 11.13 shows the simplest variant of such a transducer configuration, a so-called *optical bar* [33]. Two detuned cavities with a heavy end mirror each (EM1 and EM2) share a lightweight mirror (MR) so that MR is connected via optical springs to both end test masses. In case a GW with a frequency below the optical spring frequency is incident on the detector perpendicular to the plane spanned by the interferometer arms, one of the arms will shrink, while the other will stretch. If you think about this in terms of a system of coupled mechanical springs, then this means that one spring will be squeezed, i.e., it will push on MR, while the spring in the perpendicular arm will be stretched and pull on MR, so that as a consequence in the presence of a GW the mirror denoted MR will move in its local frame. This local movement of MR can then be read out relative to a local reference. The trick of the optical bar topology is that this

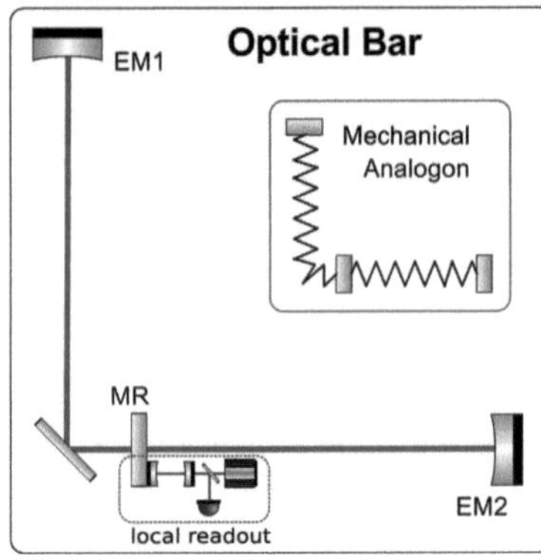

Fig. 11.13 Simplified optical layout of an optical bar. A lightweight mirror (MR) is connected via optical springs to two heavy end mirrors (EM1 and EM2). If there occurs a change in the differential armlength (e.g., due to GW), MR is displaced with respect to its local frame. A separate local readout can then be used to readout the GW signal

concept allows to separate the GW transducer from the readout process and hence gives the possibility to optimize both parts of the optical bar system independently to minimize quantum noise in the measurement. More advanced and more complex configurations involving optical springs, such as optical levers [34, 35] and local readout configurations [36] have been proposed in the literature, but due to lack of space will not be discussed here.

11.8 Speedmeter Topologies

We have seen above that Michelson interferometers are limited by the SQL. The reason for this is that quantum theory imposes a limitation on how accurate subsequent position measurements can be carried out. This is a direct consequence of the fact that position measurements at different times $x(t)$, $x(t')$ do not commute, i.e.,

$$\left[\hat{x}(t), \hat{x}(t')\right] \neq 0. \tag{11.9}$$

However, already in the 1930s John von Neumann realized that for some systems it is possible to find observables which are not limited by an uncertainty relation [37]. For instance in systems where the momentum is conserved you can in principle measure the momentum p of an ensemble of mirrors continuously with arbitrary precision since

$$\left[p(t), p(t')\right] = 0. \tag{11.10}$$

In 1990, Braginsky and Khalili proposed this concept for application in *bar detectors*, which were the predecessors of laser-interferometric GW detectors [38]. Hence,

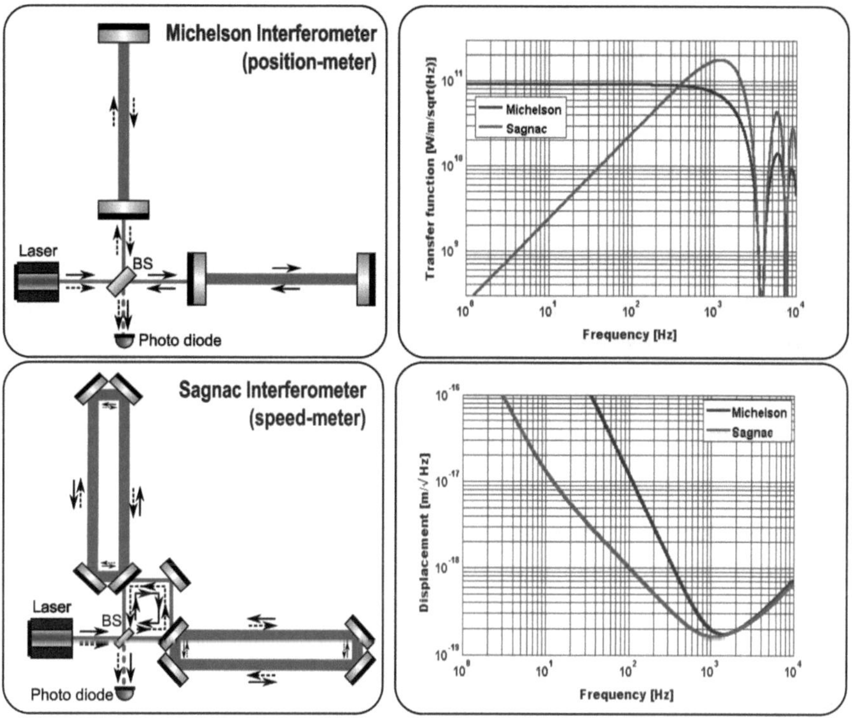

Fig. 11.14 Simplified optical layout of a Michelson interferometer with arm cavities (*top left*) and a Sagnac interferometer with arm cavities (*bottom left*). Whereas in the case of the Michelson interferometer all photons only travel through one of the arms, in the case of the Sagnac interferometer all photons travel through both arms. Transfer functions from differential arm length to signal on the main photo detector (*top right*). Quantum noise limited effective "displacement" sensitivity for a Michelson and a Sagnac interferometer (*bottom right*)

for detectors of this type, in which the velocity would be measured instead of the position of the test mass, the name "speed-meter" was coined. Later, the concept was also suggested for application in interferometric GW detectors.The first speed-meter interferometer configuration was based on a Michelson interferometer with a so-called shloshing cavity at the output port the interferometer. With the aid of this shloshing cavity the conventional Michelson interferometer is transformed into a speedmeter [39]. In the year 2003, Chen demonstrated in a theoretical work that a Sagnac interferometer naturally constitutes a speedmeter [40] .

In order to understand why a Sagnac interferometer is a natural representative of the class of speedmeters we should have a look at the optical layouts of a Michelson interferometer and a Sagnac interferometer as displayed in the two left-hand panels of Fig. 11.14. In a Michelson interferometer, depending on whether the photons in the input laser beam are transmitted or reflected, the light either travels along the horizontal or the vertical arm. It is important that when the returning photons interfere at the beam splitter, each photon has only sensed the mirror positions of a single arm. Contrasting this, in a Sagnac interferometer all photons travel along both arm cavities, i.e., they sense the positions of all relevant mirrors. Only after all photons have passed through both arm cavities the light interferes again at the output port of the beam splitter. This means that the light initially transmitted through the beam splitter measures the position of the end mirrors of the horizontal arm at a time t_1, while it measures the position of the end mirrors in the vertical arm at a slightly later time t_2. The light initially reflected at the beam splitter measures the position of the end mirrors of the vertical arm at a time t_1, while it measures the position of the end mirrors in the horizontal arm at a slightly later time t_2. So, each of the end mirrors is sensed at two different times, or, in other words, we measure the velocity of the end mirrors, but not their absolute position.

As the Sagnac speedmeter is not sensitive to absolute displacement, its response to *absolute* differential arm lengths goes down to zero. This can be seen in the top right plot of Fig. 11.14. While the Michelson has a flat response at low frequencies, the Sagnac response falls linearly towards lower frequencies. So, if the signal at low frequencies is smaller in the Sagnac compared to the Michelson Interferometer, how can the quantum noise limited sensitivity be better for the speedmeter?

Please recall that the achievable sensitivity is related to the SNR. For the Michelson the signal is flat, but the radiation pressure noise is present and increases quadratically towards lower frequency. In contrast, in the Sagnac speedmeter the signal decreases linearly towards low frequency, while the noise is flat because the radiation pressure noise is mostly cancelled, resulting in a significantly reduced and constant value of the radiation pressure noise at low frequencies. So, in conclusion the Sagnac speedmeter wins in terms of low frequency sensitivity over the Michelson.

The slight bend in the Sagnac sensitivity trace below 10 Hz (compare Fig. 11.14) originates from the presence of optical losses in the considered configuration.

11.9 Challenges Towards Sub-SQL Interferometry

It is a fabulous achievement that the advanced GW detectors such as Advanced LIGO are on the brink of reaching a sensitivity limited by the Heisenberg Uncertainty Limit for 40 kg mirrors! However, further improvements are possible since the SQL does not really pose an *ultimate limit*. Above we have introduced, and briefly discussed, a wealth of concepts available to tackle the SQL and eventually surpass it.

It is important to point out that all of the discussed concepts are not exclusive in terms of their application, but many of them can be combined to maximize the sensitivity improvement. For instance, already at the moment GEO 600 uses phase quadrature squeezing in combination with signal recycling. The baseline design for ET as well as upgrades to Advanced LIGO are likely to employ signal recycling with frequency-dependent squeezing, while at the same time speedmeters with squeezing and/or signal recycling have been proposed as alternatives to the well-established Michelson interferometers.

So far pure phase squeezing is the only example of a technique discussed in this chapter that has already been demonstrated in a km-scale GW detector. All other concepts have so far only been analyzed theoretically. Careful experimental verification of these techniques in table-top experiments and in suspended low-noise prototypes is urgently required to have these technologies available for installation in the future GW detectors, to outwit quantum mechanics and measure with sensitivities below the Standrad Quantum Limit.

Acknowledgments The author acknowledges the support from the Science and Technology Facilities Council (STFC), UK, and the European Research Council (ERC-2012-StG: 307245). The author thanks Kenneth Strain, Stefan Danilishin, and Farid Khalili for many fruitful discussions. Christian Gräf and Neil Gordon have provided many useful comments on this manuscript. All interferometer sketches in this chapter have been drawn using Alexander Franzen's ComponentLibrary.

References

1. V.B. Braginsky, F.Y. Khalili, *Quantum Measurement*, (Cambridge University Press, Cambridge, 1992)
2. S. Danilishin, F.Y. Khalili, Quantum measurement theory in gravitational-wave detectors. Living Rev. Relativ. **15**, 5 (2012). http://www.livingreviews.org/lrr-2012-5
3. H. Müller-Ebhard et al., Review of quantum non-demolition schemes for the Einstein Telescope, (2009) ET-note, https://tds.ego-gw.it/itf/tds/file.php?callFile=ET-010-09.pdf
4. H. Miao et al., Comparison of Quantum Noise in 3G Interferometer Configurations, (2012). https://dcc.ligo.org/cgi-bin/private/DocDB/ShowDocument?docid=78229
5. R. Abbott et al., AdvLIGO Interferometer Sensing and Control Conceptual Design, Technical note LIGO-T070247-01-I (2008)
6. GWINC website, http://lhocds.ligo-wa.caltech.edu:8000/advligo/GWINC
7. V.B. Braginsky, F.Y. Khalili, Quantum nondemolition measurements: the route from toys to tools. Rev. Mod. Phys. **68** (1996)
8. Y. Chen, Advanced interferometer configurations, in *Presentation at Australian-Italian Workshop on GW Detection*, Oct 2005, https://dcc.ligo.org/LIGO-G050517-x0
9. Y. Chen, Various ways to beat the standard quantum limit, in *Presentation at Aspen Workshop* 2004, https://dcc.ligo.org/LIGO-G040210-x0
10. C.M. Caves, Quantum-mechanical noise in an interferometer. Phys. Rev. D **23**, 1693–1708 (1981)
11. M. Mehmet et al., Squeezed light at 1550 nm with a quantum noise reduction of 12.3 dB. Opt. Express **19**(25), 25763 (2011)
12. H. Vahlbruch, S. Chelkowski, K. Danzmann, R. Schnabel, Quantum engineering of squeezed states for quantum communication and metrology. N. J. Phys. **9**(10), 371 (2007)
13. S. Chelkowski et al., Coherent control of broadband vacuum squeezing. PRA **75**, 043814 (2007)
14. K. McKenzie et al., Squeezing in the audio gravitational-wave detection band. PRL **93**, 161105 (2004)
15. J. Abadie et al., A gravitational wave observatory operating beyond the quantum shot-noise limit. Nat. Phys. **7**(12), (2011)

16. H. Grote et al., First long-term application of squeezed states of light in a gravitational-wave observatory. PRL **110**, 181101 (2013)
17. J. Abadie et al., Enhanced sensitivity of the LIGO gravitational wave detector by using squeezed states of light. Accepted for publication in Nature Photonics, (2013). doi:10.1038/NPHOTON.2013.177
18. R.L. Ward et al., DC readout experiment at the Caltech 40m prototype interferometer. Class. Quantum Gravity **25**, 114030 (2008)
19. S. Hild et al., DC-readout of a signal-recycled gravitational wave detector. Class. Quantum Gravity **26**, 055012 (2009)
20. C. Gräf, Optical design and numerical modeling of the AEI 10 m Prototype sub-SQL interferometer, Ph.D. thesis, University of Hannover, 2013
21. S. Gossler et al., The AEI 10 m prototype interferometer. Class. Quantum Gravity **27**(8), 084023 (2010)
22. C. Gräf, S. Hild, H. Lück, S. Gossler, B. Willke, K. Strain, K. Danzmann, Optical layout for a 10 m Fabry-Perot Michelson interferometer with tunable stability. Class. Quantum Gravity **29**(7), 075003 (2012)
23. H.J. Kimble, Y. Levin, A.B. Matsko, K.S. Thorne, S.P. Vyatchanin, Conversion of conventional gravitational-wave interferometers into quantum nondemolition interferometers by modifying their input and/or output optics. Phys. Rev. D **65**, 022002 (2002)
24. S. Hild, Beyond the second generation of laser-interferometric gravitational wave observatories. Class. Quantum Gravity **29**(12) 124006 (2012)
25. M. Punturo et al., The third generation of gravitational wave observatories and their science reach. Class. Quantum Gravity **27**, 084007 (2010)
26. S. Hild et al., Sensitivity studies for third-generation gravitational wave observatories. Class. Quantum Gravity **28**(9), 094013 (2011)
27. B. Barr et al., LIGO 3 Strawman Design, Team Red, Technical Note LIGO-T1200046-v1 2012/01/30, https://dcc.ligo.org/LIGO-T1200046-v1
28. R. Adhikari, K. Arai, S. Ballmer, E. Gustafson, S. Hild, Report of the 3rd Generation LIGO Detector Strawman Workshop, Technical Note LIGO-T1200031-v3 2012/05/15, https://dcc.ligo.org/LIGO-T1200031-v3
29. T. Corbitt, D. Ottaway, E. Innerhofer, J. Pelc, N. Mavalvala, Measurement of radiation-pressure-induced optomechanical dynamics in a suspended Fabry-Perot cavity. Phys. Rev. A **74**, 021802 (2006)
30. T. Corbitt et al., An all-optical trap for a gram-scale mirror. Phys. Rev. Lett. **98**, 150802 (2007)
31. H. Rehbein et al., Double optical spring enhancement for gravitational-wave detectors. Phys. Rev. D **78**, 062003 (2008)
32. A. Buonanno, Y. Chen, N. Mavalvala, Quantum noise in laser-interferometer gravitational-wave detectors with a heterodyne readout scheme. Phys Rev D **67**, 122005 (2003)
33. V.B. Braginsky, F.Y. Khalily, Nonlinear meter for the gravitational wave antenna. Phys. Lett. A **218**, 167–174 (1996)
34. F.Y. Khalili, The optical lever intracavity readout scheme for gravitational-wave antennae. Phys. Lett. A **298**, 308–314 (2002)
35. S.L. Danilishin, F.Y. Khalili, Practical design of the optical lever intracavity topology of gravitational-wave detectors. Phys. Rev. D **73**, 022002 (2006)
36. H. Rehbein et al., Local readout enhancement for detuned signal-recycling interferometers. Phys. Rev. D. Am. Phys. Soc. **76**, 062002 (2007)
37. J. von Neumann, *Mathematische Grundlagen der Quantenmechanik* (Springer, Berlin, 1932)
38. B. Braginsky, F.Y. Khalili, Gravitational wave antenna with QND speed meter. Phys. Lett. A **147**, 251–256 (1990)
39. P. Purdue, Y. Chen, Practical speed meter designs for quantum nondemolition gravitational-wave interferometers. Phys. Rev. D **66**, 122004 (2002)
40. Y. Chen, Sagnac interferometer as a speed-meter-type, quantum-nondemolition gravitational-wave detector. Phys. Rev. D **67**, 122004 (2003)

Chapter 12
The Parametric Instability in Advanced Gravitational-Wave Interferometers

Pierre-François Cohadon and Slawomir Gras

Abstract The Parametric Instability (PI) consists in the coherent ringing of a mechanical mode of a cavity mirror triggered by the circulating optical power, which may significantly jeopardize the stable operation of high-power advanced gravitational-wave interferometers. We present a model and a simple physical picture of radiation pressure effects and relevant experiments in the single-cavity configuration. We discuss the reasons why the PI is an essential feature of the advanced interferometers, how it can be computed using finite-element modeling, and possible schemes or techniques to overcome it.

12.1 Radiation-Pressure Effects

The high-power operation of the advanced gravitational-wave interferometers currently being built, with incident optical power at the 100 W level and above, comes at multiple costs, such as thermal lensing or scattered light issues. We investigate in the following another one: the radiation-pressure induced dynamics of the suspended mirrors, which may lead to an optomechanical instability [1], with the potential to threaten the interferometer locking and stable operation.

P.-F. Cohadon (✉)
Laboratoire Kastler Brossel, UPMC-ENS-CNRS,
4 place Jussieu, 75252 Paris cedex 05, France
e-mail: cohadon@lkb.upmc.fr

S. Gras
LIGO Laboratory, Massachusetts Institute of Technology,
185 Albany St, Cambridge, MA 02139, USA
e-mail: sgras@ligo.mit.edu

M. Bassan (ed.), *Advanced Interferometers and the Search for Gravitational Waves*, Astrophysics and Space Science Library 404, DOI: 10.1007/978-3-319-03792-9_12,

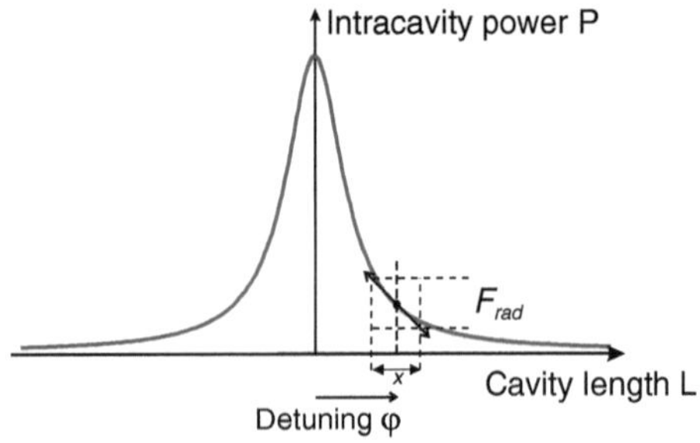

Fig. 12.1 Intracavity power P as a function of the cavity length L. Around a steady-state detuning φ (defined in Eq. 12.2), the force F_{rad} applied by the intracavity field upon the moving mirror is proportional to the *slope* of the Airy *peak* and to the displacement x

12.1.1 Optomechanical Dynamics

Light has been commonly used to probe mechanical displacements for decades, but the reverse coupling *field → mechanical motion* has gained experimental interest only recently. We present here a simple theory of these *optomechanical effects* in a single cavity.

At frequencies Ω close to a resonance frequency Ω_{m} of the mirror, its dynamics can be approximated as the one of a single harmonic oscillator, with resonance frequency Ω_{m}, mass M, and mechanical quality factor Q. Note that this is adequate to describe all relevant degrees of freedom such as pendulum modes, violin modes, or internal modes of the test masses, once the appropriate parameters (and especially the effective mass M) are used.

As the intracavity power P depends on the cavity length, any displacement x of the resonator induces a variation of the intracavity radiation pressure $F_{\text{rad}} \propto P$ (see Fig. 12.1). As a consequence, the intrinsic stiffness $M\Omega_{\text{m}}^2$ of the resonator is balanced by the radiation pressure force: for a positive detuning, the displacement creates a negative linear force and thereby an additional binding force, increasing the effective stiffness, whereas for a negative detuning, the force corresponds to a softening of the oscillator. These effects obviously cancel at resonance, but are maximum halfway of the optical cavity resonance, and proportional to the incident power. Such *optical spring* effects (see also Chap. 3) have been first demonstrated on a suspended mirror [2] or on a specially-designed flexure oscillator [3].

The previous argument is however only valid when the mechanical frequency Ω_{m} is negligible compared to the cavity bandwidth Ω_{c}. Otherwise, the cavity storage time has to be taken into account to evaluate the resonator dynamics. Linearizing the cavity equations around the corresponding steady-state detuning, one gets the radiation pressure force F_{rad} Fourier component at frequency Ω [4]:

$$F_{\text{rad}}[\Omega] = -2\frac{\varphi\varphi_{\text{NL}}}{\Delta} M\Omega_{\text{m}}^2\, x[\Omega], \tag{12.1}$$

with

$$\varphi \equiv 4\pi L/\lambda\gamma \pmod{2\pi} \tag{12.2}$$

$$\varphi_{\mathrm{NL}} = \frac{8\pi}{\lambda\gamma c}\frac{P}{M\Omega_{\mathrm{m}}^2} \tag{12.3}$$

$$\Delta = (1 - i\Omega/\Omega_{\mathrm{c}})^2 + \varphi^2, \tag{12.4}$$

where L is the cavity length, λ the laser wavelength, φ the detuning normalized to the cavity damping rate γ, and φ_{NL} the normalized phase-shift of the cavity induced by the static recoil effect of the mirror.

In the framework of linear response theory, the mirror responds to an external force F (superimposed to the radiation-pressure force exerted by the intracavity light field) with a displacement $x[\Omega]$ given by:

$$x[\Omega] = \chi_{\mathrm{m}}[\Omega]\left(F[\Omega] + F_{\mathrm{rad}}[\Omega]\right), \tag{12.5}$$

with the mechanical susceptibility $\chi_{\mathrm{m}}[\Omega]$:

$$\chi_{\mathrm{m}}[\Omega] = \frac{1}{M\left(\Omega_{\mathrm{m}}^2 - \Omega^2 - i\Omega\Omega_{\mathrm{m}}/Q\right)}. \tag{12.6}$$

Using (12.1) to include cavity-induced radiation pressure, the mirror responds now to F with an effective mechanical susceptibility $\chi_{\mathrm{eff}}[\Omega]$:

$$\chi_{\mathrm{eff}}[\Omega]^{-1} = \chi_{\mathrm{m}}[\Omega]^{-1} + 2\frac{\varphi\varphi_{\mathrm{NL}}}{\Delta}M\Omega_{\mathrm{m}}^2. \tag{12.7}$$

In the limit of a mechanical quality factor $Q \gg 1$, $\chi_{\mathrm{eff}}[\Omega]$ still has a lorentzian shape, but with effective resonance frequency and mechanical quality factor given by:

$$\Omega_{\mathrm{eff}} = \Omega_{\mathrm{m}}\left(1 + \mathrm{Re}\frac{\varphi\varphi_{\mathrm{NL}}}{\Delta(\Omega_{\mathrm{m}})}\right) \tag{12.8}$$

$$1/Q_{\mathrm{eff}} = 1/Q_{\mathrm{m}}\left(1 - 2Q\,\mathrm{Im}\frac{\varphi\varphi_{\mathrm{NL}}}{\Delta(\Omega_{\mathrm{m}})}\right). \tag{12.9}$$

The in-quadrature component of the radiation pressure F_{rad} (with respect to the displacement x) yields a damping/anti-damping force, which can significantly alter the mechanical Q for $\Omega_{\mathrm{m}} \simeq \Omega_{\mathrm{c}}$.

12.1.2 Experimental Results: Optical Damping

After the first optical spring demonstrations [2, 3], a number of related table-top experiments have demonstrated this optomechanical damping behavior, in the context

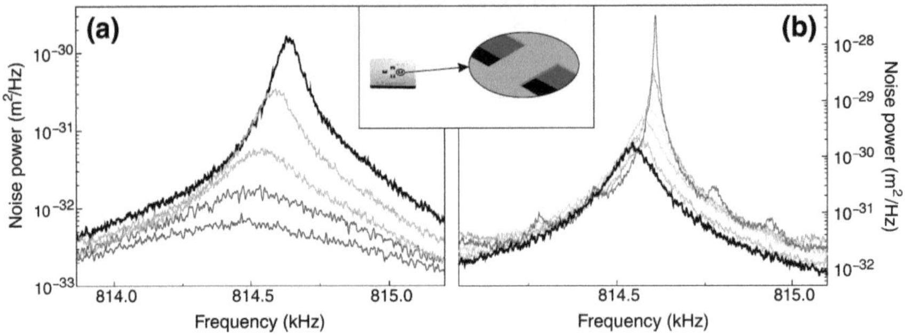

Fig. 12.2 Thermal noise spectra, normalized as micro-resonator displacements. *Black thick curves* correspond to the resonant case $\varphi = 0$. **a** *Lighter color curves* are obtained for negative detunings φ from -0.1 to -0.6 and for an incident power of 5 mW. **b** *Lighter color curves* are obtained for positive detunings φ from 0.03 to 0.13 and for an incident power of 2.5 mW. The damping and anti-damping effects are demonstrated by the area reduction or increase of the spectra. Note the related shift of the mechanical resonance frequency. The insert shows a picture of the silicon micro-mirror, with a close-up on the actual mechanical resonator used

of radiation-pressure cooling of macroscopic mechanical resonators, a subject which has gained momentum since 2006.

Figure 12.2 shows the thermal noise spectra obtained in a table-top experiment at Laboratoire Kastler Brossel, with a silicon micro-mirror [4] (see the insert in the figure), both for negative (a) and positive detunings (b). In both cases, the thick curve corresponds to the resonant situation $\varphi = 0$. Both damping and anti-damping effects are evident on the noise spectra. For a negative detuning, the spectra are both widened and drastically decreased at their resonance frequencies. The decrease of the area of the curves, which is directly related to the effective temperature by the equipartition theorem, is a strong indication of the simultaneous effective temperature reduction of the mirror mechanical mode [4]. The situation is opposite for positive detunings, where the spectra are strongly narrowed and increased at resonance. Note in both cases the clear frequency shift ($\Omega_{\text{eff}} - \Omega_{\text{m}}$) of the mechanical resonances.

The simple theoretical model can be further tested by looking at the dependence of the effective damping ratio Q/Q_{eff} with respect to the detuning φ. Figure 12.3 presents our experimental results, for five optical powers from 0.5 to 3.2 mW, along with the theoretical fits deduced from Eqs. (12.3) and (12.9). The insert shows the intracavity power at resonance P, single free parameter derived from the preceeding fits, as a function of the incident power P_{in}. It exhibits the expected linear dependence, with a slope in agreement at the 1 % level with the one deduced from cavity and mode-matching losses. This clearly shows that the observed damping/anti-damping effects are solely due to radiation pressure.

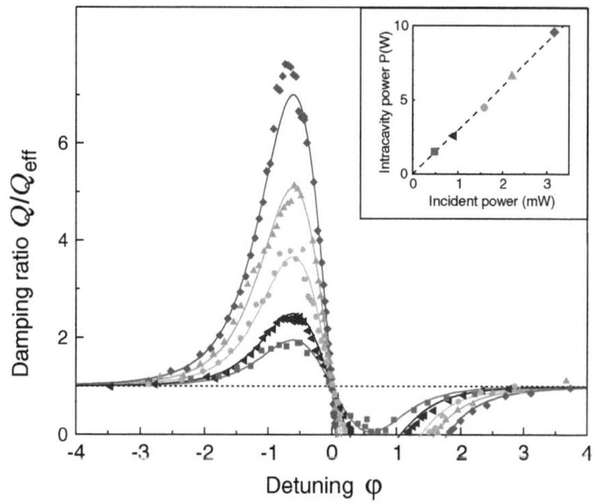

Fig. 12.3 Evolution of the cavity damping/anti-damping effects with respect to the detuning. Damping ratio Q/Q_{eff} for five values of incident power, from 0.5 to 3.2 mW (*red*). Points are experimental results and full lines are fits. The region where $1/Q_{\text{eff}}$ vanishes is the PI region. Insert: Evolution of the adjusted intracavity power P with the incident power. The *dashed line* is the linear dependence expected from cavity and mode-matching losses

12.1.3 Optomechanical Instability

At some point, for positive detunings, the effective damping cancels: the resonator becomes unstable and starts to oscillate at its effective resonance frequency, with a net energy flow from the optical field to the mechanical motion. Such an effect has been first demonstrated in a fused silica micro-toroidal resonator [5]. This is the *Parametric Instability*, well-known in the microwave domain and first discussed by Braginsky in the gravitational-wave interferometry context [1].

12.2 Three-Mode Parametric Effects

Figure 12.4a presents an alternative interpretation of the damping/anti-damping processes as a scattering of photons (at angular frequency ω_0) by the mechanical phonons of the mirror (at Ω_{m}), giving rise to Stokes (at $\omega_0 - \Omega_{\text{m}}$) and anti-Stokes (at $\omega_0 + \Omega_{\text{m}}$) motional sidebands.

For a resonant cavity, both sidebands have equal amplitudes: they can be used to infer the mirror motion (as a phase modulation of the reflected field) but have no net mechanical effect. As the cavity optical resonance curve acts as an effective density of output modes, detuning the cavity promotes the scattering of photons to one side-band with respect to the other. For negative detunings for instance, more photons are scattered at the $\omega_0 + \Omega_{\text{m}}$ frequency than at $\omega_0 - \Omega_{\text{m}}$ and the net effect of the optomechanical coupling is a nonzero energy transfer from the mechanical motion to the optical field, i.e., damping. Anti-damping is obtained as well for positive detunings. The optimal efficiency is very simple to state in the *resolved sideband regime* ($\Omega_{\text{m}} \gg \Omega_{\text{c}}$) (Fig. 12.4a, dark curve) as the damping/anti-damping processes are

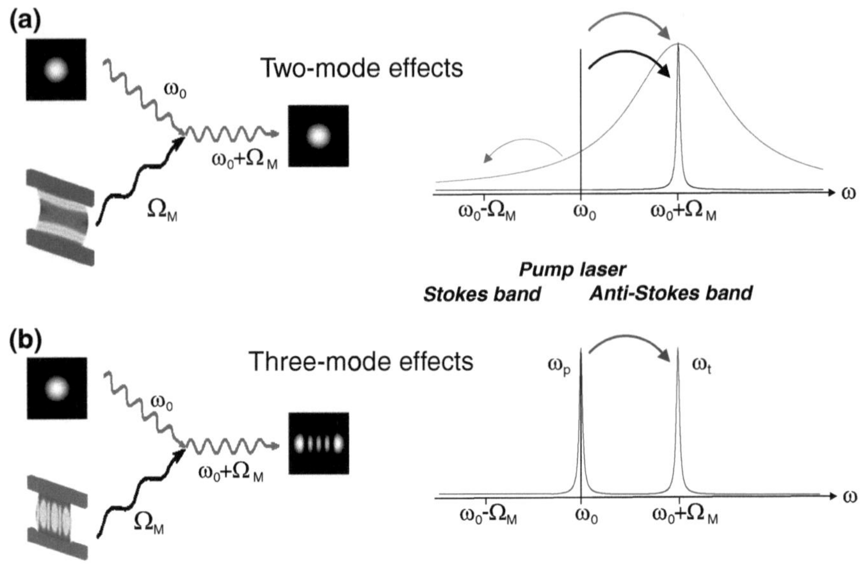

Fig. 12.4 Radiation-pressure damping seen as a scattering process. **a** *Two-mode damping*. The photon is scattered to the same optical mode, with a frequency offset Ω_m with respect to the incident photon. The cavity is detuned in order to favor the anti-Stokes process, and the pump laser is out of resonance. **b** *Three-mode damping*. The photon is scattered from a *pump mode* ω_p to a *target mode* ω_t of the cavity. Both the pump laser and the anti-Stokes band can be simultaneously resonant with the cavity. The vibration profile of the mechanical mode must somehow match the output mode

resonant for a detuning matching the mechanical resonance frequency Ω_m. Figure 12.4a presents the case where the anti-Stokes process is resonant, both in the non-resolved and resolved sideband regimes.

12.2.1 From Two-Mode to Three-Mode Effects

As the driving laser is strongly out of resonance, the process is somehow not efficient. But the situation is different in the case with two optical modes involved, one resonant with the driving laser, the other with the (anti-)Stokes band (see Fig. 12.4b). In that case, the anti-Stokes process corresponds to an effective scattering of one photon of the pump mode ω_p to the target mode ω_t, possibly of a different optical transverse profile, by absorption of mechanical phonons ω_m. Such a process actually involves three modes (one mechanical and two optical) and naturally embodies the physics of the radiation-pressure effects. It is resonant close to an optical frequency mismatch:

$$\Delta\omega = \omega_t - \omega_p = \pm\Omega_m \qquad (12.10)$$

(− for the Stokes process, + for the anti-Stokes).

In such a configuration, close to a Stokes resonance for instance, the effective mechanical Q_{eff} of the mechanical resonator is altered:

$$Q/Q_{\mathrm{eff}} = 1 - R, \tag{12.11}$$

with the parametric gain R (neglecting the nonresonant anti-Stokes effect):

$$R = \frac{16}{\pi \lambda c} \frac{\mathscr{F}_{\mathrm{p}} \mathscr{F}_{\mathrm{t}} P_{\mathrm{in}} Q \Lambda}{M \Omega_{\mathrm{m}}^2} \left(\frac{\Omega_{\mathrm{t}}^2}{\Omega_{\mathrm{t}}^2 + (\omega_p - \Omega_{\mathrm{m}} - \omega_{\mathrm{t}})^2} \right), \tag{12.12}$$

where $\mathscr{F}_{\mathrm{p}}$ and $\mathscr{F}_{\mathrm{t}}$ are the optical finesses for the pump and target modes of the cavity, Ω_{t} the related optical bandwidth, P_{in} the incident power, and Λ the spatial overlap between mechanical mode u_{m} and optical modes ϑ_{p} and ϑ_{t}:

$$\Lambda = \left(\int \mathrm{d}r_{\perp} \, u_{\mathrm{m}}(r_{\perp}) \, \vartheta_{\mathrm{p}}(r_{\perp}) \, \vartheta_{\mathrm{t}}(r_{\perp}) \right)^2, \tag{12.13}$$

where $r_{\perp}$ is the transverse position upon the mirror surface. To avoid the unstable regime, the gain R must never exceed one. At this point, it is worth noting that three-mode effects are only possible with the internal vibration degrees of freedom of the mirrors as for center-of-mass motion (such as for a suspension pendulum mode), the corresponding flat vibration profile $u_{\mathrm{m}}(r_{\perp}) = u_{\mathrm{m}}$ cancels the overlap Λ due to the orthogonality of the TEM modes $\vartheta_{\mathrm{p}}(r_{\perp})$ and $\vartheta_{\mathrm{t}}(r_{\perp})$.

12.2.2 Experimental Demonstration of Three-Mode Coupling

Three-mode optomechanical effets have been first demonstrated by Zhao et al. in a suspended 80-m cavity, pumped by a high-power TEM_{00} laser beam [6]. In that experiment, the resonance condition $\Delta\omega = \Omega_{\mathrm{m}}$ of the anti-Stokes process has been demonstrated through the variation of output intensity of the TEM_{01} mode. The mirror is first driven into motion by an electrostatic drive at the resonance frequency of a mechanical mode whose vibration profile gives a significant overlap Λ with the optical modes. The mode frequency difference $\Delta\omega$ is tuned by heating a compensation plate which changes the g-factor (see Chap. 3) of the cavity. Three-mode parametric effects are demonstrated with the measured TEM_{01} power. As the optical spacing is swept through the resonance condition (12.10), the power detected on the output quadrant photodiode reaches a peak value (see Fig. 12.5b).

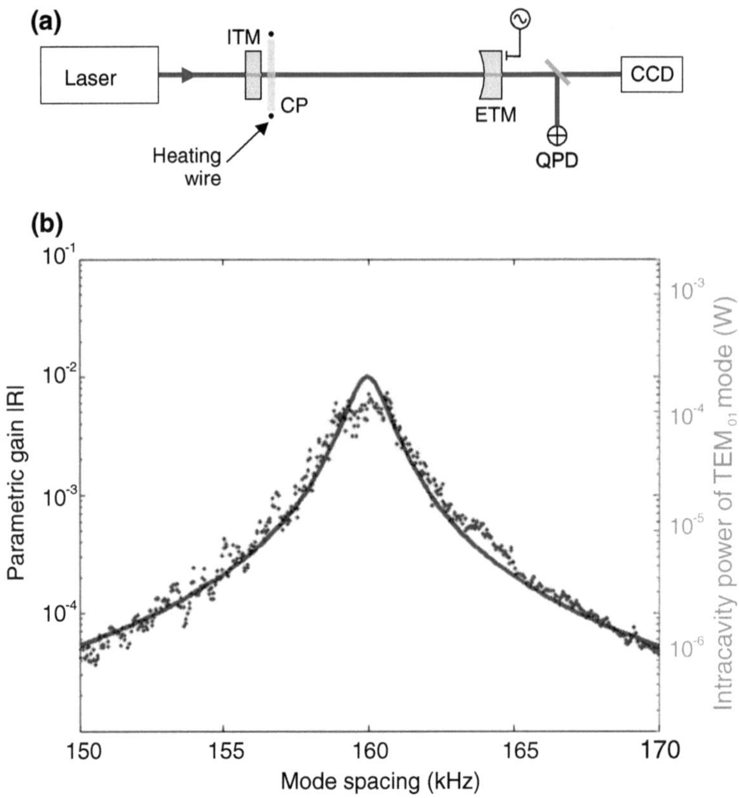

Fig. 12.5 Experimental demonstration of three-mode parametric coupling. **a** Setup. The Nd:YAG laser is locked to a TEM_{00} mode of a 80-m suspended-mirror cavity. The end test mirror ETM is electrostatically driven at a mechanical resonance frequency. The frequency difference between the TEM_{00} and TEM_{01} modes is tuned by heating the compensation plate CP. The cavity tuning and the TEM_{01} mode power are measured with the CCD device and quadrant photodiode QPD. **b** Measured power of the TEM_{01} mode (*dots*) as a function of the mode spacing between the TEM_{00} and TEM_{01} modes. The solid line is a fit. Adapted from Ref. [6]

12.2.3 Toward Table-Top Demonstrations of the PI

Yet, with only 4 W of optical power available at the cavity input mirror ITM, the maximum detected power only corresponds to $|R| \simeq 10^{-2}$ as the high Q and optical finesses are insufficient to balance the large mass of the suspended mirror. A number of experiments have therefore aimed at demonstrating the PI with microoptomechanical systems, for which incident powers at the mW level should be sufficient.

In the experiment performed at Laboratoire Kastler Brossel, the moving mirror is a $1\,\text{mm} \times 800\,\mu\text{m} \times 30\,\mu\text{m}$ silicon doubly-clamped beam, very similar to the one already used in [4]. A mechanical mode with a resonance frequency close to

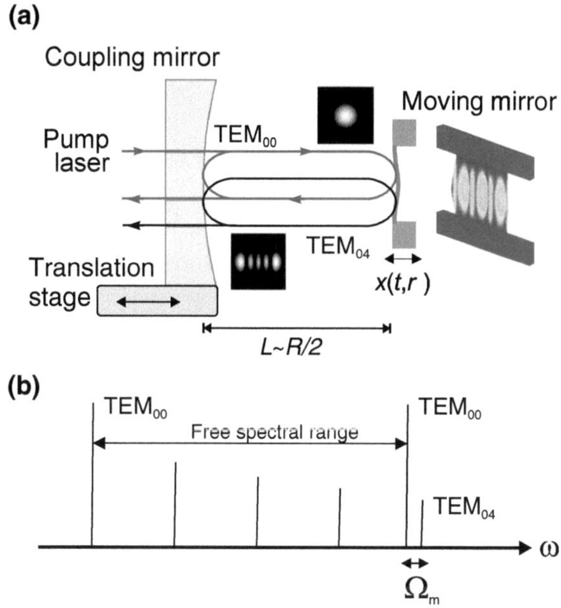

Fig. 12.6 Experimental implementation of three-mode coupling. **a** Three-mode opto-mechanical system. A TEM_{00} pump laser is sent into a resonant cavity. Its photons are scattered to the TEM_{04} mode by the moving mirror. The cavity length is set in order to set one of the motional sidebands at resonance. **b** Mode spectrum of the cavity close to the semi-confocal working point. The cavity length is set such that the TEM_{04} mode associated with a given longitudinal mode of the cavity is almost resonant with the subsequent TEM_{00}

5 MHz and a significant overlap Λ with the optical modes is used (see its profile on Fig. 12.6). Gaussian modes TEM_{00} and TEM_{04} are used as pump and target modes.

As the mechanical frequency is in the MHz range, the frequency resonance conditions (12.10) are reached with a cavity close to the semi-confocal configuration, with the moving mirror cavity as a plane end mirror. The input mirror is concave, with a 50-mm radius of curvature (see Fig. 12.6). The cavity length is finely tuned to set alternatively both TEM_{00} motional sidebands at resonance with the TEM_{04} mode. The cavity has a finesse $\mathscr{F} \simeq 20{,}000$ for the TEM_{00} and TEM_{04} modes. As demonstrating three-mode effects requires the frequency difference $\Delta\omega$ to be much lower than the cavity bandwidth Ω_{04}, cavity length has to be controlled at the 100-nm level.

One finds a three-mode mechanical amplification close to the Stokes process resonance $\Delta\omega = -\Omega_m$ and a three-mode damping close to the anti-Stokes resonance $\Delta\omega = \Omega_m$. Results are displayed in Fig. 12.7 for one specific mechanical mode, but similar results have been obtained for two adjacent modes as well. Even with an incident power up to 5 mW, the three-mode parametric instability regime (corresponding to $R > 1$) has however eluded observation, the maximum gain obtained being limited to $|R| \simeq 0.5$.

A related experiment is under progress at UWA, with a thin dielectric membrane embedded inside a high-finesse optical cavity [8].

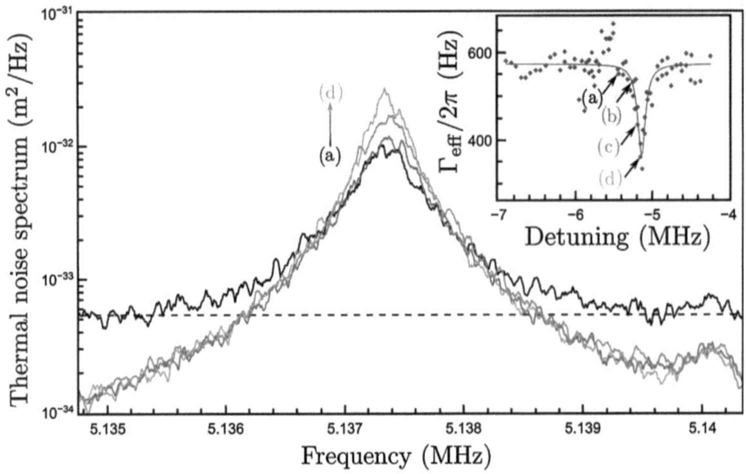

Fig. 12.7 Experimental demonstration of anti-damping by the Stokes mechanism. Calibrated thermal noise spectra measured for the (0,6) mode close to the Stokes resonance condition. Insert, **a–d**: Effective mechanical damping of the (0,6) mechanical mode, deduced from the noise spectra. The *solid line* is a fit by Eq. (12.12)

12.3 Computing the PI in Large-Scale Interferometers

The three-mode instability we have just discussed is likely to play a major role in the advanced gravitational-wave interferometers. Both Advanced Ligo and Advanced Virgo, with their substantial intracavity powers and the high optical finesses of the compound cavities, are indeed favorable playgrounds for this kind of effects, all the more since they present a high mechanical mode density in the tens of kHz band, which is the typical spacing between optical transverse modes as well.

It is therefore of utmost importance to fully understand and eventually to control these instabilities, which are likely to drive the mechanical modes over very large amplitudes, resulting in a loss of the interferometer lock, with typical timescales

$$\tau \propto \frac{Q}{\Omega_{\mathrm{m}}(R-1)}, \tag{12.14}$$

on the order of a few seconds to reach the lock-breaking amplitude. Unstable modes will therefore prevent stable operation of the interferometer even though they are out of the detection window.

12.3.1 Finite-Element Computation of the Parametric Gain

After Braginsky's pioneering works over a single cavity [1], a number of groups have modeled a complex interferometer such as AdvLigo or AdvVirgo, with dual recycling. David Blair's group at the University of Western Australia for instance has shown that three-mode coupling may involve as many as dozens of optical modes for a given mechanical one. The parametric gain then has to be summed over all possible Stokes and anti-Stokes processes, even far from the resonance condition:

$$R = \frac{8P_{\text{in}}Q}{\lambda c M\Omega_{\text{m}}^2}\sum_p\left(\frac{\mathcal{F}_p\,\Lambda_p\Omega_{\text{c}p}^2}{\Omega_{\text{c}p}^2+(\omega_0-\omega_p-\Omega_{\text{m}})^2}-\frac{\mathcal{F}_p\,\Lambda_p\,\Omega_{\text{c}p}^2}{\Omega_{\text{c}p}^2+(\omega_0-\omega_p+\Omega_{\text{m}})^2}\right), \tag{12.15}$$

where the summation is on all optical longitudinal and transverse modes p.

Results by the UWA group [9, 10] have allowed a reliable estimate of the number of potentially unstable modes, even if they are out of the detection window. Estimating the chances to reach the PI threshold requires a detailed modeling of the mechanical modes (resonance frequency, effective mass, quality factor Q, vibration profile), which can be performed by Finite-Element Modeling, for example with ANSYS.

A MatLab code then computes successively:

- the mechanical Qs, taking into account the mechanical losses due to the optical coating of the mirrors (see typical results on Fig. 12.8)

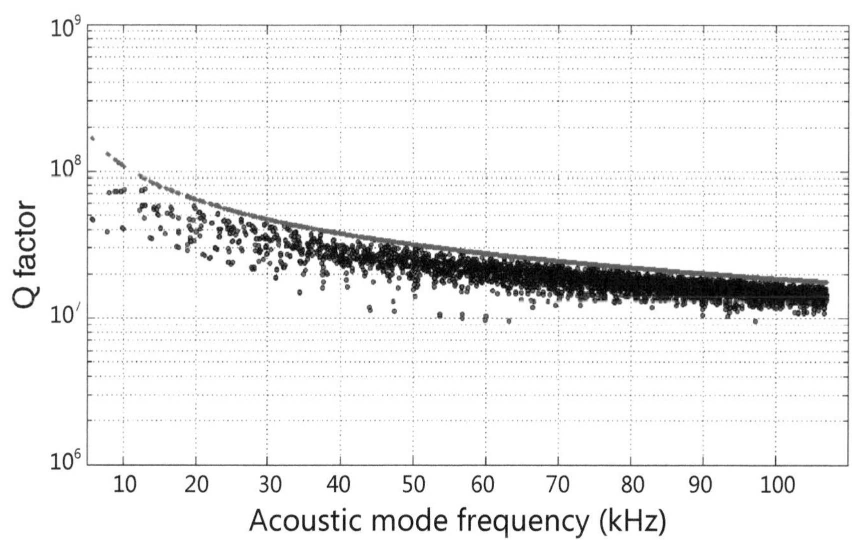

Fig. 12.8 Computed Q-factors for the mechanical modes of an Advanced Virgo test mass, up to 107 kHz. *Upper curve* Q-factors due to the substrate losses. *Scattered dots* Q-factors due to the coating and substrate losses

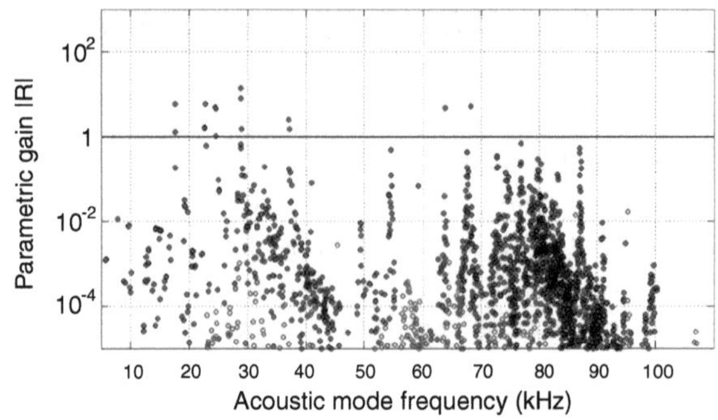

Fig. 12.9 Parametric gain $|R|$ for the mechanical modes of an Advanced Virgo test mass, in its standard configuration at full power (125 W). *Full dots* indicate modes with a positive gain, open dots modes with a negative one. The $R = 1$ *horizontal line* corresponds to the Parametric Instability threshold

- the optical high-order transverse modes of the interferometer cavities (resonance frequencies, losses and field profiles)
- the spatial overlap Λ_p for all triplets of modes (mechanical mode, TEM_{00} pump mode, and target mode p)
- at last, the mechanical gain R for each mechanical mode, by direct summation of Eq. (12.15).

The FEM modeling takes into account the details of the test mass geometries such as the flats, but ignores up to now suspension-induced mechanical losses, which may significantly alter the Q and hence lower the gain R. Optical coating $\lambda/4$ layers are modelled as an effective medium, taking into account both SiO_2 and Ta_2O_5 mechanical properties and their respective total thicknesses.

The whole computation can be performed for different configurations: single cavity, interferometer, power-recycled or dual-recycled interferometer. Figure 12.9 presents a typical plot obtained with that simulation tool: parametric gain R for every mechanical mode of an Advanced Virgo test mass, up to 107 kHz, for a dual-recycled interferometer at full power (125 W). In such a configuration, about 15 modes (per test mass) are likely to be unstable, but this is an upper limit, as Qs are expected to be somehow limited by clamping losses.

12.3.2 Feedback Approach and Monte-Carlo Estimate of the Gain

An alternative approach by Evans et al. is based on a feedback description of the PI: a small excitation of a given mechanical mode scatters light from the TEM_{00} pump mode to higher-order modes, the interferometer responds to the scattered field and radiation pressure couples back the field to the mirror motion, thereby closing the feedback loop. This approach is suitable for matrix methods to compute the corresponding parametric gain. It gives similar results as the previous one [11].

As already mentioned, the parametric gain R and hence the number of unstable modes are extremely sensitive to details of the interferometer design and configuration: exact geometry of the test masses, computed mechanical resonance frequencies, mechanical Q, optical finesse of high-order modes. For example, the sole manufacturing tolerances of the radii of curvature of the test masses or the FEM computation inaccuracies (at the 1 % level) are sufficient to alter significantly the Gouy phase in the arm cavities (see Chap. 3), or to change the resonance conditions. As a consequence, computing the exact number of unstable modes in a planned configuration is just impossible.

To get a tractable evaluation of the gain for a given mechanical mode, a technique is to compute it for a (large) number of plausible neighbouring interferometer configurations. In practice, varying all parameters in an independant way is both untractable (because heavily time-consuming) and unnecessary as the gain variations are mainly driven by the relative frequencies of the mechanical modes and the optical resonances in the compound cavitites [11].

Evans et al. limited accordingly their analysis to Gouy phase variations of 5×10^{-3} rad in the arm cavities and a few degrees in the recycling cavities. After 120,000 trials, they define a "worst case" situation by the value of R greater than 99 % of the results and establish that AdvLigo faces the possibility of a couple of unstable modes (per test mass).

12.4 Mitigation Strategies

Thermal tuning of the Gouy phase through the radius of curvature has been proposed to find a *sweet spot* where the number of unstable modes is the lowest possible [10]. Yet, not even considering the thermal time scale issues that might jeopardize the efficiency of such a method, it is unlikely to find a configuration without any unstable modes at all.

Mitigation strategies have therefore to be planned to ensure the stable operation of the advanced interferometers. Such strategies can be either passive- primarily targeted Q reductions to lower the parametric gain- or active, feedback loops that can use either local or global readout (at the interferometer dark fringe) to monitor the PI and cancel it by real-time actuation of the mirror. We present here a few strategies currently investigated.

12.4.1 Passive Damping

Reducing the mechanical quality factors is an obvious solution to the PI problem as the gain is proportional to Q. The main issue is the frequency band where the sensitivity is thermal-noise limited, as a global (i.e., frequency-independent) increase of the mechanical losses should directly impact the interferometer sensitivity. A targeted reduction is however possible, taking advantage of the wide frequency gap between the mechanical resonance frequencies (above 10 kHz, where the parametric gain has to be lowered) and the science frequency band.

Fig. 12.10 Acoustic Mode Damper on a LIGO test mass at LASTI (MIT). Photograph by S. Gras

A first proposal was to coat the barrels of the test masses with a thin strip of a lossy material (such as Al_2O_3 or gold). It is indeed possible to reduce the Q by a factor >10 for most modes, with a limited thermal noise increase penalty. Yet, the damping strongly depends on the relative location of the strip with respect to the mechanical mode and it turns out difficult to achieve sufficient damping for all potentially unstable modes [12].

Another promising technique has been recently studied. It consists in resonant dampers made of a piezoelectric resonator shunted by a resistor (see Fig. 12.10). If located close to an antinode of a given mechanical mode, such a damper can lower significantly its Q and its parametric gain. A preliminary experiment at LASTI (MIT) has confirmed the damping effect [13].

Open issues are the thermal noise increase due to the epoxy bonding between the dampers and the test masses and the required number of dampers to be glued to the test mass. Further investigations are under way at MIT.

12.4.2 Active Damping with an Electrostatic Drive

A possible mitigation strategy would be to monitor the PI in real-time on the interferometer dark fringe and feed the signal back to the mirror, taking advantage of the dielectric behavior of the test mass. Actuation could be performed with electrostatic drives, deposited on the surface of the AdvLigo reaction masses for control purpose (see Fig. 12.11) [14]. With a 800 V bias voltage, the typical available force is on the order of 200 μN while only 10 μN are required at most to damp a typical internal vibration mode. Unlike passive damping, such a technique doesn't introduce any additional thermal noise.

Fig. 12.11 Electrostatic Drive actuator implemented on an Advanced Ligo prototype reaction mass. From Ref. [14]

The practical implementation of such a technique is however far from straightforward. The dark fringe signal at mechanical resonance frequencies should be split up to individually address each unstable mode of each test mass. A PI signal would then require some commissioning to link it to a specific test mass, an amount of time which could turn problematic as the power is ramped up and more and more instabilities appear.

12.4.3 Active Damping with Optical Feedback

The generation of intracavity optical higher-order modes is part of the PI process and its feedback effect. An elegant way to circumvent the PI would then be to monitor the target optical mode in real time and use that signal to inject a field with a similar transverse profile but out of phase to kill the mechanical amplification process *in situ* by destructive interference. Such a proof-of-principle experiment has been successfully performed at the High Power Facility in Gingin [15] (see setup on Fig. 12.12).

It is in principle applicable to advanced interferometers. Yet, the technique seems rather difficult to implement. It would not only require a dedicated commissioning to assign each detected instability to a given test mass (as for the electrostatic actuation) but also to generate a complete set of phase-locked higher-order modes to inject into the interferometer cavities (up to 60 modes are usually taken into account for the PI computation, though a couple should be enough to suppress the PI).

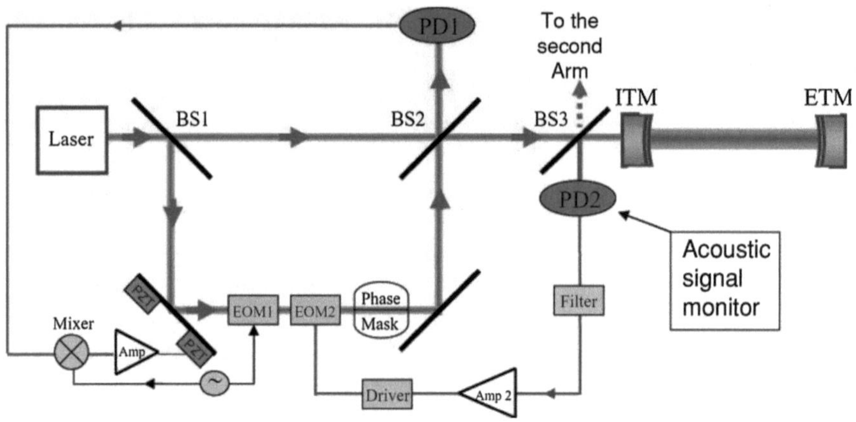

Fig. 12.12 Experimental setup to demonstrate the optical injection feedback technique. The main laser beam enters the cavity at beamsplitter BS3. The mirror displacement signal is read out on photodiode PD2. An auxiliary laser beam is picked off at BS1, converted to the transverse optical mode by the phase mask, and has a sideband at the mechanical resonance frequency Ω_m created by the electro-optical modulator EOM2. Its phase is controlled by EOM1. Depending on the loop phase, the injected sidebands will interfere either constructively or destructively with the optical field created by the three-mode parametric process, possibly suppressing the parametric instability. From Ref. [15]

12.5 Conclusion

The Parametric Instability is a serious threat to the stable operation of advanced interferometers once they will be operated at full optical power. We have presented a simple model of the PI in the simplest optical configuration, and discussed its relevance based on optical and mechanical characteristics of the currently built large-scale projects. Nevertheless, a number of PI-suppression schemes are envisioned and currently being investigated.

Acknowledgments Thanks are due to M. Evans, P. Fritschel, D. Blair, L. Ju, C. Zhao, A. Heidmann, T. Briant, S. Deléglise, F. Bondu, and J. Marque for discussion. Parts of this work have been made possible by the FP7 STREP Minos of the European Union.

References

1. V.B. Braginsky, S.E. Strigin, S.P. Vyatchanin, Phys. Lett. A **287**, 331 (2001)
2. A. Di Virgilio et al., Phys. Rev. A **74**, 013813 (2006)
3. B.S. Sheard, M.B. Gray, C.M. Mow-Lowry, D.E. McClelland, S. Whitcomb, Phys. Rev. A **69**, 051801(R) (2004)
4. O. Arcizet, P.-F. Cohadon, T. Briant, M. Pinard, A. Heidmann, Nature **444**, 71 (2006)
5. H. Rokhsari, T. Kippenberg, T. Carmon, K.J. Vahala, Opt. Expr. **13**, 5293 (2005)
6. C. Zhao et al., Phys. Rev. A **78**, 023807 (2008)

7. C. Molinelli et al., in preparation
8. J.D. Thompson et al., Nature **452**, 72 (2008)
9. C. Zhao et al., Phys. Rev. Lett. **94**, 121102 (2005)
10. S. Gras, C. Zhao, D.G. Blair, L. Ju, Class. Quantum Grav. **27**, 205019 (2010)
11. M. Evans, L. Barsotti, P. Fritschel, Phys. Lett. A **374**, 665 (2010)
12. S. Gras, D.G. Blair, L. Ju, Phys. Lett. A **372**, 1348 (2008)
13. S. Gras, M. Evans, P. Fritschel, LIGO Doc.: LIGO-G1001023 (2010).
14. John Miller et al., Phys. Lett. A **375**, 788 (2011)
15. Z. Zhang, C. Zhao, L. Ju, D.G. Blair, Phys. Rev. A **81**, 013822 (2010)

Chapter 13
A Third Generation Gravitational Wave Observatory: The Einstein Telescope

Michele Punturo, Harald Lück and Mark Beker

Abstract The first decade of the second millenium has seen the realization and the operation of the initial generation of large interferometric gravitational wave detectors, like Virgo and LIGO; these detectors demonstrated the capability of reaching their design sensitivity, which due to the novelty of their design was quite a challenging task. Achieving the target of the detection of gravitational waves still requires a large improvement in sensitivity. This is promised by the operation of the advanced detectors that are dominating the gravitational wave scene in the second decade of this century. But, in order to open the era of routine gravitational wave astronomy a new (third) generation of gravitational wave observation instruments will be needed. Will the third generation (3G) of gravitational wave observatories be the core of the gravitational astronomy in the third decade of this century? An overview of the technological progress needed to realize a 3G observatory, like the Einstein Telescope (ET), and a possible evolution scenario are discussed in this chapter.

13.1 Introduction

In the first decade of this century, the first generation of interferometric gravitational wave (GW) detectors (GEO600 [1], LIGO [2], TAMA [3], Virgo [4]) has been completed, commissioned and a series of scientific data taking runs has been performed;

M. Punturo (✉)
Istituto Nazionale di Fisica Nucleare—Sezione di Perugia, Via Pascoli 06123 Perugia, Italy
e-mail: michele.punturo@pg.infn.it

H. Lück
Institut für Gravitationsphysik, Leibniz Universität Hannover, Callinstr. 38, 30167 Hannover, Germany
e-mail: harald.lueck@aei.mpg.de

M. Beker
Nikhef—National Institute for Subatomic Physics, Sciencepark 105, 1098 XG Amsterdam, The Netherlands
e-mail: m.beker@Nikhef.nl

M. Bassan (ed.), *Advanced Interferometers and the Search for Gravitational Waves*, Astrophysics and Space Science Library 404, DOI: 10.1007/978-3-319-03792-9_13,

this not only demonstrated the capability of the scientific collaborations, that are operating these complex machines, to reach the design sensitivity of the detectors but also confirmed the viability of the working principle [5].

Even achieving the design sensitivity with the largest initial detectors (LIGO and Virgo) didn't resulted in the detection of a GW signal in these data taking runs, as the detection rate, in agreement with the expectations from existing GW sources [6], was too low. Although amazing, the sensitivity of the initial GW interferometers yielded a too short detection range (up to few tens of Megaparsec) to include enough potential GW sources in the detection volume. The initial GW interferometers are evolving to the so-called Advanced or Second Generation Detectors ("Advanced LIGO" [7] and "Advanced Virgo" [8]); the aim of this evolution is to gain a factor of ten in sensitivity, corresponding to about a factor of thousand in accessible volume and therefore in detection rate. The expectation is to have a few tens of GW signal detections per year from coalescing binary neutron star systems, with signal-to-noise ratios (SNR) larger than eight [6]. Advanced detectors are implementing rather well established technologies (some tested in reduced scale prototypes, some already implemented in GEO 600) in the same infrastructures hosting the initial detectors. The installation of the advanced detectors will take up the first half of the current decade, whereas the commissioning and scientific data taking will be done in the second half.

Although the detection of a GW signal is expected in the first years of operation of the advanced detectors, the expected SNR is generally too low for detailed astronomical investigations of the GW sources, for pinpointing the source, and finding the electromagnetic counterpart with optical and X-ray telescopes. To open the era of routine precision gravitational wave astronomy, roughly another factor of ten in sensitivity gain is needed, in order to transform the simple detection of a GW signal into an observation and measurement of the physical parameters of a GW astronomical source. These considerations pushed the GW scientific community into the investigation of a third generation (3G) of GW observatories, where the technological limitations of the advanced interferometers are overcome; in particular, in the Einstein Telescope (ET) [9, 10] conceptual design study, a special focus has been put on the design of a new infrastructure, capable to host the new generations of future GW observatories. Probably, at the end of the third decade of this century, the 3G GW observatories will play a new and crucial role in the observation of the Universe.

13.2 Scientific Targets of ET

The scientific objectives of a 3G GW observatory like ET have been discussed in detail in the ET design study document [9] and more recently by B. Sathyaprakash et al. [11]. A detailed review of relevant sources, accessible to the advanced detectors, is given in Chap. 2. Hereafter a short and not exhaustive overview of the scientific potential of ET is presented.

The main ingredients of the ET observatory are the factor of 10 gain in sensitivity and an even bigger improvement in the low frequency range (1−10 Hz) with respect to second generation (2G) detectors. This will enable ET to detect many loud events with SNR > 50; in these conditions it is possible to perform detailed studies of single source systems, like black-holes (BH) quasi-normal modes and neutron stars (NS) cores and of systems characterized by strong gravitational fields.

The structure of a NS and the role of its magnetic field are still unknown, but the signature of its equation of state (EoS) is present in almost any GW emission of a NS system. For example, the GW emission by a coalescing binary neutron star system (BNS), beyond the simple point-particle approximation, is expected to contain [12] characteristics sensitive to the NS EoS; ET is able to distinguish between different models of EoS [9]. According to the model described in [13], ET is able to measure the radius of a NS for a 1.35–1.35$M_{\odot}$ BNS at 100 Mpc with an error of $\pm 0.5 - 1.0$ km, with a fractional incertitude in pressure of 5–10 %. The result of a BNS coalescence could be a hyper-massive remnant object oscillating and emitting GWs for relatively long time; the EoS of the coalescing NS is affecting the presence and the characteristics of these oscillation. According to the models in [14], ET should be able to see that signal and determine the corresponding physical mechanisms. Furthermore, the observation of glitches and r-mode instabilities [15] in the GW emitted by isolated NS could provide a probe of the NS dynamics.

Gamma-ray bursts (GRBs) are the most luminous explosions in the Universe; they are usually subdivided, according to the duration Δt of the impulse, into short-GRB ($\Delta t < 2$ s) and long-GRBs ($\Delta t > 2$ s); core collapse Supernovae are thought to be the progenitors of long-GRBs [16, 17], whereas BNS and NS-BH coalescences are expected to be the progenitors of short-GRB [18]. First generation GW detectors (LIGO and Virgo) performed several searches of GW emissions in coincidence with GRB [19–24], but the detection distance of initial GW interferometers was insufficient in view of the cosmological distance of the GRBs detected by the satellites. In the second generation of GW interferometers, the detection of the GW emitted in short-GRBs is expected to be possible [24], although with a low expected rate (3 year^{-1} for NSBH progenitors and 0.3 year^{-1} for BNS progenitors) [25]. ET promises to detect a BNS up to a luminosity distance of 17 Gpc corresponding to a red-shift of about 2 and a NSBH system up to a red-shift $z \simeq 2 - 5$, matching the expected distance of the emitted short-GRBs ($z < 1$) with an expectation of BNS mergers detected with electromagnetic (EM) counterpart of the order of 10^2 events per year [26].

As demonstrated by B.Schutz [27], the GW signal from compact binary systems can be seen as a cosmological standard candle or better *standard siren*, the frequency being in the audio band. In fact, the amplitude of the GW signal is completely given by the chirp mass and by the luminosity distance D_L, with no need to introduce complex astrophysical models; through the reconstruction of the coalescence GW signal it is possible, hence, to determine both of these properties and therefore to measure the luminosity distance of the binary system. Actually, for cosmological distances, the GW signal will be red-shifted ($f(t) \rightarrow f(t)/(1+z)$) causing a wrong reconstruction of the masses ($M \rightarrow (1+z)M$) and of the distance ($D_L \rightarrow (1+z)D_L$). As shown by B. Sathyaprakash et al. [26], the simultaneous detection of the BNS coalescence GW signal (in ET) and of the short-GRB flash (in an EM telescope) will allow us to measure both the luminosity distance and the red-shift of the source; detecting about 500 events of this kind (about 5 years of data taking in ET) it is possible to constrain the cosmological model of the Universe with a good precision (e.g., computing the dark-energy equation of state parameter with an error of percent [11, 26]). More recent studies are investigating the possibility to use tidal effects to better reconstruct the NS masses (and then overcome the red-shift ambiguity) [28] or to use a population of BH-BH binaries [29].

Understanding the mechnisms of Supernovae (SN) explosions is another target of ET. The details of the explosion mechanism of core-collapsing SN is still unknown, despite a long series of modeling attempts. The collapsing core cannot be investigated by means of optical telescopes, because the surrounding stellar envelope is opaque to the EM emissions. GW emissions instead bring information directly from the core of the collapsing massive star and their observation could constrain [30] the plethora of the proposed shock-revival mechanisms. Advanced detectors have a marginal probability to detect SN explosions because a few events per century are expected within their detection volume; on the other hand, within the sight distance of ET (up to 5 Mpc) a few events per decade are expected, but it will still be possible to gain insights into the SN mechanisms. Furthermore, the simultaneous detection of EM waves, GWs, and neutrino emissions of a SN explosion (if enough sensitive neutrino detectors will be available) would be a wonderful multimessenger framework for the study of such a complex phenomenon.

13.3 The Einstein Gravitational Wave Telescope Project

We need two key ingredients for the ET project:

- The design of an infrastructure that can host the future evolutions of GW detectors, without being or becoming a limiting factor to their sensitivities;
- the identifications of the technologies needed to overcome the current limitations of the advanced detectors, aiming at a gain in sensitivity of a factor of 10 with respect to them.

These two ingredients are closely related and the best way to understand how to get them is to start from the expected sensitivity of a second generation GW detector, as shown in Fig. 13.1. Considering only the *fundamental* noises, the dominating sources are:

- At very low frequency (below 4–5 Hz) the seismic noise and the Newtonian noise (see Sect. 13.3.1); these noises are mainly related to the characteristics of the site hosting the observatory.
- Below 300 Hz, the thermal noise of the optics suspension system (see Sect. 13.3.3), the thermal noise of the suspended mirrors (mainly the coating contribution, see Sect. 13.3.3) and the radiation pressure component of the quantum noise (see Sect. 13.3.4.3).

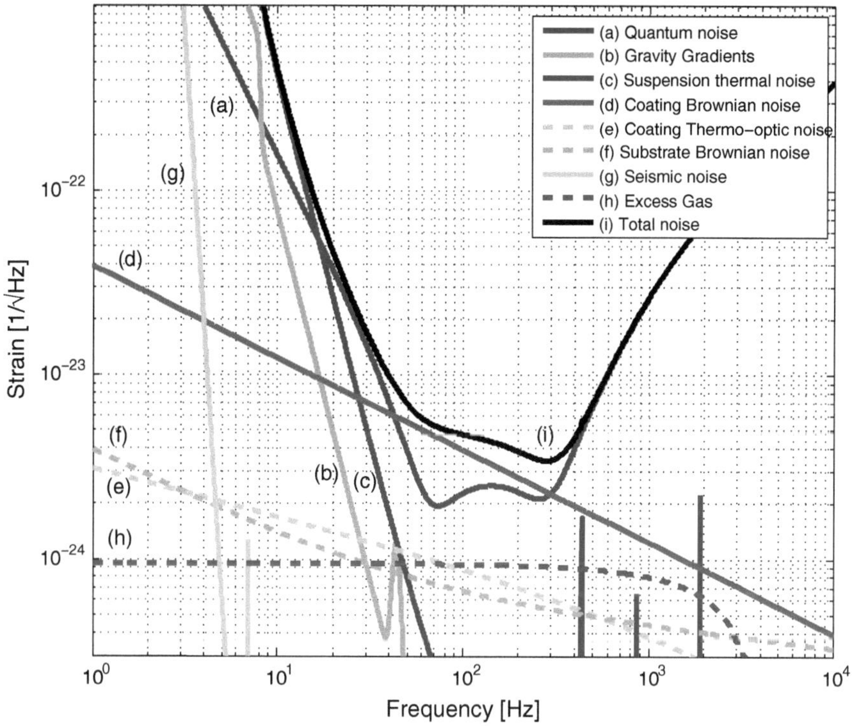

Fig. 13.1 Sensitivity model of an advanced GW detector (in particular this is a model quite close to the expected sensitivity of the advanced Virgo detector [31] with 125 W of injected power). At very low frequency, the noises mainly related to the site characteristics (seismic and Newtonian noises) are dominating. The remaining part of the spectrum is instead dominated by noises related to the detector characteristics: thermal noises, related to the suspension systems and to the mirrors properties, and the radiation pressure component of the quantum noise play a crucial role at low and intermediate frequencies; at high frequencies the shot noise component of the quantum noise dominates

- Above 300 Hz, the shot noise component of the quantum noise (see Sect. 13.3.4.1).

The last two components of the sensitivity are mainly related to the design and to the technologies adopted in the detector.

The initial design of the Einstein Telescope will be based on the principles of the advanced detectors, which have reached a very high level of maturity and have proven their viability in the existing detectors. The capability of these techniques will be expanded to allow reaching the sensitivity aimed for. The main extensions are the following:

- Increasing the interferometer arm length to dilute displacement noises[1]

[1] Indicating with displacement noises the pile-up of all the noise sources that are causing an effective movement of the test masses, like seismic noise, thermal noise, …

- Build the observatory at an underground site to minimize the influence of seismic motion
- cool parts of the interferometer optics and supensions to cryogenic temperatures to minimize thermal noises
- Use silicon as optics substrates for the cryogenic optics to optimize their performance
- Use high circulating light powers of up to 3 MW to cut down shot noise in the high frequency regime
- Use frequency dependent input squeezing to lower shot noise and radiation pressure noise

The conflicting requirements of high light power (required for low shot noise at high frequencies but unfortunately accompanied by high levels of absorbed light) and cryogenic operation of the mirrors (required to cut down thermal noises in the low frequency range) are resolved by using two interferometers each dedicated to a different frequency range, the so called Xylophone concept as described in Sect. 13.4.

13.3.1 Seismic Noise

Seismic noise is the continuous and nonstationary soil vibration; it is due both to natural sources, like volcanic and geoseismic activities, weak or remote earthquakes, ocean or sea waves, atmospheric activities like wind and rain, Moon tides, and to human disturbances like transportation, industries, towns, agricultural activities. Each site has its own seismic signature, given by the geological properties of the soil, the natural and human activities and this signature fluctuates with the environmental conditions. A worldwide characterization of the seismic noise spectral properties has been done by J. Peterson [32]; measuring the seismic noise in several sites, distributed around the world, he has been able to present an envelope for the seismic noise power spectrum density (PSD) bounded by the NLNM (New Low Noise Model) and NHNM (New High Noise Model) curves in Fig. 13.2. Natural disturbances usually dominate below 1 Hz, whereas human activities dominate above 1 Hz. The very low frequency (few $\times 10^{-5}$ s) motion is produced by tidal effects of the Moon, the region between 0.01 Hz and 1 Hz (the so-called microseism) is dominated by wind and sea activities; in particular the peaks visible in many spectra in that frequency range are generated by oceanic [33] (and local sea [34]) waves. In the ET project, several sites (mainly in Europe) have been investigated measuring the seismic noise in underground and surface locations [9, 35]; driven by the pioneer experience of the CLIO [36] experiment in Japan, now continued with the construction of the KAGRA [37] detector (former LCGT project), the attention has been focused on possible underground sites, where the contribution to the seismic noise given by the surface waves is reduced, as highlighted by the comparison in Fig. 13.3 of the seismic noise level at the CLIO site with the corresponding spectra measured in the sites hosting the other GW detectors in the World. As result of the sites survey, operated in the ET project, a list of possible candidates has been produced; in Fig. 13.4 the best three underground candidates

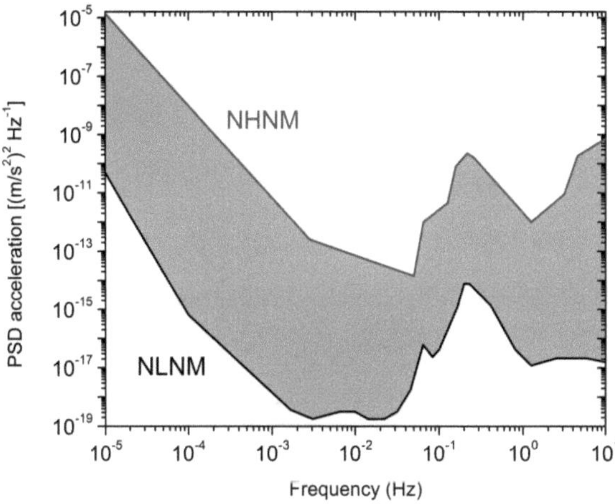

Fig. 13.2 Peterson Seismic Noise Model [32]. The NLNM (New Low Noise Model) and NHNM (New High Noise Model) curves are a composite of station spectra obtained in different World-wide distributed stations, acquired with heterogeneous instruments and different conditions. These curves delimit the coloured area, where a new seismic noise measurement usually should be found, although it is always possible to find noisier periods/sites because of seismic storms or intensive human activities

(Hungary-Gyöngyösoroszi, Spain-Canfranc Underground Laboratory and Italy in Sardinia) are compared with a good surface site (in the Netherlands).

Ground-based GW interferometers are and will be limited, in the low frequency range, roughly to 1–10 Hz, by soil vibration acting on the suspended test masses through two mechanisms:

- shaking the hanging point of the suspension chain,
- shaking directly the test masses of the interferometer via the mutual attraction forces, given by Newton's universal law of gravitation, between these masses and the soil layers which are perturbed by the seismic waves (Newtonian noise, see Sect. 13.3.2).

13.3.1.1 Seismic Noise Filtering

As it has been shown in the previous section, the seismic noise is peculiar of each site and depends on several factors. However, an over-simplified model, in terms of displacement amplitude spectral density (ASD), can be used to introduce the filtering issues. As mentioned in Chap. 7.3, the seismic displacement spectral amplitude x_{seism} can roughly be expressed, above 1 Hz, by:

$$x_{\mathrm{seism}}(f) \simeq \frac{\alpha}{f^2} \tag{13.1}$$

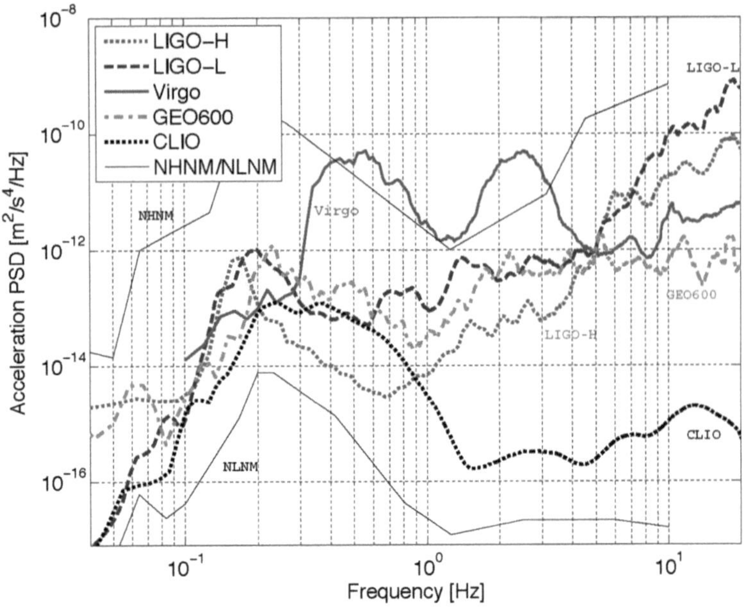

Fig. 13.3 The seismic noise (acceleration) measured at the CLIO underground site in Japan is compared with the corresponding spectra measured at the sites hosting the other GW detectors in the World. The advantage to realize a site underground is quite evident. In this plot the Virgo site seismic noise is affected by a microseism excess because of the bad weather conditions during the data taking

where the amplitude α usually ranges between 10^{-8} and 10^{-6} m$\cdot$Hz$^{3/2}$, depending on the quietness of the detector site (more refined models for the displacement spectral amplitude at the Virgo site can be found in [31] and in Fig. 4.8b of [38]). From Eq. 13.1 it is clear that, in order to have a reasonable sensitivity at a few Hertz, the seismic vibration must be filtered by the suspension chain before it reaches the test masses. This has been true for the initial and advanced detectors and it is clearly even more true for the 3G observatories. If the selected location for the 3G observatory is seismically quiet, like the underground sites shown in Fig. 13.4, the technologies adopted in the advanced detectors are still valid. For example it has been shown in [39] that a passive seismic filtering (*à* la Virgo [40]) approach is possible in ET if the chain of the Virgo Super-Attenuator (SA) is prolonged up to about 17 m (with respect to the current $\approx$ 9 m) and if the site has a seismic noise similar to that measured at the KAGRA [37] site. Braccini and coll. in [39] showed that in these conditions it is possible to meet the ET required sensitivity above 3 Hz. Probably, ET will implement an evolution of the Virgo SA, based on the Geometric Anti-Spring

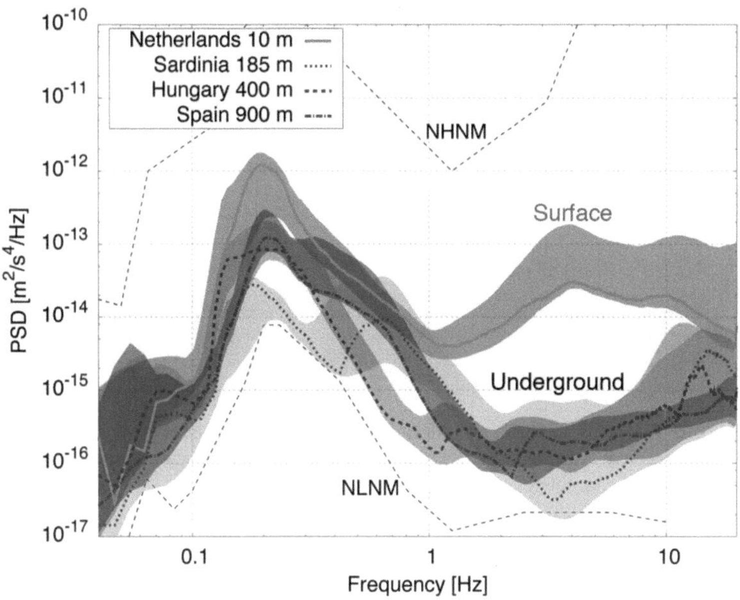

Fig. 13.4 The seismic noise (acceleration) measured at three underground sites in Europe compared to the same measurement at the surface (in the Netherlands, courtesy of the Nikhef-ET group). It is well evident, in the spectrum acquired in Sardinia, the peak at 0.5–0.7 Hz, a typical feature of the Mediterranean sites

(GAS) [41, 42] technology (as KAGRA is doing), allowing a filtering chain more compact and tunable with respect to the Magnetic Anti-Spring technology used in Virgo.

13.3.2 Newtonian Noise

Each object in the neighborhood of the suspended test masses is exerting a mutual attraction on them given by the Newtonian gravitational force. Therefore, the GW interferometer is affected by any variation of the local mass distribution. This effect is known as Newtonian noise or gravity gradient noise [43]. The major contributors to this noise source are density perturbations in the surrounding media as a result of seismic motion or atmospheric effects. In addition, Newtonian noise from objects, such as people, cars, and tumbleweeds, moving in the vicinity of the test masses has also been studied [44, 45]. While some of these sources can be controlled, others will provide technical challenges for the low-frequency sensitivity of future generations of ground-based detectors.

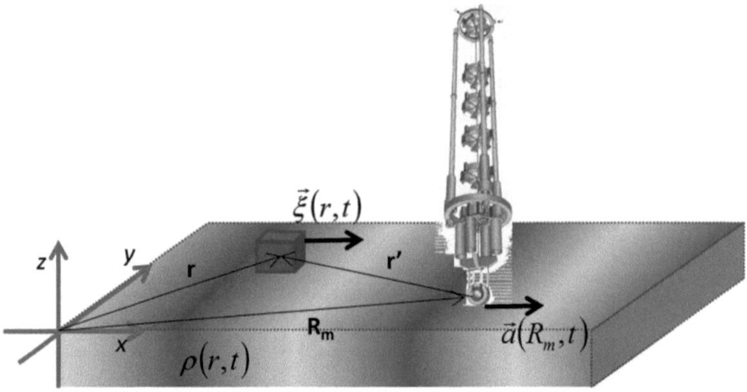

Fig. 13.5 Basic scheme of the Newtonian noise: the fluctuation of the gravitational attraction between the mass element at position **r** and the mirror at position $\mathbf{R_m}$, due to the seismic displacement ξ of the mass element, accelerates the mirror itself

13.3.2.1 Seismic Newtonian Noise

Fluctuating gravitational forces due to seismic density perturbations will interact with the suspended test masses, producing Newtonian noise. It is evident that the seismic filtering chain doesn't play any role in the suppression of this noise source. Being a gravitational force, the test masses cannot be shielded from it in any way. In [38] it is possible to find a clear mathematical description of the Newtonian noise and hereafter only the conclusions of that description are included. The acceleration variation $\delta \mathbf{a}_{NN}$ caused by the displacement $\overrightarrow{\xi}$ of the element dV at position **r** (see Fig. 13.5) is given by:

$$\delta \mathbf{a}_{NN}(\mathbf{R_m}, t) = G\rho_0 \int_V \left[\overrightarrow{\xi}(\mathbf{r}, t) \cdot \nabla \right] \mathbf{k}\, \mathrm{d}V \tag{13.2}$$

where ρ_0 is the equilibrium density (considering a homogeneous medium) and $\mathbf{k} = (\mathbf{r} - \mathbf{R_m}) / |\mathbf{r} - \mathbf{R_m}|^3$.

The Newtonian noise for surface GW detectors has been modeled in the past by several authors [43, 46, 47]; a well-established model is due to Hughes and Thorne [47], where the differential mirror displacement ASD due to the Newtonian noise is given by

$$S_{NN}(\omega) = 4\pi \frac{G\rho_0}{\omega^2} \beta S_{\mathrm{seism}}(\omega) \tag{13.3}$$

where S_{seism} is the displacement ASD of the seismic motion and β is a dimensionless parameter accounting for the coupling of the various seismic wave modes to Newtonian noise.

In the initial GW interferometers the low-frequency sensitivity of the detector was too poor to be affected by the Newtonian noise. However, in the advanced

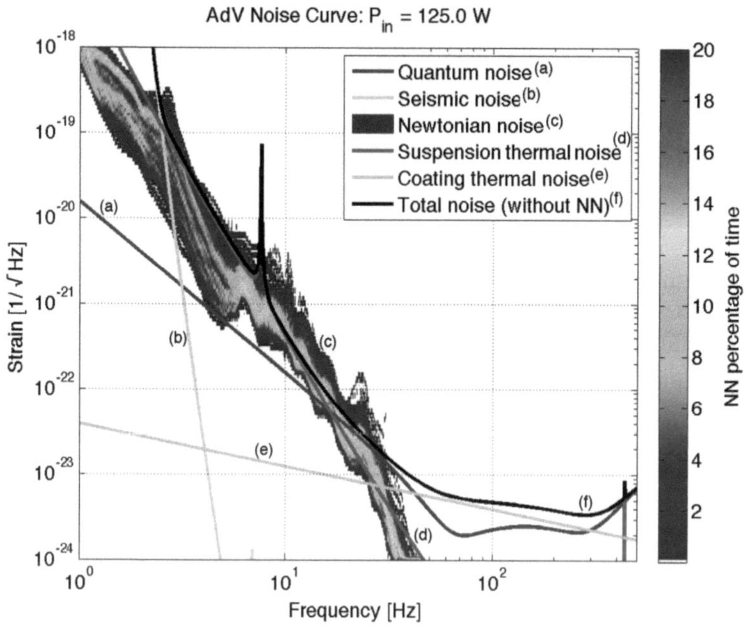

Fig. 13.6 Newtonian noise estimated in [38] for the Advanced Virgo detector. Estimation is based on formula 13.3 and on one week of seismic data taking; in the figure it is indicated the amount of time spent in each level of noise amplitude spectral density

detectors it approaches the overall noise level (see Fig. 13.1) and in the case of high seismic activity it is expected to dominate the low frequency noise budget as shown in Fig. 13.6.

In the third generation of gravitational wave detectors, the more stringent requirements in terms of sensitivity at low frequencies will enhance the importance of this noise source. Using the results obtained during the ET site survey presented in Fig. 13.4, combined with estimates of the Newtonian coupling to the test masses and the desired level of sensitivity, a limit could be set on the acceptable levels of seismic motion for ET. The requirement sets a limit below which the seismic motion, averaged over several days, should remain and is parameterized by the average displacement amplitude spectral density given in Eq. (13.1) with $A_0 = 5 \times 10^{-10}\mathrm{m} \cdot \mathrm{Hz}^{3/2}$ for frequencies between 1 and 10 Hz. Note that this corresponds to a acceleration power spectral density (PSD) requirement that is constant in frequency and given by $P_{aa}(f) = 4 \times 10^{-16}\ (\mathrm{m}^2/\mathrm{s}^4)/\mathrm{Hz}$.

As was already discussed above, seismic noise in the frequency range of interest to third generation gravitational wave detectors (>1 Hz), originates predominately from human activity on the surface. The largest fraction of the seismic energy from these sources is coupled to surface waves that propagate along the surface with amplitudes reducing with depth. Therefore, it is proposed that ET is constructed at an underground and / or remote location where seismic noise was found to be lower and more stable [38]. Despite such a low seismic noise environment however, the ET sensitivity (in Fig. 13.9) will still be limited by Newtonian noise at frequencies below 6 Hz. To improve low-frequency sensitivity even further, subtraction techniques will be implemented. These monitor the seismic wave fields with arrays of sensors and subsequently estimate and subtract the Newtonian signal from the interferometer output.

13.3.2.2 Atmospheric Newtonian Noise

Density perturbations in air will produce fluctuations in the local gravitational field. Such perturbations are caused by a number of mechanisms. Saulson [43] was the first to consider the effects of pressure or infrasound waves. These act on a gravitational wave detectors in an analogous way to seismic pressure waves but were estimated to contribute roughly an order of magnitude less Newtonian noise than their seismic counterparts in the frequency band of interest to second generation detectors. Other types of atmospheric Newtonian noise were studied by Creighton [45] and include: density perturbations caused by temperature variations that are transported past the test masses by wind, vortices around detector buildings, and shockwaves from, for example, sonic booms.

For third generation observatories, these noise sources need to be re-evaluated, in particular to include the significance of an underground location. Sources related to features with length scales much smaller than the detector depth, such as wind and vortices, are expected to have Newtonian contributions vanishing with depth. Large-scale phenomena, such as infrasound, needs to be studied carefully as little is known about low frequency infrasound waves and the reduction of their Newtonian effects with depth.

13.3.2.3 Newtonian Noise Subtraction

To construct an optimal Newtonian noise filter algorithm, we make use of the fact that the seismic Newtonian noise will be correlated with the seismic wave field but not with the gravitational wave signal. In this way, we can adopt a Wiener filter approach to develop filter coefficients for seismic sensors in a 2 or 3D array around each test mass. Consider the gravitational wave signal H that is corrupted by the Newtonian noise N, such that the interferometer output is $Y = H + N$. The frequency domain signal of the i-th seismic sensor is given by $X_i = U_i + \Sigma_i$ where U_i is the the seismic displacement signal and Σ the instruments self noise. Assuming that the seismic noise is stationary and has no nonlinear effects we can then construct filters $\alpha_i(\omega)$ to produce an estimate of the Newtonian noise signal $\hat{N}$ and minimize an error function defined by

$$\begin{aligned} E(\omega) &= Y(\omega) - \hat{N}(\omega) \\ &= Y(\omega) - \sum_i \alpha_i(\omega) X_i(\omega). \end{aligned} \tag{13.4}$$

It can then be shown [38] that in order to minimize the power of this error function the filter coefficients must be chosen such that

$$\alpha_j(\omega) = \sum_i \langle X_i(\omega)^* X_j(\omega)\rangle^{-1} \langle X_i(\omega)^* Y(\omega)\rangle, \tag{13.5}$$

where $\langle\cdot\rangle$ denote expectation values. We note that the above filters can be derived from measurable quantities. The term $\langle X_i(\omega)^* X_j(\omega)\rangle$ is the cross correlation matrix between all of the sensors in the array, while the last term $\langle X_i(\omega)^* Y(\omega)\rangle$ is a vector containing the cross spectra between each sensor and the interferometer output. We define a Newtonian noise subtraction residual ρ^2 as the ratio between the power of the error signal and the interferometer output. It is be given by

$$\rho^2 = \frac{\langle E(\omega)^* E(\omega)\rangle}{\langle Y(\omega)^* Y(\omega)\rangle} = 1 - \frac{\langle X_i(\omega)^* Y(\omega)\rangle \langle X_i(\omega)^* X_j(\omega)\rangle^{-1} \langle Y(\omega)^* X_j(\omega)\rangle}{\langle Y(\omega)^* Y(\omega)\rangle}. \tag{13.6}$$

In the case that the interferometer output is dominated by Newtonian noise and the subtraction is perfect then the residual will approach zero. Numerical and analytical simulations of seismic motion with virtual arrays of sensors have been used to evaluate the effectiveness of subtraction schemes and various array topologies [38, 48, 49]. For example, numerical simulations based on seismic noise characteristics measured at the Virgo site showed that significant subtraction factors are possible with a few hundred sensors in a regular grid around each test mass. The results of these simulations for a surface "Advanced Virgo"-like detector are presented in Fig. 13.7. The sensors measure vertical seismic displacement are positioned in a regular 2D grid around each test mass. We see that with just four sensors close to the test mass roughly half of the Newtonian noise could already be subtracted. Additional sensors decrease the residual up to 98 % subtraction, reached with a thousand sensors. These studies also revealed that more careful placement of the sensors will lead to improved subtraction efficiency.

Another important notion when considering seismic Newtonian noise is its apparent relation to the amount of seismic correlation. For low seismic correlation, or short coherence lengths, the Newtonian noise integration process described in Eq. (13.2) will take place over a larger number of incoherently vibrating volumes of soil, that partially cancel each other out. High coherence lengths, on the other hand, produce Newtonian contributions over larger coherently moving volumes, producing a stronger level of net Newtonian forces. With respect to subtraction techniques, long coherence lengths means sensors will need to be distributed over larger areas. Short coherence lengths will require more densely distributed sensors in closer proximity to the test masses. This is an important consequence for third generation detectors as

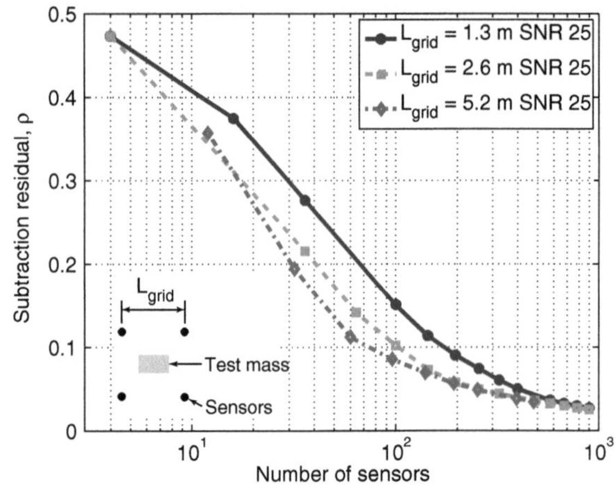

Fig. 13.7 Newtonian noise subtraction residual for a grid of sensors around each test mass for a surface detector. Each curves present different separation distances between sensors

seismic correlation over lengths of tens of meters is generally high for frequencies around a few Hz but drops quickly toward higher frequencies [38].

We have discussed Newtonian noise subtraction techniques from a seismic Newtonian noise point of view. Similar techniques can be used to address atmospheric Newtonian noise issues. Both tasks will be challenging for third generation detectors as the required levels of measurement sensitivity for both seismic and infrasound sensors border on that of currently available technology.

13.3.3 Thermal Noise

Thermal noise in GW interferometers is due to all processes that modulate the phase of the light on its optical path, coupling it to Brownian fluctuations, i.e., to stochastic fluctuations of the temperature field in the optical components. Thermal noise plays an important role both in initial and in advanced GW detectors; in fact, its component due to the last stage suspension mechanics competes with seismic and quantum (radiation pressure) noises at low frequencies whereas its component due to the mirrors (coatings) competes with quantum (shot noise) at medium frequency (see Fig. 13.1 as example). A detailed description of thermal noise and its many ways to couple to the detector has been given in Chap. 8.

Due to its importance, thermal noise effects on GW detectors have been modeled in detail (see [50] for a review and for an outlook on the perspectives for 3G interferometers) and mitigation strategies have been implemented in advanced detectors. The amount of thermal energy in each vibrational degree of freedom is clearly defined by the equipartition theorem and, since initial and advanced interferometers are operating at room temperature, the only handle available to reduce the effect of thermal noises on the detector's sensitivity is to *confine* the influence of the thermal energy to frequencies around the resonance frequencies of mechanical modes of the mirrors or suspensions, making it less prejudicial in the detection frequency band of interest. Indications on how to do it are given by the fluctuation-dissipation theorem (see Chap. 8 and [51]) relating the PSD of the fluctuation of a system in thermal equilibrium to the dissipation mechanisms, described by its mechanical impedance. It is possible to generalize the equations introduced in Chap. 7 for the thermal noise of the last stage of the Advanced Virgo suspension,using the matrix formalism developed in [31, 52] , and to relate the vector $\widetilde{\mathbf{F}}$ of the spectral amplitudes of the thermal stochastic forces acting on the payload elements to the fluctuation velocity vector $\widetilde{\dot{\mathbf{X}}}$ (describing the motion of the various payload bodies) through the impedance matrix $\mathbf{Z}$:

$$\widetilde{\mathbf{F}} \equiv \mathbf{Z}\widetilde{\dot{\mathbf{X}}} \quad ; \quad \mathbf{Z} = \frac{1}{i\omega}\left[\mathbf{M}\cdot(\boldsymbol{\Omega} + i\omega\boldsymbol{\Gamma} - \omega^2\mathbf{I})\right] \tag{13.7}$$

where $\mathbf{M}$ is the mass diagonal matrix, containing the masses of the elements (Mirror, Reference Mass, Marionette, ...) of the payload:

$$\mathbf{M} = \begin{pmatrix} m_1 & 0 \\ 0 & m_2 \end{pmatrix} \tag{13.8}$$

$\boldsymbol{\Omega}$ is the vibrational mode frequencies matrix,

$$\boldsymbol{\Omega} = \begin{pmatrix} \omega_1^2 & -\omega_1^2 \\ -\frac{\mu_1}{\mu_2}\omega_1^2 & \frac{\mu_1\omega_1^2+\omega_2^2}{\mu_2} \end{pmatrix} \tag{13.9}$$

$\boldsymbol{\Gamma}$ is the (viscous) dissipation coefficients matrix

$$\boldsymbol{\Gamma} = \begin{pmatrix} \gamma_1 & -\gamma_1 \\ -\frac{\mu_1}{\mu_2}\gamma_1 & \frac{\mu_1\gamma_1+\mu_2\gamma_2}{\mu_2} \end{pmatrix} \tag{13.10}$$

and $\mathbf{I}$ is the identity matrix. The Eqs. 13.8, 13.9, 13.10 are referred to a payload composed by the mirror (1) and the marionette (2), like in Advanced Virgo. The Virgo design, described in Chap. 7, includes, instead, also a third body, the reference mass. In these equation is $\mu_j = m_j / \sum_j m_j$ and the frequencies ω_j contain an immaginary part,

$$\omega_j^2 = \frac{g}{L_j}\left[(1 + D_j) + iD_j\phi_j\right] \tag{13.11}$$

to take into account the structural dissipation (L_j is the suspension wire length and D_j is the geometrical dilution factor [53] related to the geometry of the suspension). The computation of the relevant pendulum thermal noise (the fluctuation of the mirror), taking into account the dissipative contribution of the other elements of the payload, is, then, formally given by:

$$S_{p,1}(\omega) = \frac{4k_BT}{\omega^2}\Re\left\{\left(\mathbf{Z}^{-1}\right)_{11}\right\}$$
$$= \frac{4k_BT}{\omega^2}\Re\left\{\left[\left[\frac{1}{i\omega}\left[\mathbf{M}\cdot\left(\boldsymbol{\Omega}+i\omega\boldsymbol{\Gamma}-\omega^2\mathbf{I}\right)\right]\right]^{-1}\right]_{11}\right\} \tag{13.12}$$

The minimization of the dissipation processes (in $\boldsymbol{\Gamma}$ and $\boldsymbol{\Omega}$) confines the thermal energy around the internal modes frequencies ($\boldsymbol{\Omega}$). In advanced detectors, the mechanical dissipation mechanisms are minimized adopting a monolithic fused silica design [54, 55] in the suspension system, special low losses fused silica in the mirror substrates and optimizing the design [56] and the chemical composition [57, 58] of the reflective dielectric coatings on the mirrors.

In order to have a significant decrement of the thermal noise, in 3G detectors it is necessary to reduce also the amount of thermal energy available in the system, reducing the temperature of the main optics down to cryogenic values. This strategy is anticipated by the KAGRA detector, that will adopt cooling technologies (cryo-coolers, radiation shields, ...) of clear interest for ET. The choice to operate at cryogenic temperature implies relevant novelties in 3G detectors. The first important change with respect to 2nd generation detectors is related to the realization of the optics and of the suspension last stage; the use of fused silica, crucial in initial and advanced detectors for the test mass substrates and suspension realization is impossible in 3G detectors both because of a well-established dissipation peak [59] and because of the poor thermal conductivity [60] exhibited by that material at cryogenic temperature. New materials are investigated for cryogenic GW detectors: KAGRA focused its interest on Sapphire suspension fibers and test masses [37], ET is investigating the possibility to use Silicon. Both these crystalline materials show an excellent loss angle at low temperature (see [61] for a complete review), good thermal conductivity (reference values are $\kappa_{Al_2O_3} \simeq 2330\,\mathrm{W\,m^{-1}K^{-1}}$ at about 10 K, $\kappa_{Si} \simeq 1200\,\mathrm{W\,m^{-1}K^{-1}}$ at about 12 K), crucial in order to extract the heat deposited by laser beams impinging on the suspended test masses), relatively high Young's modulus[2] ($Y_{Al_2O_3} \simeq 400\,\mathrm{GPa}$, $Y_{Si} \simeq 132\,\mathrm{GPa}$), high tensile strength.[3] The thermal expansion coefficient of Silicon shows a peculiar behavior, vanishing at about 120 K and 20 K, allowing a efficient suppression, around these temperatures, of the thermoelastic noise both in the suspensions wires [62] or ribbons [63] and in the mirror substrates. Furthermore it is known [64] that Silicon is suitable for chemical bonding procedures (*Hydroxide Catalysis Bonds*) allowing to realize monolithic all-Silicon suspensions (recent unpublished and preliminary results [65] are suggesting the possibility to reach a similar achievement with all-Sapphire suspensions). It is a fact that, anyways, the fabrication of a monolithic suspension with these crystalline materials is still to be realized.

Silicon and Sapphire are candidate materials in the realization of the mirror substrates in future GW detectors; the mechanical properties (high thermal conductivity,

---

[2] The higher the Young's modulus of a body, the less deformation is generated by the same thermal energy, thus less thermal noise.

[3] Useful to realize high dilution factor suspensions.

low loss angle, high Young's modulus, and low thermal expansion coefficient) indicating them as excellent materials for the realization of the monolithic suspensions are obviously still valid in the selection of a low thermal noise material for the test masses, but in this case the optical properties play a crucial role. Sapphire is transparent to the laser wavelength ($\lambda = 1064\,\text{nm}$) adopted in initial and advanced interferometer, allowing an easy re-use of the optoelectronics developed for these detectors; it shows birefringence effects [66] (AL_2O_3 is an uni-axial crystal and the light propagating along the C-axis does not change its polarization state, but imperfections in the direction of the C-axis could give rise to birefringence effects) and it suffers of high levels of optical absorption, 30–60 ppm/cm [67], quite high with respect to silica at room temperature (and affecting the design of KAGRA). Silicon is not birefringent, it is available in quite large pieces, but it is not transparent at 1064 nm, forcing the adoption of a longer wavelength. Current studies are focused on the optical properties of Silicon at $\lambda = 1550\,\text{nm}$ and recent results [68] confirmed the relationship between free-carrier concentration and optical absorption, identifying a high resistivity ($10^4\,\Omega\text{cm}$) n-doped Silicon sample having an optical absorption at room temperature lower than 5 ppm/cm. Still to be understood is the excess of absorption found in [68] at cryogenic temperature, that contrary to expectations, for some samples, has twice the optical absorption measured at room temperature.

Mechanical losses in the high reflectivity dielectric coating deposited on the Fabry-Perot cavities mirrors are the main noise source for the advanced detectors in the intermediate frequency range (see Fig. 13.1); mitigation strategy (see Chap. 8) adopted in advanced detectors (like doping the Tantala layers with Titanium atoms [58] or optimizing the amount of high dissipation material (Ta_2O_5) while keeping the same reflectivity [56]) are needed in 3G interferometers, but further developments are needed. Several are the R & D activities in progress in this framework; *grating*[4] *reflectors* [70–72] are reaching high reflectivity values and promising a reduction of thermal noise in case of Silicon mirrors [73]; *crystalline coatings*, based on AlGaAs [74] or AlGaP [75] structures, are showing high reflectivity, low thermal noise [76, 77] but still too high absorption with respect to the amorphous multilayer coatings used in advanced detectors.

In order to reduce the mirror thermal noise, another mechanism is under investigation for ET; it is well known that a larger beam spot on the cavity mirrors reduces the thermal noise effects because, naively, the beam is *averaging* over a larger fluctuating surface of the mirror. But because of the limited diameter of the mirror in ET and because of the stringent constrains in clipping losses,[5] it will be difficult to have in ET beams substantially larger than the beams in the advanced detectors. Having,

[4] Currently used coatings are based on alternate layers of dielectric materials (see Chap. 8 for a detailed description); the high refraction index material Ta_2O_5 is dissipative and generates the largest fraction of the thermal noise. Grating reflectors are based on resonant waveguide grating nanostructures, grown on the surfaces of the mirror substrates. With this design it is possible to obtain high reflectivity with a single layer of high refraction index material or even without the deposition of Ta_2O_5 on a Silicon test mass. See [69] for an introduction.

[5] Optical losses due, roughly, to the finite size of the mirrors with respect to the nominally infinite radial extension of a Gaussian beam.

instead, a beam with an intensity distribution "flatter" than the Gaussian one, can generate a thermal noise reduction in the interferometer (see Sect. 3.2). For this reason the possibility to use high-order Laguerre-Gauss modes has been introduced [78, 79] and several aspects are currently under evaluation about the generation of these modes [80] and the mitigation of the effects of the mirror surface imperfections on their cross-coupling [81, 82].

13.3.4 Quantum Noise

The initial phase design of the Einstein Telescope relies on well-explored *conventional* technologies, defined by their maturity, either proven in the first generation, planned for the advanced generation gravitational wave detectors, or tested in prototypes. Consequently, the configurations described in the following sections are a challenging application of the technologies and of the principles described in Chap. 3; for this reason, a limited level of detail has been adopted.

In the high frequency range, all existing and planned advanced GW detectors will be limited by shot noise, which is one facet of quantum noise. The other side of quantum noise manifests itself in radiation pressure noise which, due to the frequency dependent mechanical susceptibility of the test masses, becomes dominant in the low frequency region. Although the product of amplitude noise and phase noise, resulting in radiation pressure noise and shot noise, is limited by Heisenberg's uncertainty principle, both noise contributions can be traded against each other using squeezed states. An introduction to Quantum Noise and means to overcome some limitations is given in Chap. 11, while a comprehensive description can be found, e.g., in [83] or [84] and references cited therein.

The design chosen for the optical layout of initial ET in the conceptual design study is similar to the one for the advanced detectors. In addition to the advanced detectors topology, the Einstein Telescope will use frequency dependent squeezed vacuum injected to the output port. The technological feasibility of alternative topologies like speed meters (see Chap. 11.8), which promise advantages especially in the low frequency range will be investigated in the close future, but until they have reached a sufficient level of maturity, the base plan for the Einstein Telescope will be a dual recycled arm cavity Michelson Interferometer with frequency dependent input squeezing as depicted in Fig. 13.8. In the following, we will shortly list how these techniques lower the quantum noise.

Quantum noise arises from the zero point (vacuum) fluctuations of the electromagnetic field, resulting from its quantization. These vacuum fluctuations enter the interferometer through each open port. The interferometer arm end mirrors ETM1 and ETM2 have got very low transmission (for quantum noise considerations these ports can be regarded as closed); that leaves only the laser port and the detection port where vacuum fluctuations can enter. Mostly for technical reasons all interferometers are operated close to the so-called *dark port*, an operation point where the arm lengths are adjusted such that the laser carrier fields returning from the inter-

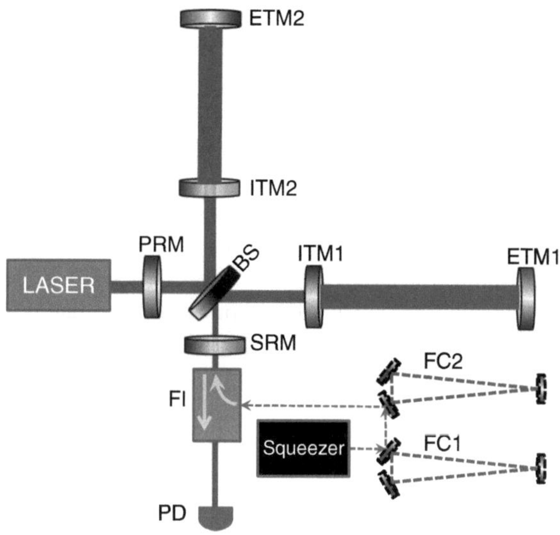

Fig. 13.8 Schematic views of the optical topology of the Einstein Telescope: a dual recycled Michelson with arm cavities and filtered input squeezing. The optical elements are: ETM1 and ETM2: highly reflective arm end mirrors; ITM1 and ITM2: partly transmissive cavity coupling mirrors; BS: the beam splitter; SRM: Signal Recycling mirror; PRM: Power Recycling mirror. FC1 and FC2 are filter cavities. FI is a Faraday Isolator acting as an optical diode and PD is the photo detector

ferometer arms destructively interfere toward the detection port of the beam splitter and consequently constructively interfere toward the laser. Under these conditions, the interferometer acts like a mirror, meaning it reflects light back if shone into one of these two ports. Quantum fluctuations entering the laser port get reflected back to it and do not (or only to a very small amount) reach the output. Vacuum fluctuations entering the detection port of the interferometer also get reflected and reach the photo detector. Fluctuations in the amplitude quadrature (relative to the carrier light exiting through the output port) will affect the power detected by the photo detector and appear as noise—the shot noise. One of the goals of reducing quantum noise is to reduce the relative amount of the fluctuations in relation to the signal. The most obvious way to do so is to enhance the signal.

13.3.4.1 Increasing the Light Power

Increasing the signal can be done by using more light in the interferometer arms where the GW converts carrier light into signal sidebands. High power in the interferometer arms is achieved by three different means:

- High laser power
- Arm Cavities
- Power Recycling

High Laser Power

The most obvious way to increase the light power inside the interferometer is to use a powerful laser. All first and second generation interferometric GW detectors (see Chap. 4) use lasers with a wavelength of 1064 nm. During the construction and operation of the first generation detectors, the available laser power increased from about 10–20 W in the first generation to 200 W available for the second generation. According to current plans, solid state 1 kW lasers will be available for the third generation. The use of high laser power gives rise to thermal problems resulting from the light absorbed in auxiliary optics like modulators, Faraday rotators or polarizers. The absorbed light will lead to a local temperature increase which, through the temperature dependence of the refractive index, will lead to thermal lensing. This in turn can lead to a reduced quality of mode matching to the optical resonators. Although the absorption levels of fused silica have been cut down in the past to a level of 0.25 ppm/cm for bulk absorption and to the sub-ppm level for coating absorption, thermal problems will still pose a problem for the second generation of detectors. Elaborate thermal compensation systems, discussed in Chap. 9, are needed to compensate the effect.

As stated in Sect. 13.3.3, the possibility to use Silicon as test mass material in 3G detectors is pushing the development of a new generation of lasers; in fact Silicon is opaque at the light at wavelenghts around $\lambda = 1064$ nm (adopted in the current GW detectors), while it promises extremely low absorption values at the wavelength of $\lambda = 1550$ nm [68, 85]. High power fiber lasers are currently being developed but have not yet reached the performance level of solid state lasers.

Arm Cavities

Placing Fabry-Perot cavities in the interferometer arms (see Fig. 13.8) resonantly enhances the light in the individual arms and with it the signals created by the GWs. The amount of power enhancement is limited by the fact that the line width of the optical resonator decreases with increasing buildup until relevant parts of the signal frequency spectrum fall outside of the resonance of the cavity and the GW sidebands do not get simultaneously enhanced with the carrier light any more. The low transmission of the input mirror then causes a reduction of the signal amplitude for high frequencies signals at the output port, i.e., a reduced sensitivity. The possible power enhancement in the interferometer arms is hence a compromise between a low relative shot noise level and bandwidth. This effect can partly be *undone* with the Signal Recycling technique.

Power Recycling

The carrier light as well as all common fluctuations returning from the interferometer arms constructively interfere toward the input port where they can be reflected with

Table 13.1 Typical light powers and mirror masses for different generations of GW interferometers

Detector generation	1st	2nd	3rd
Laser power	10 W	200 W	1 kW
Light power @ beam splitter	200 W	2 kW	20 kW
Light power inside ifo arms	15 kW	800 kW	3 MW
Mirror mass	10 kg	40 kg	200 kg

a mirror located in the input, called the Power Recycling mirror (see Fig. 13.8). The interferometer together with this mirror then forms an optical resonator where the light power can resonantly be enhanced. This technique is called Power Recycling and essentially has the same effect as directly increasing the laser power injected into the interferometer. All detectors of the first and second generation used and will use Power Recycling. With a laser power of about 200 W the second generation will reach power levels inside the arm cavities of around 800 kW. The Einstein Telescope is aiming at a light power in the interferometer arms of 3 MW. With the promising prospect of 1 kW high-power-lasers available for the third generation there will be no need to increase the Power Recycling factor beyond what already planned in the second generation.

Approximate numbers for power requirements for the 1st to the third generation are given in the upper part of Table 13.1.

13.3.4.2 Signal Recycling

The second generation of interferometric gravitational-wave detectors (in this sense, we regard GEO600 as a precursor second generation instrument) uses an additional technique to enhance the GW signal: Signal Recycling ([86, 87]). The signals being generated by the GW (with a phase difference of π) in the interferometer arms constructively interfere toward the output port. If these signals are sent back toward the interferometer with a recycling mirror (see Fig. 13.8), they get reflected back to the output port by the interferometer, similar to the laser power being reflected back to the input port. Together with the interferometer this mirror forms an optical resonator, the Signal Recycling cavity. The signals created from the carrier field inside this resonator (in the interferometer arms) are resonantly enhanced, resulting in an increased signal even after being transmitted through the Signal Recycling mirror. By microscopic adjustment of the position of the Signal Recycling mirror, the resonance of this cavity can be tuned to any desired frequency. Depending on this tuning the bandwidth of the interferometer can either be narrowed (called Signal Recycling in the case where the carrier frequency is close to a resonance) or widened (called Resonant Sideband Extraction in the anti resonant case). The tuning changes the resonance conditions of the fields in the signal recycling cavity and with it the phase relation between the signal sidebands and the carrier at the output port and inside the interferometer arms. Signal Recycling enhances the GW Signal but does not enhance the vacuum fluctuations reaching the photo detector.

Beside the optical resonance, detuned Signal Recycling produces another opto-mechanical resonance, typically at lower frequencies (see Fig. 13.1), which gives an increase in sensitivity and, through the correlations that it introduces between amplitude and phase fluctuations, even allows to surpass the Standard Quantum Limit [88–91].

13.3.4.3 Radiation Pressure Noise

While increasing the power in the interferometer lowers the relative readout noise, i.e., the shot noise, it increases the radiation pressure noise. The impulse transfer of the photons onto the mirrors upon reflection causes a force acting onto the mirrors, the radiation pressure $p = 2I/c$, where I is the light power and c the speed of light. The phase quadrature of the vacuum fluctuations entering the output port gets converted into differential amplitude fluctuations in the interferometer arm in the same way as the phase modulation caused by the GW in the interferometer arms get converted into detectable amplitude modulation at the output port. Here, when saying "phase quadrature of the vacum fluctuations," we define the quadrature by comparing the phase of the vacuum fluctuations, after being reflected from the interferometer and arriving at the photo detector, with the phase of the carrier light emerging from the interferometer, also at the location of the photo detector; "amplitude quadrature" we call in-phase light and "phase quadrature" we call the 90° shifted quadrature. In the interferometer arms, these amplitude fluctuations beat with the strong carrier field and cause fluctuating radiation pressure forces on the mirrors.

In this way, the amplitude fluctuations on the light in the interferometer arms get converted into mirror motion causing the so-called radiation pressure noise. This radiation pressure noise increases with increasing light power in the interferometer arms.

The mechanical susceptibility (displacement/applied force) of a free mass, or of a pendulum mass well above its resonance frequency, is $(M\Omega)^{-2}$, were M is the mass of the test mirror and Ω the frequency of interest. Radiation pressure therefore is more important at low frequencies. The worsening of the Quantum Noise at frequencies below about 20 Hz shown in Fig. 13.1 is due to this effect.

Increasing the mirror mass decreases the mechanical susceptibility and hence cuts down the effect of radiation pressure on test mass movement. Consequently the mirror mass will increase from the first to the third generation as indicated in Table 13.1. With the Xylophone strategy (see Sect. 13.4) of the Einstein Telescope the radiation pressure in the low frequency interferometers due to the lower power of only 18 kW is much less. Still due to the increasing suceptibility of the mirrors to applied forces toward low frequencies the mirror mass needs to be as high as for the high-power (3 MW), high-frequency interferometers.

13.3.4.4 Squeezing

The other option to increase the ratio of GW signal to vacuum-fluctuation-produced noise is to lower the vacuum fluctuations. The fluctuations in the different quadratures of the vacuum field entering the output port can be traded against each other, still fulfilling the requirements of Heisenberg's uncertainty relation. So, as discussed in Chap. 11, the noise in one quadrature, e.g., the amplitude quadrature, can be lowered while at the same time raising the noise in the phase quadrature. If such a squeezed state with lowered amplitude noise (as defined in Sect. 13.3.4.3) is injected into the output port, the shot noise of the light registered by the photo-detector will be lowered. At the same time, the radiation pressure contribution will increase. The fact that shot noise and radiation pressure noise are dominant in disjunct frequency ranges allows to use amplitude squeezed vacuum in one frequency range and phase squeezed vacuum in the other. By appropriately rotating the phase angle of the squeezing as a function of frequency, which can be achieved by reflecting the squeezed state from a filter cavity [92–95], the shot noise (at high frequencies) and the radiation pressure noise (at low frequencies) can be simultaneously reduced. In case of using squeezing together with a detuned interferometer, i.e., not tuned to zero frequency, which means that the laser carrier frequency is not resonant in the Signal Recycling cavity, the rotation of the squeezing ellipse with respect to the light coming from the interferometer (due to the dispersion of the Signal Recycling cavity) causes an increased amount of Quantum Noise outside the Signal Recycling resonance due to the anti-squeezed quadrature of the injected squeezing [95]. This effect can be avoided by filtering the injected squeezing with a second filter cavity.

Squeezing is already implemented in the current generation of interferometric gravitational-wave detectors in the case of GEO-600 [96], aiming at a squeezing level of 6 dB over the entire frequency range of interest. 6 dB effective squeezing at the photo-detector lowers the shot noise by a factor of two corresponding to a light power increase of a factor of 4. An overall gain from squeezing of about 10 dB, i.e., a factor of 3 in noise amplitude spectral density, is foreseen for the Einstein Telescope and seems realistic with the squeezing levels already achieved today [97].

13.3.4.5 Parametric Instabilities

The use of high power in the resonant Fabry-Perot cavities may show a crucial drawback: the parametric instabilities [98], discussed in detail in Chap.12. This is a process in which the coupling between optical modes of resonant cavities and acoustic modes in the cavity mirror substrates creates a combined resonance, which when excited would spoil the cavity's performance, disrupting controlled operation (unlock) of the cavity. These parametric instabilities may be a serious problem for the advanced interferometers [99–101]. Studies for LCGT [102] (now called KAGRA) have shown that the problem is less severe in this interferometer because of the different geometry and materials of the mirrors, since the radius of curvature of the mirror and the sound velocity at the operating temperature play a relevant role in determining the number of unstable modes.

Hence, for a third generation gravitational wave detector, the evaluation of the relevance of the parametric instabilities issue strongly depends on the selected design options and for this reason the current ET design parameters are hereafter used to describe the possible situation. It has been computed [103] that the number of the instable modes of the ET cavity is a few hundreds times larger than that of the KAGRA cavity and it is likely that a method to suppress the parametric instabilities is needed. The reduction of the mechanical Q factor of elastic modes is a promising method currently investigated in advanced detectors and LCGT: a mechanical loss is deliberately added to the mirror, in order to damp the elastic modes. If the mechanical loss is concentrated far away from the optical beam, it will not significantly contribute to the mirror thermal noise. A possible implementation of this solution is realized by coating the barrel surface with a dissipative layer of Ta_2O_5. This method seems promising for ET if a Sapphire substrate would be used; on the other hand, in the case of a Silicon substrate, the excess of thermal noise could affect the expected sensitivity. These studies suggest that the parametric instabilities may be an issue for the Einstein telescope, but there are technical solutions available that will evolve with the progress of dedicated R&D.

13.4 Infrastructures, Geometries, and Topologies

As described in the previous sections, the site hosting a future 3G observatory like ET will most likely have to be underground; this obviously requires a new infrastructure to be built. The ET design study [9] was mainly addressed to a feasibility study for such an infrastructure and to the understanding of the scientific potential of the observatory. The proposed infrastructure is an underground (100–200 m) laboratory, in a site having the low seismic noise characteristics cited in Sect. 13.3.1, allowing a length of the interferometers arms of the order of 10 km, capable to host more than one detector. Each detector should be able to span a different frequency interval, all in the range 2–3 Hz to 10 kHz, guaranteeing an improvement, in terms of sensitivity, of roughly a factor of ten with respect to advanced detectors. The wide detection frequency range requirement heavily affects the design of a 3G detector; the request to have an excellent low-medium frequency sensitivity, achieved by reducing the thermal noise with cryogenic suspensions and cryogenic test-masses, conflicts with the request of an excellent high frequency sensitivity, achieved by increasing the light power circulating in the Fabry–Perot cavities up to 3 MW.[6] In the ET design

[6] The optical power deposited by the ligh impinging on the mirror surfaces, is extracted through the suspension fibers; the amount of heat to be extracted constrains the minimal diameter of the suspension fibers to a value that conflicts with the requirement to have very thin suspension fibers imposed by the request to have a high dilution factor in order to minimize the thermal noise.

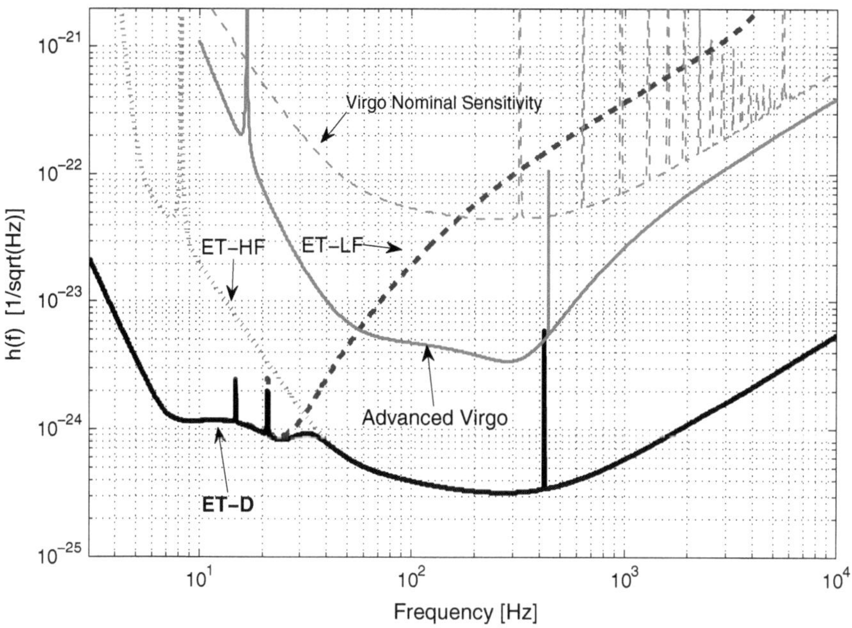

Fig. 13.9 Sensitivity model (named ET-D, *black and thick curve*) of a single ET detector, having 10 km *L-shaped* (90° opening angle) arms, dual-recycled Michelson layout with Fabry-Perot arm cavities. The ET sensitivity is composed by the overlap of the sensitivities of the low-frequency interferometer (ET-LF, *blue dashed curve*) and high-frequency interferometer (ET-HF, *red dotted curve*). Parameters and specifications of these interferometers are described in Table 13.1 of [105]. The ET detector sensitivity is compared in the figure to the reference Advanced Virgo sensitivity (*gray curve*) and to the nominal Virgo sensitivity (*green dashed line*)

this conflict has been solved with the so-called Xylophone [104] design: each ET detector is composed of two interferometers:

- **ET–LF**, devoted to the low-frequency range, implementing a state-of-the-art suspension system, cryogenic ($\simeq$10 K) silicon test masses and cryogenic suspensions, low laser power at $\lambda = 1550$ nm;
- **ET–HF**, devoted to the high-frequency range, operating at room temperature, using fused silica test masses and high laser power at $\lambda = 1064$ nm.

With the implementation of such a configuration and the technologies described in the previous paragraphs, the ET observatory will reach a sensitivity [105] as shown in Fig. 13.9.

As previously stated, the ET observatory will be composed of more than one detector, resulting in several advantages, already largely analyzed for the similar configuration of the LISA low frequency space observatory [106] :

- The capability to fully reconstruct the two polarizations of the GW [11], through virtual interferometry compositions;

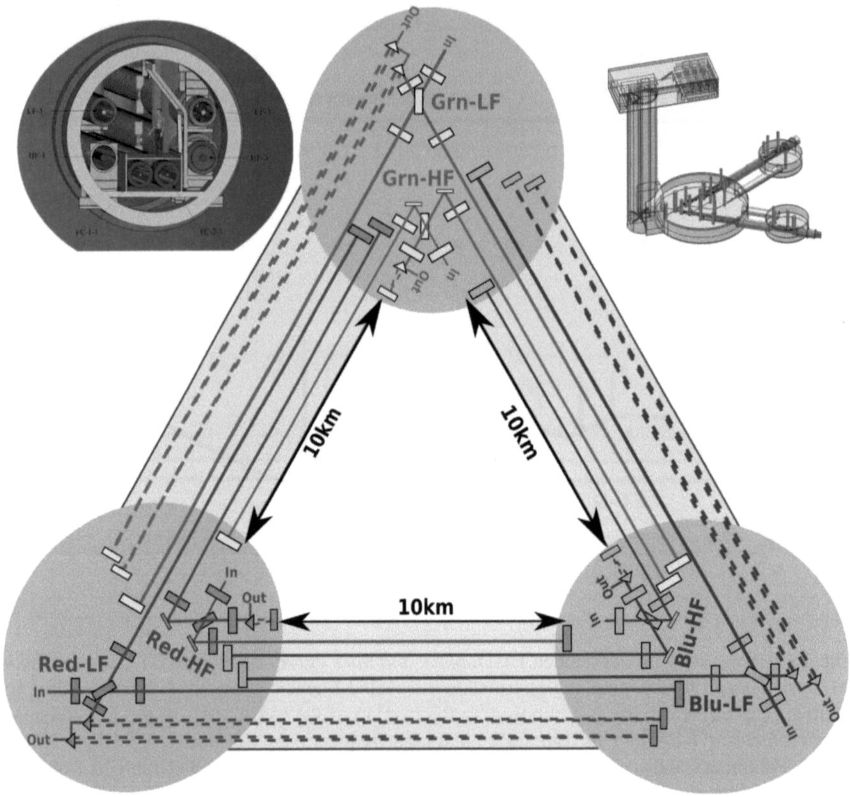

Fig. 13.10 Schematic implementation of the ET layout; in the triangular geometry the three (*color coded*) detectors are visible, each composed by two interferometers (Xylophone configuration); in the *top–left* small image a possible distribution of the beam pipes is shown; in the *top–right* image, a schematic view of a corner station is shown. All the pictures are derived from [9]

- The possibility to construct a null-data-stream [107], in which by design no GW signal is present. Signals that appear in this data stream can be used for vetoing noisy events;
- High duty cycle for the observatory is guaranteed by the redundancy of the configuration: even if one detector completely fails, the two remaining can assure full sky coverage, even if at the loss of polarization information.

It is evident that the capability to host more than one detector, has a cost implication. In an underground installation, the largest cost is the excavation of the underground tunnels and halls; the ET design study revealed that a triangular geometry optimises the construction cost of the infrastructure with respect to the number of interferometer composing the observatory. The implementation of the ET observatory is schematized in Fig. 13.10. It is worth to note that the use of a triangular geometry has a minimal impact on the sensitivity of the observatory; the loss in sensitivity due to the 60° angle between the arms is compensated by the conbined use of the tree detectors outputs, resulting in a loss in sensitivity ranging between 0 and 6 % [107].

Acknowledgments This work has been performed with the support of the European Commission under the Framework Programme 7 (FP7), project *ELiTES* (Grant Agreement 295153), http://www.et--gw.eu/descriptionelites.

References

1. H. Grote, L.S. Collaboration, Class. Quantum Gravity **25**(11), 114043 (2008)
2. B. Abbott et al., LIGO Scientific Collaboration. Reports on Progress in Physics **72**(7), 076901 (2009)
3. K. Arai, the TAMA Collaboration. J. Phys.: Conf. Ser. **120**(3), 032010 (2008)
4. F. Acernese et al., Class. Quantum Gravity **25**(11), 114045 (8pp) (2008)
5. J. Hough, S. Rowan, J. Opt. A: Pure Appl. Opt. **7**(6), S257 (2005)
6. J. Abadie et al., Class. Quantum Gravity **27**(17), 173001 (2010). doi:10.1088/0264-9381/27/17/173001
7. D.S. for the Advanced LIGO Team. Advanced ligo reference design (2009). https://dcc.ligo.org/cgi-bin/DocDB/ShowDocument?docid=1507
8. The Virgo Collaboration, Advanced virgo technical design report (2012). https://tds.ego-gw.it/ql/?c=8940. VIR-0128A-12
9. The ET Science Team, Einstein gravitational wave telescope conceptual design study (2011). https://tds.ego-gw.it/ql/?c=7954. ET-0106C-10
10. M. Punturo et al., Class. Quantum Gravity **27**(8), 084007 (2010)
11. B. Sathyaprakash et al., Class. Quantum Gravity **29**, 124013 (2012)
12. L. Baiotti, B. Giacomazzo, L. Rezzolla, Phys. Rev. D **78**, 084033 (2008). doi:10.1103/PhysRevD.78.084033, http://link.aps.org/doi/10.1103/PhysRevD.78.084033
13. J.S. Read et al., Phys. Rev. D **79**, 124033 (2009). doi:10.1103/PhysRevD.79.124033
14. M. Shibata, Phys. Rev. Lett. **94**(20), 201101 (2005). doi:10.1103/PhysRevLett.94.201101
15. N. Andersson, Class. Quantum Gravity **20**(7), R105 (2003). http://stacks.iop.org/0264-9381/20/i=7/a=201
16. S. Woosley, Astrophys. J. **405**, 273 (1993)
17. K. Iwamoto et al., Nature **395**, 672 (1998). http://dx.doi.org/10.1038/27155
18. J.S. Bloom et al., Astrophys. J. **654**(2), 878 (2007). http://stacks.iop.org/0004-637X/654/i=2/a=878
19. B. Abbott et al., Phys. Rev. D **72**, 042002 (2005). doi:10.1103/PhysRevD.72.042002, http://link.aps.org/doi/10.1103/PhysRevD.72.042002
20. B. Abbott et al., Phys. Rev. D **77**, 062004 (2008). doi:10.1103/PhysRevD.77.062004, http://link.aps.org/doi/10.1103/PhysRevD.77.062004
21. F. Acernese et al., Classical and Quantum Gravity **24**(19), S671 (2007). http://stacks.iop.org/0264-9381/24/i=19/a=S29. (Virgo Collaboration)
22. F. Acernese et al., Class. Quantum Gravity **25**(22), 225001 (2008). http://stacks.iop.org/0264-9381/25/i=22/a=225001. (Virgo Collaboration)
23. B.P. Abbott et al., Astrophys. J. **715**(2), 1438 (2010). http://stacks.iop.org/0004-637X/715/i=2/a=1438. (The LIGO Scientific Collaboration and the Virgo Collaboration)
24. J. Abadie et al., Astrophys. J. **760**(1), 12 (2012). http://stacks.iop.org/0004-637X/760/i=1/a=12
25. B.D. Metzger, E. Berger, Astrophys. J. **746**(1), 48 (2012). http://stacks.iop.org/0004-637X/746/i=1/a=48
26. B.S. Sathyaprakash, B.F. Schutz, C.V.D. Broeck, Class. Quantum Gravity **27**(21), 215006 (2010). http://stacks.iop.org/0264-9381/27/i=21/a=215006

27. B.F. Schutz, Nature **323**, 310 (1986). doi:10.1038/323310a0
28. C. Messenger, J. Read, Measuring a cosmological distance-redshift relationship using only gravitational wave observations of binary neutron star coalescences. (2011). http://arxiv.org/pdf/1107.5725v2.pdf
29. W. Del Pozzo, Inference of the cosmological parameters from gravitational waves: application to second generation interferometers. (2012). http://arxiv.org/abs/1108.1317
30. C.D. Ott, Class. Quantum Gravity **26**(6), 063001 (2009). http://stacks.iop.org/0264-9381/26/i=6/a=063001
31. M. Punturo, Advanced virgo sensitivity document (2012). https://tds.ego-gw.it/ql/?c=8877. VIR-0073D-12
32. J. Peterson. Observations and modeling of seismic background noise. Open-File report 93–322 (1993). http://earthquake.usgs.gov/regional/asl/pubs/files/ofr93-322.pdf, http://earthquake.usgs.gov/regional/asl/pubs/files/ofr93-322.pdf
33. A. Friedrich, F. Krüger, K. Klinge, J. Seismolog. **2**(1), 47 (1998). doi:10.1023/A:1009788904007
34. E. Marchetti, M. Mazzoni, Evidence of oceanic microseism as a source of low frequency seismic signal recorded at virgo. Techical Report Virgo (2004). https://tds.ego-gw.it/ql/?c=1502. VIR-NOT-FIR-1390-261
35. M.G. Beker et al., J. Phys.: Conf. Ser. **363**(1), 012004 (2012). http://stacks.iop.org/1742-6596/363/i=1/a=012004
36. S. Miyoki et al., Class. Quantum Gravity **23**(8), S231 (2006). http://stacks.iop.org/0264-9381/23/i=8/a=S29
37. K. Somiya, Class. Quantum Gravity **29**(12), 124007 (2012). http://stacks.iop.org/0264-9381/29/i=12/a=124007
38. M.G. Beker, Low-frequency sensitivity of next generation gravitational wave detectors. Ph.D. Thesis, Vrije Universiteit, 2013
39. F. Acernese et al., Astropart. Phys. **33**(3), 182 (2010)
40. S. Braccini et al., Astropart. Phys. **23**(6), 557 (2005)
41. A. Bertolini et al., Nucl. Instrum. Methods Phys. Res. **435**(435), 475 (1999)
42. A. Stochino et al., Nucl. Instrum. Methods Phys. Res. Sect. A: Accelerators, Spectrometers, Detectors and Associated Equipment **580**(3), 1559 (2007). doi:10.1016/j.nima.2007.06.029, http://www.sciencedirect.com/science/article/pii/S0168900207012302
43. P.R. Saulson, Phys. Rev. D **30**(4), 732 (1984)
44. K.S. Thorne, C.J. Winstein, Phys. Rev. D **60**, 082001 (1999). doi:10.1103/PhysRevD.60.082001
45. T. Creighton, Class. Quantum Gravity **25**(12), 125011 (2008). http://stacks.iop.org/0264-9381/25/i=12/a=125011
46. M. Beccaria et al., Class. Quantum Gravity **15**, 3339 (1998)
47. S.A. Hughes, K.S. Thorne, Phys. Rev. D **58**(12), 122002 (1998). doi:10.1103/PhysRevD.58.122002, http://link.aps.org/doi/10.1103/PhysRevD.58.122002
48. M. Beker et al., Gen. Relativ. Gravit. **43**, 623 (2011)
49. J.C. Driggers, J. Harms, R.X. Adhikari, Phys. Rev. D **86**, 102001 (2012). doi:10.1103/PhysRevD.86.102001, http://link.aps.org/doi/10.1103/PhysRevD.86.102001
50. R. Nawrodt, S. Rowan, J. Hough, M. Punturo, F. Ricci, J.Y. Vinet, Gen. Relativ. Gravit. **43**(2), 593 (2011). doi:10.1007/s10714-010-1066-5, http://dx.doi.org/10.1007/s10714-010-1066-5
51. H.B. Callen, T.A. Welton, Phys. Rev. **83**, 34 (1951)
52. F. Piergiovanni, M. Punturo, P. Puppo, The thermal noise of the virgo+ and virgo advanced last stage suspension (the ppp effect). (2009). https://tds.ego-gw.it/ql/?c=2199. VIR-015E-09
53. G. Gonzalez, Class. Quantum Gravity **17**(21), 4409 (2000). http://stacks.iop.org/0264-9381/17/i=21/a=305
54. P. Amico et al., Rev. Sci. Instrum. **73**(9), 3318 (2002)
55. A.V. Cumming et al., Class. Quantum Gravity **29**(3), 035003 (2012). http://stacks.iop.org/0264-9381/29/i=3/a=035003

56. A. Juri et al., Adv. Thin-Film Coatings Opt. Appl. III **6286**(1), 628608 (2006)
57. G.M. Harry et al., Class. Quantum Gravity **19**(5), 897 (2002)
58. G.M. Harry et al., Class. Quantum Gravity **24**(2), 405 (2007)
59. R.E. Strakna, Phys. Rev. **123**, 2020 (1961). doi:10.1103/PhysRev.123.2020, http://link.aps.org/doi/10.1103/PhysRev.123.2020
60. Y.S. Touloukian, C.Y. Ho, in *Conductivity–Nonmetallic Solids*, vol. 2 (Plenum, New York, 1970)
61. R. Nawrodt et al., Mechanical loss in bulk materials (2013). https://tds.ego-gw.it/ql/?c=9409. ET-0002A-13
62. M. Alshourbagy et al., Rev. Sci. Instrum. **77**(4), 044502 (2006)
63. S. Reid, G. Cagnoli, D. Crooks, J. Hough, P. Murray, S. Rowan, M. Fejer, R. Route, S. Zappe, Phys. Lett. A **351**(4–5), 205 (2006)
64. N.L. Beveridge, Characterisation of silicon-silicon hydroxide catalysis bonds for future gravitational wave detectors. Ph.D. Thesis, University of Glasgow (2012). http://theses.gla.ac.uk/id/eprint/3526
65. R. Douglas, Sapphire bonding work (2013). Talk at the ELiTES-WP2 teleconference
66. F. Benabid, M. Notcutt, L. Ju, D. Blair, Phys. Lett. A **237**(6), 337 (1998). doi:10.1016/S0375-9601(97)00818-9, http://www.sciencedirect.com/science/article/pii/S0375960197008189
67. E. Hirose, K. Watanabe, N. Mio, Sapphire for the kagra project (2011). http://gwdoc.icrr.u-tokyo.ac.jp/JGWDoc/. JGW-T1100626-v1
68. J. Degallaix et al., Opt. Lett. **38**(12), 2047 (2013). doi:10.1364/OL.38.002047, http://ol.osa.org/abstract.cfm?URI=ol-38-12-2047
69. S. Kroker et al., Grating reflectors for future gwd (2012). https://tds.ego-gw.it/ql/?c=9328. Talk at the 4th ET general meeting
70. F. Brückner, D. Friedrich, T. Clausnitzer, O. Burmeister, M. Britzger, E.B. Kley, K. Danzmann, A. Tünnermann, R. Schnabel, Opt. Express **17**(1), 163 (2009). doi:10.1364/OE.17.000163, http://www.opticsexpress.org/abstract.cfm?URI=oe-17-1-163
71. F. Brückner, D. Friedrich, T. Clausnitzer, M. Britzger, O. Burmeister, K. Danzmann, E. Kley, A. Tünnermann, R. Schnabel, Phys. Rev. Lett. **104**, 163903 (2010)
72. S. Kroker, T. Kasebier, S. Steiner, E.B. Kley, A. Tunnermann, Appl. Phys. Lett. **102**(16), 161111 (2013). doi:10.1063/1.4802883, http://link.aip.org/link/?APL/102/161111/1
73. S. Kroke et al., Thermal noise in grating reflectors (2013). Talk at the GWADW 2013 conference
74. G.D. Cole, S. Groblacher, K. Gugler, S. Gigan, M. Aspelmeyer, Appl. Phys. Lett. **92**(26), 261108 (2008). doi:10.1063/1.2952512, http://link.aip.org/link/?APL/92/261108/1
75. A.C. Lin, M. Fejer, J.S. Harris, J. Cryst. Growth **363**, 258 (2013). doi:10.1016/j.jcrysgro.2012.10.055, http://www.sciencedirect.com/science/article/pii/S0022024812007701
76. G.D. Cole et al., Tenfold reduction of brownian noise in optical interferometry (2013). http://arxiv.org/abs/1302.6489. ArXiv:1302.6489 [physics.optics]
77. A.C. Lin et al., Prospects and progress in crystalline coatings: AlGaP (2013). Talk at the GWADW 2013 conference
78. S. Chelkowski, S. Hild, A. Freise, Phys. Rev. D **79**(12), 122002 (2009). doi:10.1103/PhysRevD.79.122002
79. B. Mours, E. Tournefier, J.Y. Vinet, Class. Quantum Gravity **23**, 5777 (2006)
80. L. Carbone, C. Bogan, P. Fulda, A. Freise, B. Willke, Phys. Rev. Lett. **110**, 251101 (2013). doi:10.1103/PhysRevLett.110.251101, http://link.aps.org/doi/10.1103/PhysRevLett.110.251101
81. R.A. Day, G. Vajente, M. Kasprzack, J. Marque, Phys. Rev. D **87**, 082003 (2013). doi:10.1103/PhysRevD.87.082003, http://link.aps.org/doi/10.1103/PhysRevD.87.082003
82. G. Vajente, R.A. Day, Phys. Rev. D **87**, 122005 (2013). doi:10.1103/PhysRevD.87.122005, http://link.aps.org/doi/10.1103/PhysRevD.87.122005
83. T. Corbitt, Quantum noise and radiation pressure effects in high power optical interferometers. Ph.D. Thesis, MIT 2008

84. T. Corbitt, N. Mavalvala, J. Opt. B: Quantum and Semiclassical Opt. **6**(8), S675 (2004). http://stacks.iop.org/1464-4266/6/i=8/a=008
85. A. Khalaidovski, J. Steinlechner, R. Schnabel, Class. Quantum Gravity **30**(16), 165001 (2013).http://stacks.iop.org/0264-1/30/i=16/a=165001
86. B.J. Meers, Phys. Rev. D **38**, 2317 (1988)
87. A. Buonanno, Y. Chen, N. Mavalvala, Phys. Rev. D **67**(12), 122005 (2003). doi:10.1103/PhysRevD.67.122005
88. A. Buonanno, Y. Chen, Class. Quantum Gravity **18**(15), L95 (2001). http://stacks.iop.org/0264-9381/18/i=15/a=102
89. A. Buonanno, Y. Chen, Phys. Rev. D **64**, 042006 (2001)
90. A. Buonanno, Y. Chen, Phys. Rev. D **65**, 042001 (2002)
91. A. Buonanno, Y. Chen, Phys. Rev. D **67**, 062002 (2003)
92. H.J. Kimble, Y. Levin, A.B. Matsko, K.S. Thorne, S.P. Vyatchanin, Phys. Rev. D **65**(2), 022002 (2001). doi:10.1103/PhysRevD.65.022002
93. T. Corbitt, N. Mavalvala, S. Whitcomb, Phys. Rev. D **70**(2), 022002 (2004). doi:10.1103/PhysRevD.70.022002
94. F.Y. Khalili, H. Miao, Y. Chen, Phys. Rev. D **80**(4), 042006 (2009). doi:10.1103/PhysRevD.80.042006
95. A. Buonanno, Y. Chen, Phys. Rev. D **69**(10), 102004 (2004). doi:10.1103/PhysRevD.69.102004
96. H. Grote, K. Danzmann, K.L. Dooley, R. Schnabel, J. Slutsky, H. Vahlbruch, Phys. Rev. Lett. **110**, 181101 (2013). doi:10.1103/PhysRevLett.110.181101, http://link.aps.org/doi/10.1103/PhysRevLett.110.181101
97. T. Eberle, S. Steinlechner, J. Bauchrowitz, V. Händchen, H. Vahlbruch, M. Mehmet, H. Müller-Ebhardt, R. Schnabel, Phys. Rev. Lett. **104**, 251102 (2010). doi:10.1103/PhysRevLett.104.251102, http://link.aps.org/doi/10.1103/PhysRevLett.104.251102
98. V.B. Braginsky, S. Strigin, S. Vyatchanin, Phys. Lett. A **287**, 331 (2001)
99. L. Ju, S. Gras, C. Zhao, J. Degallaix, D. Blair, Phys. Lett. A **354**, 360 (2006)
100. L. Ju et al., Phys. Lett. A **355**, 419 (2006)
101. L. Ju et al., Class. Quantum Gravity **26**, 015002 (2009)
102. K. Yamamoto, T. Uchiyama, S. Miyoki, M. Ohashi, K. Kuroda, K. Numata, Journal of Physics: Conference Series **122**(1), 012015 (2008). http://stacks.iop.org/1742-6596/122/i=1/a=012015
103. K. Yamamoto, Parametric instability of a cavity of Einstein Telescope (2009)
104. S. Hild et al., Classical and Quantum Gravity **27**, 015003 (2010). http://stacks.iop.org/0264-9381/27/015003
105. S. Hild, et al., Classical and Quantum Gravity **28**(9), 094013 (2011). http://stacks.iop.org/0264-9381/28/i=9/a=094013
106. O. Jennrich, Class. Quantum Gravity **26**(15), 153001 (2009). http://stacks.iop.org/0264-9381/26/i=15/a=153001
107. A. Freise, et al., Classical and Quantum Gravity **26**(8), 085012 (2009). http://stacks.iop.org/0264-9381/26/i=8/a=085012

Chapter 14
Low Temperature and Gravitation Wave Detectors

Fulvio Ricci

Abstract The success of a precision experiment is often associated to the use of low temperature techniques. In particular, when the thermal noise is a barrier for improving the experiment sensitivity, the cryogenics is crucial for beating this limitation. This strategy was applied in the case of the resonant gravitational wave detectors (GW) and now it is proposed for the future generation of the GW interferometers. In the following we summarize the history of GW detectors and we recall some of the basic principles of the cryogenic techniques. Then, we focus on the issues of cooling the mirrors of a GW interferometer.

14.1 Introduction

In the early 1960s, Joseph Weber initiated the experimental search for cosmic gravitational wave (GW) signals using massive aluminum cylinders with a fundamental longitudinal frequency of about 1 kHz and of a motion sensor converting the vibration of the bar into an electric signal [1].

In 1969, Weber announced the detection of signals in coincidence between two identical detectors set at a distance of nearly 1,000 miles [2]. The claim stimulated the construction and operation of new room temperature detectors, which gave no evidence that GW were being seen [3–6].[1] Despite of the controversial results, the scientific challenge of the direct detection of GW signals was launched and in 1970 William Fairbank at Stanford University proposed a new, much more sensitive

[1] Here we cite just few experiments among the several ones. A more detailed bibliography can be found in [7].

F. Ricci (✉)
Department of Physics, University of Rome La Sapienza and INFN Sezione di Roma, Piazzale Aldo Moro 5, I-00185 Roma, Italy
e-mail: fulvio.ricci@roma1.infn.it

M. Bassan (ed.), *Advanced Interferometers and the Search for Gravitational Waves*, Astrophysics and Space Science Library 404, DOI: 10.1007/978-3-319-03792-9_14,

detector, a five ton bar cooled to the mK range [8], which involve the use of cryogenics and superconducting techniques for noise reduction. W. Hamilton in Louisiana [9] and E. Amaldi and G. Pizzella in Rome [10] followed the same path starting to develop cryogenic detectors. The cryogenic antennas took data in coincidence more than two decade later and today two ultra cryogenic detectors that achieved a temperature of 0.1 K [11, 12], are in continuous data taking. Although we are still waiting for the first detection, these cryogenic systems improved by four order of magnitude the energy sensitivity over Weber's original antennas.

Since several years now, the sensitivities of the kilometer scale interferometers surpassed the sensitivity limits of the cryogenic bars [13], over a larger frequency band. At present we are upgrading the interferometer configuration and we expect (on the base of robust theoretical predictions) that these instruments in their advanced configuration will detect several GW events during the next quinquennium. In this scenario, even the study of a third-generation detector was started (see Chap. 13): it was focused on the realization of a GW observatory ten times more sensitive so that, at this level of sensitivity, it will turn the detection into a routine astronomical tool. The design study of the Einstein Telescope observatory (ET) has been recently completed, setting the specifications for the site infrastructure and the necessary technologies. Cryogenics is considered again one of the main technologies to be applied for beating the thermal noise limit achieved by the advanced detectors.

In addition, the construction of a new 3-km interferometer, KAGRA, has started in Japan: it is based on the use of cryogenics in an underground laboratory. Thus, the interest in all the aspects of the use of low temperature techniques for a GW detector is grown again.

The advantages of lowering the temperature T in a GW interferometer are summarized in the following list:

1. Test mass and suspension thermal noise are reduced at low temperature (see Chap. 8), being the displacement power spectrum of the mirror $S_{xx} \propto T$.
2. Thermoelastic noise in both the mirror substrates and in coatings decrease as T^2, because of the decrease of the thermal expansion coefficient.
3. The thermorefractive noise associated with the fluctuation of the refraction index decrease as T^2.
4. The acoustic losses of several materials characterized in terms of the loss angle Φ decrease at low temperature. Thus, the noise power spectral of the mirror displacement is expected to decrease as Φ.

Finally, we stress another important advantage: the thermal lensing is drastically reduced for two reasons:

1. Refraction index variation with temperature is very small at low temperature,
2. Generally, the thermal conductivity increases and consequently reduces thermal gradients on the mirror.

In Fig. 14.1, we show the expected sensitivity curves of the advanced detectors and a tentative curve of the ET sensitivity. The grey box indicates the spectral range where

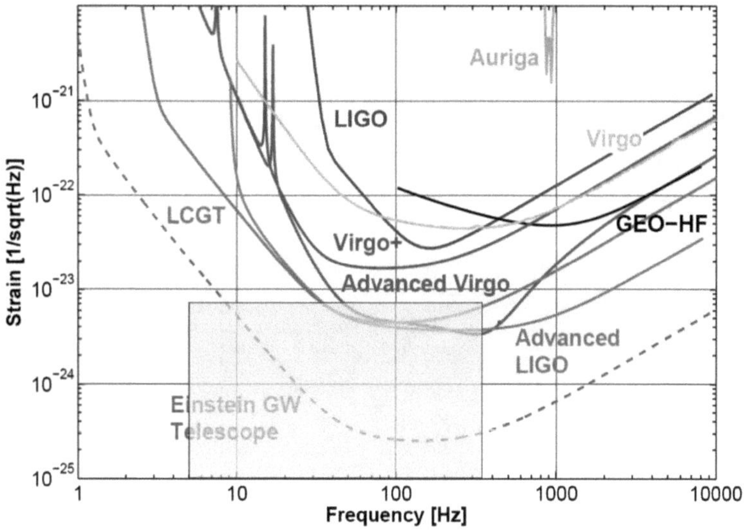

Fig. 14.1 Sensitivities of gravitational wave detectors from the first to the third generation. The *grey box* indicates the spectral range of the sensitivity dominated by the thermal noise

the dominant source noise is the thermal one. The main improvement is obtained by cooling the suspended mirror at low temperatures.

As we said before the use of low temperature for operating the GW detectors was proposed by W. Fairbank, who used to say...

Any experiment is better, if it is done at low temperature.

However, in case of a GW interferometer we are facing the challenge of cooling untouchable super-mirrors of more than 100 kg, suspended by thin crystal fibers in ultra high vacuum and in absence of any element polluting its surface. For this reason we also tend to agree with Bill Hamilton, who used to add to Fairbank's statement (Fig. 14.2):

Any experiment will be harder, if it is done at low temperature.

In the following sections, we introduce few fundamental concepts of cryogenics and then we discuss the main problem to be solved for cooling the super-mirrors of the ET detector without spoiling the stringent requirement of a GW interferometer.

14.2 Cryogenics

Research and development into the field of cryogenics began during the nineteenth century with the liquefaction of common gases as O_2 and N_2 (see Fig. 14.3). However, the great step forward, the storage of cryofluids, occurred in 1892 when sir James Dewar, a Scottish chemist and physicist, invented a double-walled container with a vacuum insulation flask. This was the crucial element that led the way to liquefy

Fig. 14.2 On the left William Fairbank in the earlier 1970s. On the right William O. Hamilton these days

Fig. 14.3 Chronological development of the cryogenic techniques up to 1 K

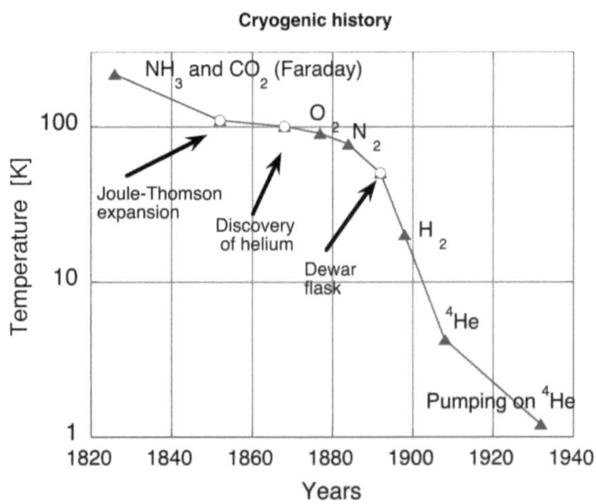

helium, a liquid with an extremely low latent heat. Kamerlingh Onnes was able to attain the goal of the first successful liquefaction of helium in 1908. In the *old days* of low temperature physics, the production of liquid helium was a painful process. The experimentalist had to start in the early morning with liquid nitrogen ($T_{N_2} = 77$ K) to produce hydrogen ($T_{H_2} = 20$ K), by means of which it was possible to obtain at the end of the day the liquid helium ($T_{He} = 4.2$ K). Then, pushed by the need to produce liquid oxygen and hydrogen used as propellants for missiles, in the 1940s commercial development of gas liquefiers began. Samuel Collins designed a machine providing helium without the need to liquefy hydrogen and even without nitrogen. Today nitrogen and helium cryogenics has progressed and a variety of liquefiers exist for small or large liquid helium production.

Cryogenics has many applications. Cryogenic liquids, such as oxygen, nitrogen, and argon, are often used in industrial and medical applications. Taking advantage of the reduction with temperature of the vapor tension of gas, the cryogenics is used for producing high vacuum. Other applications include fast freezing of some foods and the preservation of biological materials such as human blood, tissue, and embryos. Moreover, the electrical resistance of most metals decreases as temperature decreases. Certain metals lose all electrical resistance below some transition temperature and become superconductors. An electromagnet wound with a wire of such a metal can produce extremely high magnetic fields with no generation of heat and no consumption of electric power once the field is established and the metal remains cold. These metals, typically niobium and titanium alloys cooled to 4.2 K, are used for the magnets of magnetic resonance imaging (MRI) systems in most hospitals. The industrial application of cryogenics can be synthesized in various categories of applications

1. Storage of gases
2. Vacuum systems
3. Low noise detectors and superconductors
4. Biological and medical applications.

Cryogenic fluids are also used in a variety of applications for research. Among the large-scale applications of cryogenics, the most impressive is by far the cryogenic plant set for the Large Hadron Collider at CERN, a proton-proton accelerator fully equipped with cryogenic magnets cooled at 1.9 K along the entire particle ring 27 km long.

In the case of small-scale applications of cryogenics, more and more often the use of cryogenic refrigerators (cryocoolers) is preferred to cryofluids. Cryocoolers are required for a wide variety of applications, and the number of applications keeps expanding as improvements to cryocoolers technology are achieved. One of the earliest applications, appeared about 50 years ago, was for cooling infrared sensors to about 80 K for night vision capability of the military. Then, they have been in widespread use for tactical military applications to meet ground, airborne, and shipboard infrared sensor cooling needs. Today the list of targeted applications is impressively broad: it includes IR detectors, low noise amplifiers, front-end passives in communication systems, SQUIDs, MEMs and next generation nanoscale devices.

While there are several types of cryocoolers, most operate on some variation of a common process. Gas is circulated through a closed cycle to absorb heat from the interior of the device and transfer it to the outside environment. This gas may be helium or some other mixture of gases. The ability of the device to cool its interior environment depends largely on the thermodynamic properties of the gas circulating through the system. Cooling cycles in these devices begin with gas being sent to a compressor. As the compressed gas passes through a heat exchanger, it absorbs heat from the cryocooler, thereby cooling anything connected to it. When this gas absorbs heat at a constant volume in the heat exchanger, its pressure increases. It expands in volume and its pressure decreases in the next portion of the cycle. Finally, it returns to the compressor, which completes a closed loop through the cycle, and begins to circulate through the cycle again.

Common types of cryocoolers include the Joule–Thomson cooler, the Gifford-McMahon and the Stirling cooler, the pulse tube (PT) refrigerator, etc. Although less efficient than other devices, Joule–Thomson cryocoolers provide advantages in reliability and low electrical and mechanical noise levels,[2] while the more efficient coolers generate vibration due to the piston movement in the cold environment.

In the case of the mirror cooling for a GW interferometer we will analyze both methods to bring the mirror at low temperature, i.e.,

1. the first is based on the use of cryofluids
2. the second method relies on a battery of PT cryocoolers, which are more efficient than the Joule–Thomson systems and weaker source of mechanical noise than the traditional McMahon or Stirling machines.

14.2.1 The Cryofluids

The simplest way to produce low temperature is still the use of cryoliquids (e.g., nitrogen, helium). Liquid oxygen and hydrogen are rarely used in refrigeration: the former because of its high chemical reactivity, the latter again for its flammability. Both cryoliquids find also important applications in the space activity and in the industry. In the following table, we summarize some of the thermodynamical properties of cryofluids.

For our purpose, helium is the most interesting element and the most used in cryogenics after nitrogen. There exist two stable isotopes of helium, both of strong cryogenic interest (Table 14.1):

1. ^{4}He(nuclear spin $I = 0$). The ^{4}He content of near-surface atmospheric air has been determined to be 5.2204 ± 0.0041 parts per million by volume. To get it for industrial use ^{4}He is found in the natural gas field (mainly in Russia and the USA) with a concentration of $\sim$7 %.
2. ^{3}He (nuclear spin $I = \frac{1}{2}$). The ^{3}He/^{4}He ratio for atmospheric helium is established as 1.393×10^{-6}. By far the most common source of ^{3}He is the nuclear weapons program, of which it is a byproduct. In fact, the nuclear weapons program produces tritium which decays into helium-3.[3]

---

[2] The Joule–Thomson cryocoolers are based on an adiabatic expansion of the gas through an impedance orifice or a valve. This procedure is called a throttling process, and it permits to cool the gas when it is kept below its inversion temperature.

[3] During the cold war, the USA, Russia, and other countries stockpiled tens of thousands of nuclear weapons, and in doing so accumulated vast amounts of ^{3}He. Initially, the National Nuclear Security

Table 14.1 Thermodynamical properties of the two helium isotopes

Properties	^{3}He	^{4}He
Density [g/cm^3] (T = 0 K)	0.0863	0.146
Boiling temperature T_b (K)	3.191	4.215
Critical temperature T_c (K)	3.324	5.20
Critical pressure (bar)	1.165	2.289
Critical density (g/cm^3)	0.0413	0.0693
Melting pressure (T = 0 K) (bar)	29.31	25.32
Maximum superfluid transition temperature (K)	0.0025	2.177
(Gas at 300 K and 1 bar)/(Liquid at 1K) volume ratio	662	866

^{4}He plays a crucial role for achieving temperatures down to ~1 K. Below it ^{3}He or a mixture of both liquids is used.

Liquid helium is notoriously difficult to handle. If you do not use the correct equipment or techniques, you can easily lose all the liquid, because of the very low latent heat of vaporization. For example, a heat load of 1 W will evaporate 1.41 of liquid per hour. Moreover, we have to stress that the latent heat of evaporation L_{ev} and the vapor pressure p_{vap} are fundamental parameters when using a cryo-liquid in the refrigeration process. The evaporation process provides a refrigeration power W_{ev}, which depends on L_{ev} and p_{vap}. However, looking at Table 14.2, we note that the value L_{ev} = 2.56 kJ/l for ^{4}He is very small in comparison, for example, with that of hydrogen (31.8 kJ/l) or of nitrogen (160 kJ/l). A low value of L_{ev} means a small cooling power, a serious drawback when cooling by evaporation. To increase the refrigeration efficiency of the cooling process, the enthalpy of the cold gas warming from 4.2 K to room temperature is used. The curve of the ^{4}He enthalpy versus temperature is shown in Fig. 14.4.

For example, on the cooling process of a 100 g of Cu, the full use of the ^{4}He enthalpy reduces at just 0.081 the amount of the liquid helium, i.e., 1/30 times the amount needed using the ^{4}He latent heat only. In practice, if the material is cooled by cold helium gas flowing over it, the amount of liquid helium is reduced by a following theoretical factors:

1. 32 in the range 300–4.2 K
2. 10 in the range 77–4.2 K

When the liquid helium-4 and its vapor are cooled to 2.177 K, known as its λ point, a third phase forms still liquid. The new liquid *He-II* showsthe remarkable

(Footnote 3 continued)
Administration (NNSA) and its predecessor agencies, which have maintained the US tritium stockpile, used to consider the gas useless and they considered to vent it into the atmosphere. In the 1980s, however, scientists began to realize the potential of ^{3}He as a neutron detector and the price skyrocketed.

Table 14.2 Boiling and melting temperature at 1 bar of gas and their latent heat

Substance	T_b Boiling temperature at 1 bar (K)	T_m Melting temperature at 1 bar (K)	Latent heat at T_b L_{ev} (kJ/l)
H_2O	373.15	273.15	2252
NH_3	293.8	195	–
CO_2	194.6	216	–
Xe	165.1	161.3	303
Kr	119.9	115.8	279
CH_4	111.18	90.8	–
O_2	90.2	54.4	245
Ar	87.3	83.8	224
N_2	77.4	63.3	160
H_2	20.3	14	31.8
^4He	4.21	–	2.56
^4He	3.19	–	0.48

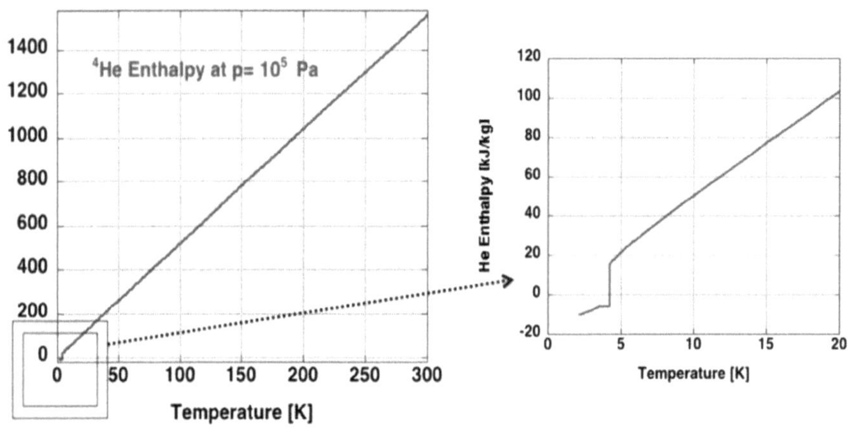

Fig. 14.4 The enthalpy of the ^{4}He gas in the range 2–300 K

property known as superfluidity.[4] It flows with practically no friction through small openings and narrow capillaries and if we plot the volume rate flowing in a capillary versus temperature we note the striking change in the curve slope crossing the λ point. The flow of helium II acts as if it were a mixture of two fluids [14–17]. One, called the superfluid, has no viscosity or entropy and can flow without dissipation through extremely narrow channels. The other, called the normal fluid, does have a finite viscosity and carries all the entropy. The liquid is pictured to be all normal fluid at the lambda transition and all superfluid at absolute zero.

Above the λ-point, ^{4}He shows transport properties similar to those of a classic fluid; it has a low thermal conductivity (a factor of 10^{-4} compared to Cu and a factor

[4] The liquid phase above the λ point is named *He-I*.

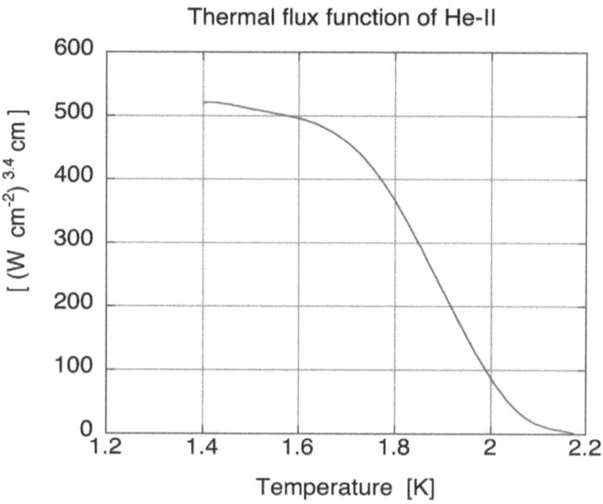

Fig. 14.5 Thermal flux function of the liquid helium $\dot{Q}^n L$ below 2.2 K versus temperature

of 10^{-1} compared to stainless steel) and boils with strong bubbling. In particular, when pumping on *He-I*, bubbles of vapor form inside the liquid, which is warmer than the surface cooled by the evaporation process.

Below 2.17 K, the properties of heat conduction in the liquid helium drastically change. The steady-state heat flux $\dot{Q}$ in a tubular conduit of length L, the ends of which are maintained at temperatures T and T_w, is expressed by the empirical relation

$$\dot{Q}^n = \frac{X(T) - X(T_w)}{L} \quad (14.1)$$

where $n = 3.4$ is the exponent value derived from the best experimental fit, and $X(T)$ is a function of temperature shown in the Fig. 14.5, physically analogue to a conductivity integral [18]. On the base of this information, it is possible to show that the highest heat flux per unit temperature is maximized near 1.9 K.

As an example, the heat flux transported by conduction between 1.9 and 1.8 K in a helium tube one meter long is ~1.2 $\mathrm{Wcm^{-2}}$, which corresponds to a specific heat flux per unit surface three orders of magnitude higher than that through a similar bar made of OFHC copper. Because of the very high thermal conductivity, the temperature gradient in the *He-II* bath is very close to zero, and for this reason ***bubbles do not form in the liquid*** during the evaporation process.[5]

The absence of boiling in *He-II* is a great advantage in a GW cryogenic detector. In fact, the long history of the cryogenic GW experiments can be told as a never ending fight for reducing the residual vibration associated to the boiling of the cryofluids. We just recall here the case of the Explorer antenna installed at CERN [19]. In the plot we show in the vertical axis the detector output filtered for enhancing the sensitivity

[5] The high thermal conductivity allows also for the propagation of temperature waves (called *second sound*), which are reflected and diffracted as standard waves, but which give rise to temperature variations in small regions.

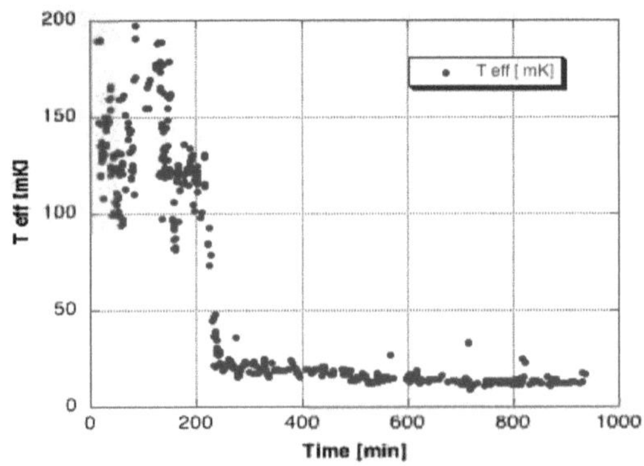

Fig. 14.6 The output of the GW detector Explorer when we were pumping on the ^{4}He bath. The sharp change is in coincidence of the λ point crossing

to the sudden change of the vibration energy of the detector. On the horizontal axis, we show the time referred to a period during which we were pumping on the liquid helium bath of ∼3000 l and decreasing its temperature from 4 K down to 1.5 K. In Fig. 14.6, we notice the striking reduction of the detector output observed during and after the λ point transition, when the boiling noise disappeared.

Noise associated with the boiling of the cryofluids was experienced by other resonant GW detectors and it can be also a potential problem for the third generation of GW interferometers. In fact, some disturbances associated to the boling process of cryofluids has been observed already during the refilling phase of the N_2 cryotraps mounted both in LIGO and VIRGO.

In a design phase of the detector, there is the need to quantify the effect of the boiling on the GW test masses. Our need is to infer the acceleration (or displacement) noise induced on the wall of the test mass container. The noise spectral density of the acceleration due to the boiling process depends on the boiling rate and on the geometry and the material of the container. For example, in the case of the GW resonant antenna Explorer, we evaluated the order of magnitude of the random displacement of the wall of the helium container, a stanley steel cylinder of ∼1.6 m diameter and ∼4 m long. The r.ms. value of the displacement noise, peaked around 900 Hz, was of the order of 10^{-10} m at 4 K when the boiling rate of the liquid helium was ∼2 l/h.

Since the power spectrum of the boiling noise is mainly concentrated in the acoustic range, it can be measured using sound level meters (see Fig. 14.7). The signal output of these microphones, proportional to the pressure change detected on the microphone surface, is calibrated to give the relative sound intensity I_r in $\mathrm{dB_{SPL}}$ units:

$$I_r\ [\mathrm{dB_{SPL}}] = 10 \times \log(I/H),$$

where I is the sound intensity (usually measured in $\mathrm{W/m^2}$) and $H = 1.0 \times 10^{-12}\ \mathrm{W/m^2}$. The quantity H is an intensity reference value corresponding to the earing threshold of a normal ear in good health. This should correspond to a hear

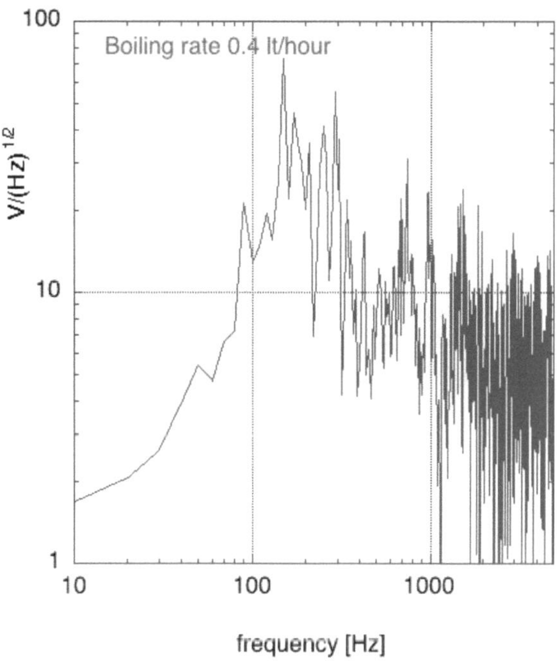

Fig. 14.7 Pressure noise spectrum of a boiling fluid in a cylindric container of 10 cm diameter and 30 cm height, made of steel. The boiling rate was ~0.4 l/h. In the *vertical axis* we show the spectral density of the amplified output voltage of the microphone and on the x-axis the frequency. The voltage/pressure factor was 6.3×10^{-2} Pa/V

pressure change $\Delta p_0 = 20\,\mu$Pa. Under this assumption, the microphone noise can be translated also in terms of a pressure change being

$$L_{SPL} = 10 \times \log(I/H) = 20 \times \log(\Delta p_1/\Delta p_0).$$

Then, the pressure fluctuation is interpreted as a stress field applied to the walls of the test mass container, provided that a correction factor, which takes into the account the distance and directionality of the microphones with respect to the container wall, is included.

Unfortunately this is not the end of the story. The companies producing audio systems calibrate often their devices in dBA units. To give an idea of the sound level in dBA, we report here a couple of common cases:

- a whispering level of noise is ranging around 20 dBA, which corresponds to $\sim 2 \times 10^{-4}$ Pa,
- a bird twitter outside at 15 m distance is quoted around 50 dBA corresponding to $\sim 6.3 \times 10^{-3}$ Pa.

The letter *A* following the abbreviation dB designate a frequency-response function that filters the sounds picked up by the sound level meter. The filter shape is not standard and it can change from microphone producer to producer. The consequence is that there is not a conversion formula for measured dBA values to sound level in

dB_{SPL} or vice versa and the lesson is that often this kind of measurements gives just the order of magnitude of the acoustic disturbance.

In practice, the most reliable approach for inferring the sound noise contribution to the noise budget of a GW detector is to put in contact with the container wall a full set of low temperature accelerometers, which monitor the random movement induced by the boiling (Fig. 14.7).

The quietest scheme for cooling the test mass of a GW detector is the *Claudet bain*: it permits to use super fluid helium at atmospheric pressure and to insure continuous refilling from the container of the helium in the normal state

A ^{4}He bath at 4 K at atmospheric pressure flows gently through a high impedance porous filter acting also as acoustic filter to a second ^{4}He bath kept in thermal contact with a cell unit cooled below the 2 K. In this way, the *He-II* bath is in a quiet hydrodynamic status well far from the boiling point and with enhanced heat removal capability. Moreover, since *He-II* is modeled as a mixture of two fluids flowing with velocities of opposite sign (the superfluid and the normal component), it should be possible to operate the cryo system in the state of zero mass flow [20]. This implies that, in absence of the vapor phase, we have

$$\rho_s v_s = -\rho_n v_n \tag{14.2}$$

where ρ and v are the density and the velocity of the normal (n) and superfluid (s) components.

In this condition, in principle the test mass can be in a direct thermal contact with a cold element filled by superfluid helium at atmospheric pressure. However, the fluid still transmits sound waves and we should limit the acoustic effect of the cooling fluid on the test masses. Therefore, this approach requires also a deeper analysis of the impact of the acoustic losses of the fluid to the dynamic behavior of the suspended test masses [21].

In conclusion, as we will discuss later, the simpler and safer approach will be to fill a copper box with He-II at atmospheric pressure and bring the refrigeration power to the mirror through its suspension wire made of a crystalline material.

14.2.2 The Cryocooler

The vapor-compression refrigerator, first developed in the latter part of the nineteenth century, in a single stage produced temperatures down to about 230 K. Since the middle of the twentieth century a steadily increasing demand for cryogenic temperatures below 120 K has developed for a wide variety of applications. Initially most of these applications required to develop very large plants with power inputs of many megawatts. Then, a need for small cryogenic refrigerators (cryocoolers) has evolved triggering an impressive research and development effort during the past 40 years (see Fig. 14.8). One such cryocooler, the pulse tube (PT) refrigerator, first conceived in the mid-1960s, was of academic interest until the mid-1980s. At the Moscow

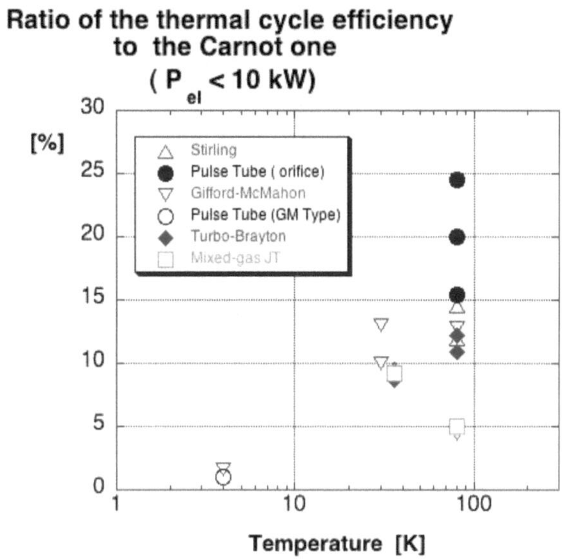

Fig. 14.8 Ratio between the thermal cycle efficiency of refrigeration and the corresponding Carnot cycle efficiency versus temperature

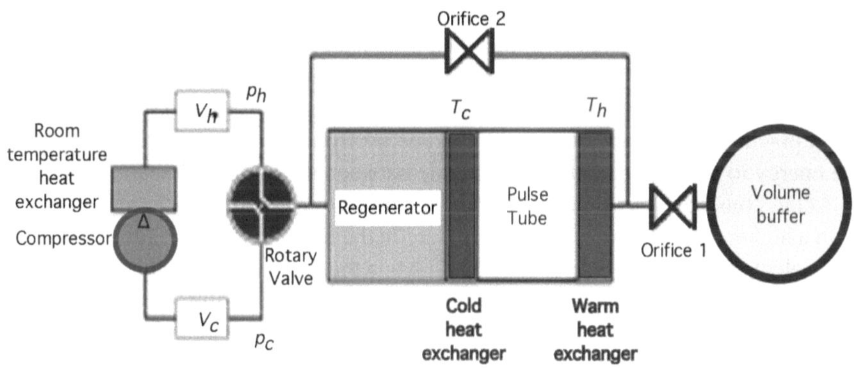

Fig. 14.9 Scheme of a pulse tube cryocooler

Bauman Technical Institute in 1984 Alexander Mikulin et al. [22] introduced an orifice inside the pulse tube near the warm end and achieved a low temperature of 105 K. Since then, improvements in its efficiency have occurred rapidly.

Unlike the Stirling or Gifford-McMahon refrigerators, the PT one has no moving parts at the cold end (see Fig. 14.9). Therefore, no expensive high-precision seals are required and the cold head can be operated without any service inspection. In this way, some of the problems associated with cryocoolers in many different applications, such as vibration and reliability are attenuated (Table 14.3).

Subsequent modifications and improved designs yielded much improved efficiencies.

Table 14.3 Progress of the pulse tube technology

Year	PT type	Final temperature (K)	P_{cool} (W)	P_{el} (W)
1993	Single stage	80	0.8	33
1994	Single stage	77	133	3400
1995	Double stage	4.2	0.5	6300
2004	2 double stages for ^4He and ^3He	1.3	0.04 (@2 K)	4500+1300

P_{cool} is the refrigeration, P_{el} the electric power of the compressor

Pushed in large part by NASA and the military, the efficiency figures, i.e., the ratio P_{cool}/P_{el} are still improving, thanks to the effort of the research laboratories and few commercial companies.

The Pulse Tube cryocooler (PT) is based on the displacement and the expansion of a gas, usually He^4. The working fluid undergoes an oscillating flow and an oscillating pressure. A typical average pressure in a PT is 25 bar, and a typical pressure amplitude range between 2 and 7 bar.

A piston compressor and a set of switching valves are used to create the pressure oscillations of the helium gas. A crucial element of the cryocooler is the PT regenerator. It is a heat exchanger, which acts as cold storage systems for the pulsed flow processes used in the cryocooler.

The piston moves periodically pushing the gas, which enters the regenerator with temperature T_h and leaves it at the cold end with temperature T_c, hence heat is transferred into the regenerator material. On its return the heat stored within the regenerator is transferred back to the gas.

In practice, the regenerator stores the energy from one stream and later transfer the energy to a second stream, for example between the out-of-phase pulses of gas.

Let us consider a gas element entering the tube with temperature T_h and leaving it with a higher temperature when the pressure in the tube is low. Later in the cycle the same element is pushed out the tube again when the pressure inside the tube is high. As a consequence, its temperature will be higher than T_h. In the heat exchanger, it releases heat and cools to the ambient temperature T_h. At the cold end of the pulse tube, there is the opposite effect: here gas elements enter the tube when the pressure is high with temperature T_c and return when the pressure is low with a temperature below T_c. The heat is extracted from the cold point, providing the desired cooling power.

The thermal environment of a gas element, that moves back and forth in the system, changes when it passes the heat exchanger. In the regenerator and in the heat exchanger, the heat contact between the gas and its surrounding material is good. Here the temperature of the gas is practically the same as that of the surrounding medium. However, in the pulse tube the gas element is thermally isolated, so, in the pulse tube, the temperature of the gas element varies with the pressure.

As we said before, Mikulin and his co-workers [22] inserted a flow resistance (the orifice) at the warm end of the pulse tube to allow some gas to pass to a large reservoir acting as buffer volume. This is typically 10 times larger than the volume of the pulse tube and its pressure is almost constant and close to the average pressure in the pulse tube. The combination of the orifice and the buffer provides a phase difference between the flow of the gas in the tube and the pressure oscillation; such phase difference is necessary for the performance of the PT.

In 1990, Zhu et al. [23] connected the warm end of the pulse tube with the main gas inlet by a tube, containing a second orifice. Thus, a part of the gas could enter the pulse tube from the warm end, bypassing the regenerator. The function of the second orifice is to reduce losses. It allows some gas to bypass the regenerator and to enter the pulse tube directly.

The quality of the regenerator material sets in practice the cryocooler performance. It consists of a porous matrix of finely divided material in the form of wire mesh, plates, or small balls. These forms maximize the surface area and minimize the thermal conduction necessary within the solid to maximize the heat transfer with the surrounding gas within the duration of a cooling cycle. The regenerator stores the heat of the gas during a half-cycle and, therefore, it must have a high heat capacity, compared to the heat capacity of the gas.

At temperatures above 10 K materials as bronze or stainless steel are often used. Below 10 K one uses rare earth materials, which are specially developed for this application.

To obtain temperatures below 20 K, the system is operated at low frequency ($\nu \sim 1$ Hz). The frequency defines the diffusion depth δ in the working gas and the regenerator material:

$$\delta = \sqrt{\frac{k}{\pi \nu C}} \tag{14.3}$$

where k is the heat conductivity, ν is the frequency and C is the volumetric heat capacity of the regenerator material. It follows that when the frequency is increased, the diffusion depth decreases, and the heat storage in the regenerator degrades. Moreover, a high operating frequency leads to a large pressure drop in the regenerator, i.e., a poor performance of the system.

A pulse tube cryocooler seems to be suitable for cooling the mirrors of the GW interferometer. However, also this kind of refrigeration system injects mechanical noise because of the gas pulse flowing in its cold head.

We measured the acceleration of the 4 K cold head of the Pulse Tube 407 of CRYOMECH and that of the Sumitomo SRP-052A (see Fig. 14.10).

It results that the noise produced by this kind of refrigerators is still too high for cooling the GW test masses so that it is necessary to design a cooling system, which can attenuate the PT cryocooler vibrations. An active system to reduce the residual vibrations associated to the gas pulse is the so-called Vibration Free Cryostat (VFC) cryostat [24]: it was conceived to host a suspended mirror cooled by a PT cryocooler.

The cryostat scheme is sketched in Fig. 14.11. It is based on the idea to attenuate the cryocooler vibrations by directly acting on it and consequently allowing a shorter heat link between the cryocooler and the mirror.

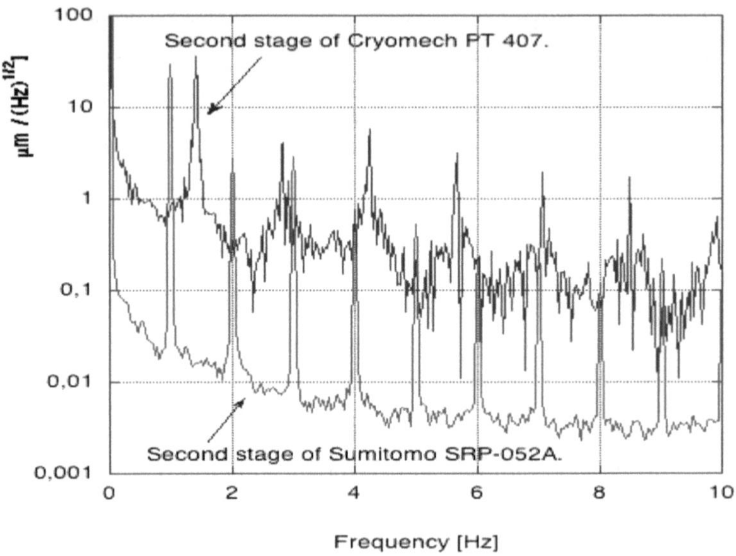

Fig. 14.10 We show the displacement noise spectrum of the coldest point of the CRYOMECH PT407 obtained by by an accelerometer for low temperature applications. The similar data taken on Sumitomo SRP-052A are obtained using an optic fiber sensor

The active control on the cold head vibration is performed by a digital feedback system in which the cold head displacement are monitored by a low temperature sensor. The sensor output provides the error signal for the correction, which is performed at room temperature by piezoelectric actuators.

The cryocooler cold head is clamped to a platform placed on dampers and it is connected to the cryostat by means of an elastic bellow designed to mechanically decouple the head and the platform from the lower part of the cryostat connected to the ground. The feedback correction signal is sent to three piezo-actuators which are loaded by the platform and can act on it from below. The attenuation provided by this system is of the order of 200. Then, in a GW detector, the cold head will be connected to the test mass via flexible heat links insuring an other signifiant insulation factor.

However, for achieving the ET sensitivity goal a R&D activity on a further reduction of the PT vibration is highly recommended. The pulsed force due to the gas deforms the tube of the cold head displacing the cold plate of several micrometers. This effect has been simulated via finite element software showing that, depending on the the cold head geometry, the longitudinal expansion of the tube can be of the same order the horizontal displacement. Thus, there is still a large margin of improvement changing the geometry of the cold head and/or enhancing both the active and passive attenuation damping of the gas pulse effects.

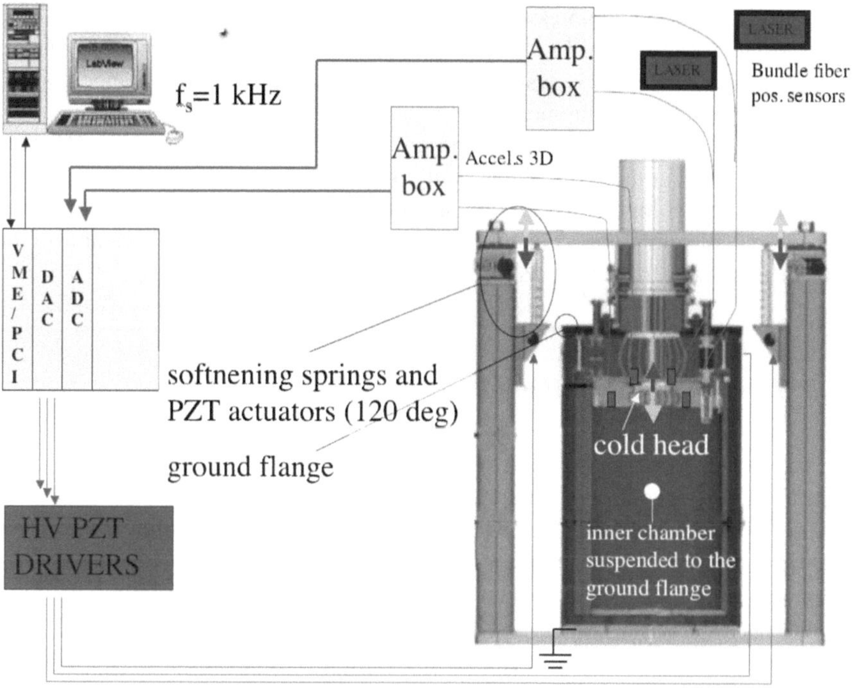

Fig. 14.11 Sketch of the vibration free cryostat

14.2.3 Cooling Strategy Comparison

The two methods for cooling the mirrors have both advantages and drawbacks. The cryocoolers insure a longer duty cycle for the GW observatory and they require less maintenance and manpower.

Moreover, the use of cryocoolers in an underground environment significantly relaxes all the safety issues associated with the use of cryofluids in an ambient with limited ventilation.

On the other hand, the main limitations of this approach are

- the low refrigeration power that implies the use of a battery of cryocoolers and associated compressors to be installed in the underground laboratory not too far from the test masses,
- the low efficiency, which requires a large amount of electric power to be distributed in the laboratory,
- the residual vibrations of the cold heads and the acoustic noise transmitted along the high pressure helium gas lines.

The use of cryofluids implies the construction of a cryoplant outside the underground laboratory surface: it will be the production center of liquid helium hosting the liquefiers and the compressors. Then, the liquid helium will be transferred to the cryostats using special low loss lines deployed from the surface to the underground interferometer. This system has a higher cost, needs more manpower and more stringent safety requirements.

The KAGRA project has already chosen the cooling strategy based on PT: they are confident that at the level of sensitivity of the advanced detectors it is possible, with present technology, to damp enough of the residual vibration of the cryocoolers. The entire GW community will follow their experience, which will orient also the final technical design of the Einstein Telescope detector.

14.3 The Cryogenic Payload

The last suspension stage, simply called the *payload*, is suspended from a long series of pendula kept at room temperature, for reducing the seismic and acoustic disturbances (the super attenuator). The payload is designed to compensate the residual seismic noise and to steer the mirror through internal forces exerted from the last element of the super attenuator. The main components of the payload are: the Marionette, the Recoil Mass, and the Mirror (see Chap. 7). The marionette is the first stage used to control the mirror position by means of coil-magnet actuators acting between the last stage upper part suspension and the marionette arms. The recoil mass[6] is used to protect and to steer the mirror by means of the coils of the electromagnetic actuators acting on four small magnets mounted on the mirror back side.

The mechanical losses of the suspended mirror determine the thermal noise contribution of interferometer sensitivity because of the well-known relation between mechanical dissipations and thermal motion in macroscopic systems (see Chap. 8). The parts in contact with the mirror are made of materials having intrinsic low mechanical dissipation; in particular, the suspension wires are attached by means of low friction mechanical clamps. Cryogenic operation introduces additional difficulties, but the benefit in thermal noise reduction can be enhanced by the selection of materials with improved properties at low temperatures. Moreover, to take advantage of the low temperature, the design of a last stage suspension should satisfy a couple of new requirements in addition to those of a room temperature payload:

1. the refrigerating power transmission to the mirror level should not deteriorate the position control performance;
2. the residual vibration transmitted to the mirror through the cryogenic system, should be well below the detector noise sensitivity curve.

As a consequence, a good mechanical isolation between the mirror and the cooler is necessary. Another important element in a cryogenic system is the thermal link

[6] The idea of using a reaction mass to control a suspended mirror has been originally pursued by the GEO group [25].

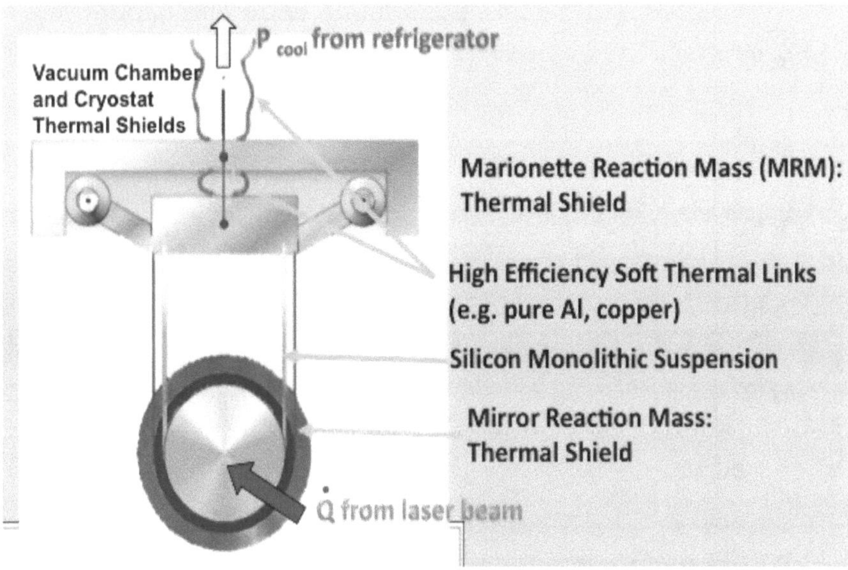

Fig. 14.12 Scheme of the cryogenic payload

with the refrigerator. Taking care of having short thermal links and good thermal couplings is important to have low refrigerating power losses. For instance, the use of a mechanical attenuator as a heat link is a good solution to damp the vibrations of the cooler, but has the disadvantage to limit the heat flow. On the other hand, the mirror suspension itself must have the required thermal conductivity to reduce thermal gradients and to optimize the cooling time.

The requirements we have just described suggest that the new system will be the result of a trade-off between the need to have a strong thermal contact with the refrigeration apparatus and the need to reduce any dissipation source due to the coupling of the suspension with the mirror. Thus, the materials used for building a cryogenic payload must have high thermal conductivity as well as good mechanical and optical properties at low temperature. Crystalline materials are preferred at low temperature for their thermal capacitance properties and in particular sapphire and silicon appears to be the best candidates, both for the mirror and for its suspensions. Indeed they have high thermal conductivity, low thermal expansion rate and very low mechanical losses at cryogenic temperatures. If test masses of large size are required, as in the ET case, then silicon is preferred.

In the Fig. 14.12 of the payload scheme, it is shown how a cryogenic suspension should work during the interferometer operation, i.e., when the payload has been cooled down and the laser is turned on. The fraction of the laser power, which is absorbed by the mirror ($\dot{Q}_{\rm laser}$), flows through suspension wires; then it crosses the points where the suspension wire clamps are set and reaches the refrigerating system.

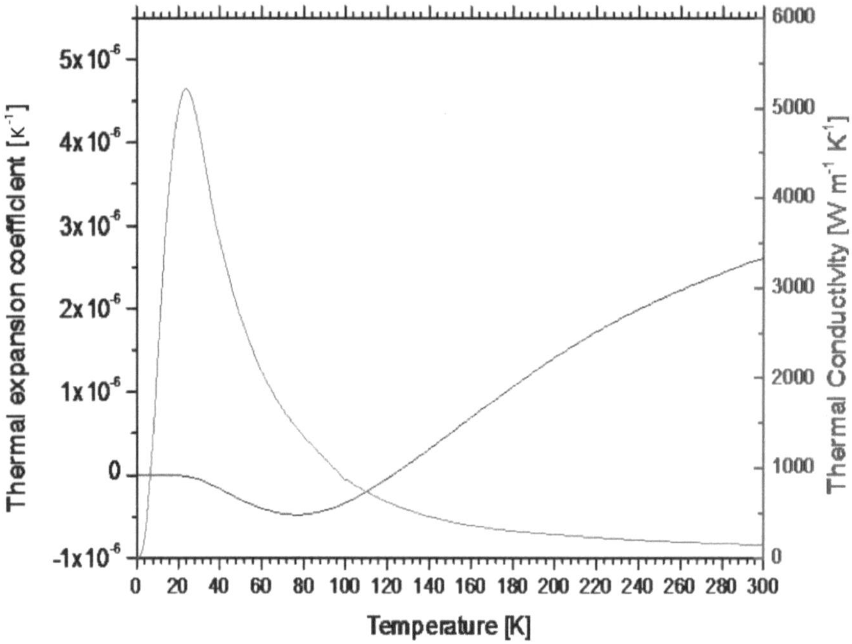

Fig. 14.13 Thermal properties of silicon. The *red curve* represents the thermal conductivity, the *black one* the thermal expansion coefficient, a crucial properties of the material for reducing the thermal lensing effect in the optical element

To this aim, the marionette has to be connected with very efficient, soft heat links to the cooling power.

In a similar way as for the mirror, the reaction mass is cooled down via its suspension wires and, if there is not extra thermal input, it reaches the thermal equilibrium and acts as a thermal shield for the mirror. The choice of the wires for the recoil mass must fulfill the request of having a good mechanical and thermal performance. For this reason, the use of a silicon suspension also for this part is the most obvious option (Fig. 14.13).

14.3.1 Payload Thermal Input and Heat Extraction

The main heat inputs into the cold mirror of the interferometer are the thermal radiation coming from the warm surface of the kilometric vacuum tube hosting the laser beam of the interferometer. To this, we have to add it the heat load due to the absorption of a small fraction of the laser light by the mirror surface (see Fig. 14.14). The latter is estimated considering that the laser power circulating in the optical resonant cavity of the interferometer. For example, let us assume 1 ppm

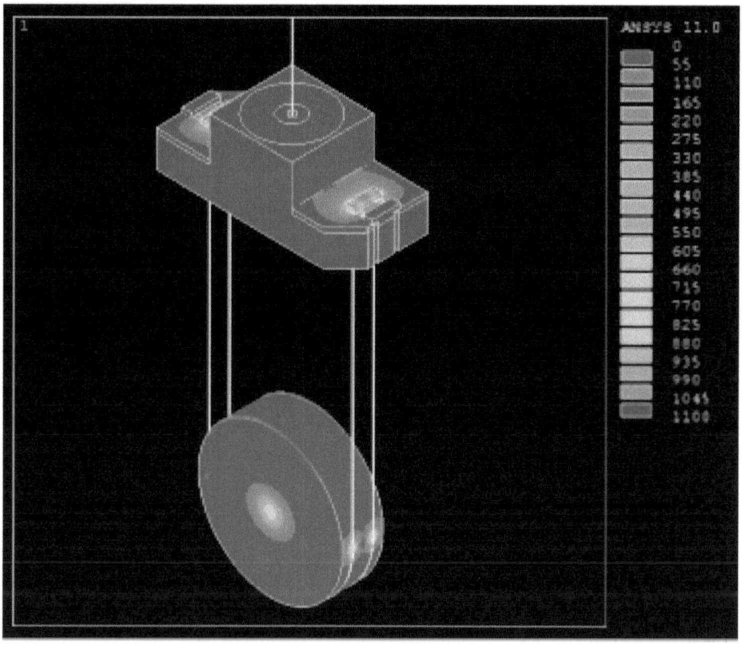

Fig. 14.14 Finite element analysis of the heat flow from the mirror to the marionette via the suspension wires, when the laser is impinging on the mirror

as a reference value for the absorption coefficient of the mirror optical coating at the laser wavelength. In this case, if ~20 kW of light power are stored in the interferometer, we have ~20 mW of absorbed laser power.

The radiation from the warm surface of the vacuum tube to the mirror is derived from the Stefan-Boltzmann equation and we have:

$$\dot{Q}_r/A_1 = \sigma F_e F_{1-2}(T_2^4 - T_1^4) \tag{14.4}$$

where $\dot{Q}_r/A_1$ is the heat transfer rate by radiation per unit area of the mirror surface, $\sigma = 5.67\,10^{-8}\,\text{W/m}^2/\text{K}^4$ is the Stefan-Boltzmann constant, F_e is the emissivity factor, and F_{1-2} is a geometric configuration factor relating the two surfaces, whose temperatures are T_2 (warm) and T_1 (cold). From Eq. 14.4 the heat flux on the mirror at $T_1 = 10\,\text{K}$, coming from the vacuum tube at $T_2 = 300\,\text{K}$ can be as high as $\dot{Q}_r/A_1 \sim 460\,\text{W/m}^2$. This huge heat flux is not compatible with the cryogenic environment and needs to be reduced by several orders of magnitude.

To limit this huge contribution, we need to reduce the solid angle under which the warm surface is seen by the cold mirror. This is obtained by providing, in the region adjacent to the mirror, cold sections of the vacuum tube, i.e., adding thermal shields (cryotraps) along the part of the vacuum tube near the cryostat hosting the mirror.

The mirror thermal equilibrium is reached when the power extracted by the cooling system is equivalent to that absorbed by the mirror ($\dot{Q}_{abs} = P_{cool}$). If this condition is met, most of the power absorbed by the mirror flows through its suspension wires (a small fraction is lost in radiation), and then is removed by the heat link directly connected to the cooling system. The equilibrium temperature of the mirror T_{mir} can be calculated by the simple analytical model relating it to the temperature of the thermal bath T_{bath} of the cooler via the power flow through the mirror wires:

$$\dot{Q}_{abs} = \frac{4\Sigma_w}{L} < K_{si} > (T_{bath} - T_{mir}) \equiv \frac{1}{Z_{therm}} \Delta T \tag{14.5}$$

with

$$< K_{si} > = \frac{1}{\Delta T} \int_{T_{bath}}^{T_{mir}} K_{si}(T)\, dT \tag{14.6}$$

and where Σ_w is the wire section, L its length and $K_{si}(T)$ is the thermal conductivity of silicon. For a given thermal input $\dot{Q}_{abs}$, the thermal impedance Z_{therm} is the relevant parameter to yield the final temperature of the mirror: this is the quantity which sets the performance of our system and influences the choice of both the material and the geometry of the mirror suspension wire.

In the design of a cryogenic payload (as an example, we show a prototype, developed in Roma, in Fig. 14.15), a few other issues must be considered. First of all, we need to limit the power dissipated in vacuum by the current flowing into the electromagnetic actuator coils used to steer the mirror: a possible solution is the use of superconducting alloys for the coil wires. In fact, despite of the significant decrease of the copper resistivity with temperature, the heat radiated in vacuum can cause a large drift of the mirror temperature because of the lower thermal capacitance of crystals at cryogenic temperatures.

During the commissioning and noise hunting phase, it can happen that we need to access the last stage of suspension for improving the detector performance. This implies that a short cooling (and warm up) time improves the flexibility of the apparatus.

To avoid any possible contamination of the mirror, the KAGRA project has chosen to cool the mirror by embedding the payload in a heavy shield thermally connected to the PT cold heads. Thin wires made of pure aluminum connect the shield to the marionette and the heat is extracted from the suspended elements of the payload both via radiation and solid conduction. This approach eliminates the danger of mirror contamination but implies to wait nearly one month to bring the 23 kg sapphire mirror at 20 K.

In the case of the proposed ET detector, the silicon mirror will have a mass with a thermal capacitance 3–4 times that of the KAGRA mirror: another approach should therefore be considered. In Fig. 14.16, we plot the typical values of the heat exchange coefficients in function of the temperature for various mechanisms of heat transfer.

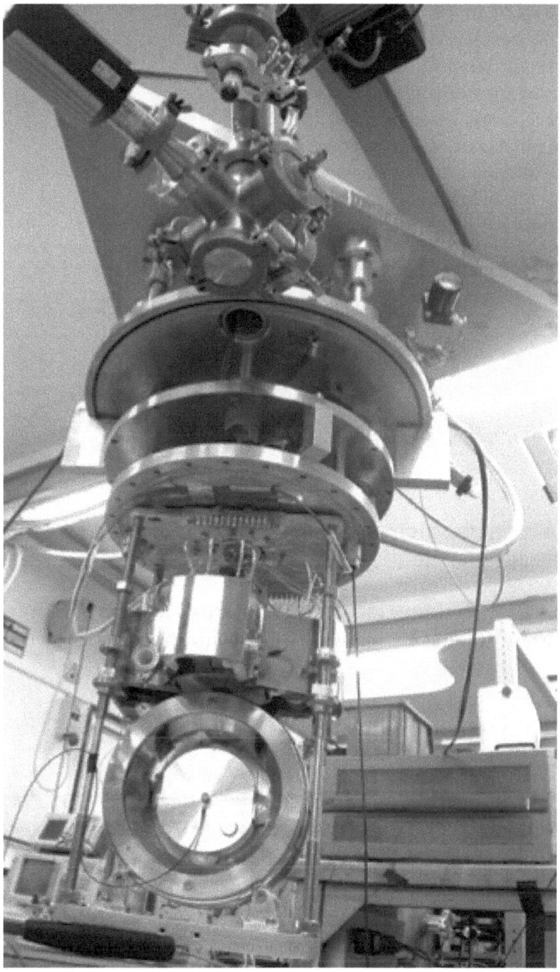

Fig. 14.15 The first prototype of cryogenic payload built in the VIRGO laboratory of the university of Rome *La Sapienza*. The payload is ready to be inserted in the inner vacuum chamber of the vibration free cryostat equipped with a PT cryocoolers

The plot shows that an improvement of the cooling time is obtained by using a low pressure helium gas, which improve the heat exchange between the cold shield of the cryostat and the payload. The drawback of this solution is a more complex design of the cryostat: it requires to transform the cold shield in a second vacuum chamber hosting the payload. During the cooling phase, the inner chamber is filled with helium gas insuring a more efficient heat extraction. Then, in the low temperature state, the gas is pumped out.[7] This approach implies to solve two problems at least:

- insure the vacuum tightness of the inner vacuum chamber, which contains the payload suspended vertically via a cable to the room temperature super attenuator and which is connected horizontally to the kilometric vacuum tube via the cold traps,

[7] This technique is applied also in the cooling process of all the GW resonant antennas.

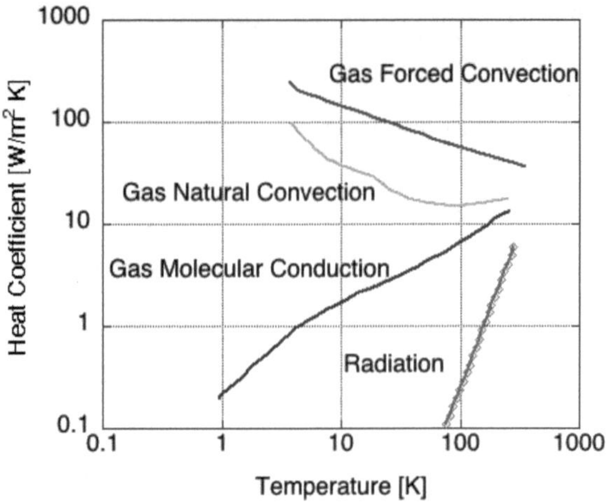

Fig. 14.16 Typical curves of the heat transfer coefficients versus temperature for the various mechanisms of heat exchange. We notice that the radiative exchange is poor, while the gas heat exchange is a way to reduce the cooling time significantly

- limit the high risk of mirror contamination by adding a purification system for the gas flowing in and out the inner vacuum chamber.

Although a few technical solutions have been proposed, at present none of them is yet validated by an experimental test.

14.4 Conclusion

The third generation of GW detectors is renewing the alliance between low temperature physics and gravitational wave astronomy. One of the most challenging goals in the construction of a new generation gravitational wave interferometers is cooling the mirrors at low temperature. Here, we presented the alternative strategies for cooling the mirrors in the case of the Einstein Telescope trying to focus on those aspects which require R&D activity and experimental tests.

At present, the construction of the Japanese detector KAGRA has started. This 3 km-arm interferometer has an innovative design and implements the key technologies of a third generation gravitational wave observatory. The KAGRA project already made few crucial choices to overcome several difficulties associated to the use of cryogenics in a GW interferometer. The common aspects of the KAGRA and ET detectors are pushing for a tighter collaboration between the Japanese and European scientists, and it will influence the final design of the ET cryogenic detector.

References

1. J. Weber, Detection and generation of gravitational waves. Phys. Rev. **117**, 306 (1960)
2. J. Weber, Evidence for discovery of gravitational radiation. Phys. Rev. Lett. **22**, 1320 (1969)
3. M. Lee, D. Gretz, S. Steppel, J. Weber, 1979 Gravitational radiation detector observations in 1973 and 1974. Phys. Rev. **D19**, 893 (1979)

4. J.L. Levine, B. Garwin, New negative results for gravitational wave detection and comparison with reported detector. Phys. Rev. Lett. **33**, 794 (1974)
5. D.H. Douglass, R.O. Gram, J.A. Tyson, R.W. Lee, Two-detector-coincidence search for bursts of gravitational radiation. Phys. Rev. Lett. **35**, 480 (1975)
6. B.W.P. Drever, J. Hough, B. Bland, G.W. Lessnoff, Search for short bursts of gravitational radiation. Nature **246**, 340 (1973)
7. O.D. Aguiar, Res. Astron. Astrophys. **11**, 1–42 (2011)
8. W. Fairbank, The use of low-temperature technology in gravitational wave experiments, in *Proceedings of the International School on Experimental Gravitation* (Italy, Varenna, 1972), p. 280
9. E. Mauceli, Z.K. Geng, W.O. Hamilton, W.W. Johnson, S. Merkowitz, A. Morse, The Allegro gravitational wave detector: data acquisition and analysis. Phys. Rev. D **54**, 1264 (1996)
10. P. Astone, M. Bassan, P. Bonifazi, P. Carelli, M.G. Castellano, G. Cavallari, E. Coccia, C. Cosmelli, V. Fafone, S. Frasca, E. Majorana, I. Modena, G.V. Pallottino, G. Pizzella, P. Rapagnani, F. Ricci, M. Visco, Long-term operation of the Rome "Explorer" cryogenic gravitational wave detector. Phys. Rev. **D47**, 362 (1993)
11. P. Astone, M. Bassan, P. Bonifazi, P. Carelli, E. Coccia, C. Cosmelli, V. Fafone, S. Frasca, A. Marini, G. Mazzitelli, Y. Minenkov, I. Modena, G. Modestino, A. Moleti, G.V. Pallottino, M.A. Papa, G. Pizzella, P. Rapagnani, F. Ricci, F. Ronga, R. Terenzi, M. Visco, L. Votano, Astropart. Phys. **7**, 231 (1997)
12. M. Cerdonio, M Bonaldi, D. Carlesso, E. Cavallini, S. Caruso, A. Colombo, P. Falferi, G. Fontana, P.L. Fortini, R. Mezzena, A. Ortolan, G.A. Prodi, L. Taffarello, G. Vedovato, S. Vitale, J.P. Zendri, The ultracryogenic gravitational-wave detector AURIGA. Class. Quant. Grav. **14**, 1491 (1997)
13. F. Acernese et al., First joint gravitational wave search by the AURIGA EXPLORER NAUTILUS Virgo Collaboration. Class. Quant. Grav. **25**, 205007 (2008)
14. L. Tisza, C. R. Acad. Sci. **207**, 1035, 1186 (1938)
15. F. London, Phys. Rev. **54**, 947 (1938)
16. L.D. Landau, J. Phys. (USSR) **5**, 7 (1941)
17. L.D. Landau, J. Phys. (USSR) **11**, 91 (1947)
18. G. Bon Mardion, G. Claudet, P. Seyfert, Practical data on steady state heat transport in superfluid helium at atmospheric pressure. Cryogenics **19**, 45 (1979)
19. P. Astone, M. Bassan, P. Bonifazi, E. Coccia, C. Cosmelli, Noise behaviour of the EXPLORER gravitational wave antenna during the l transition to superfluid phase. Cryogenics **32**, 668 (1992)
20. P. Puppo, F. Ricci, Cryogenics and Einstein telescope. Gen. Relativ. Gravit. **43**, 657–669 (2011)
21. F. Ricci, Low temperature and future gravitational wave experiments, ed. by J. Dumarchez, J.T.T. Van, in *Proceedings of the XLIInd Rencontres de Moriond, Gravitational Waves and Experimental Gravity*, vol. 177 (The Gioi Publishers, 2007)
22. E.I. Mikulin, A.A. Tarasov, M.P. Shkrebyonock, Low temperature expansion pulse tubes. In *Advances in Cryogenic Engineering*, vol. 29 (Plenum Press, New York, 1984), pp. 629–637
23. S. Zhu, P. Wu, Z. Chen, Double inlet pulse tube refrigerators: an important improvement. Cryogenics **30**, 514 (1990)
24. S. Caparrelli, E. Majorana, V. Moscatelli, E. Pascucci, M. Perciballi, P. Puppo, P. Rapagnani, F. Ricci, Vibration-free cryostat for low-noise applications of a pulse tube cryocooler. Rev. Sci. Instrum. **77**, 095102 (2006)
25. M.V. Plissi, K.A. Strain, Aspects of the suspension system for GEO600. Rev. Sci. Instrum. **69**, 3055 (1998)